Power Line Worker

Level Two: Distribution

Trainee Guide

PEARSON

Boston Columbus Indianapolis New York San Francisco Upper Saddle River
Amsterdam Cape Town Dubai London Madrid Milan Munich Paris Montreal Toronto
Delhi Mexico City São Paulo Sydney Hong Kong Seoul Singapore Taipei Tokyo

National Center for Construction Education and Research
President: Don Whyte
Director of Product Development: Daniele Stacey
Power Line Worker Project Manager for Distribution: Rob Richardson
Production Manager: Tim Davis
Quality Assurance Coordinator: Debie Ness
Desktop Publishing Coordinator: James McKay
Production Specialist: Heather Griffith-Gatson
Editor: Chris Wilson

Writing and development services provided by Topaz Publications, Liverpool, NY
Lead Writer/Project Manager: Thomas Burke
Desktop Publisher: Joanne Hart
Art Director: Megan Paye
Permissions Editors: Andrea LaBarge, Alison Richmond
Writers: Thomas Burke, Troy Staton, Gerald Shannon, Darrell Wilkerson, John Tianen, Patricia Vidler

Pearson Education, Inc.
Editorial Director: Vernon R. Anthony
Executive Editor: Alli Gentile
Senior Product Manager: Lori Cowen
Operations Supervisor: Deidra M. Skahill
Art Director: Jayne Conte
Director of Marketing: David Gesell
Executive Marketing Manager: Derril Trakalo
Marketing Manager: Brian Hoehl
Marketing Coordinator: Crystal Gonzalez

Composition: NCCER
Printer/Binder: Document Technology Resources, Fredericksburg, VA
Cover Printer: Document Technology Resources, Fredericksburg, VA
Text Fonts: Palatino and Univers

www.pearsonhighered.com

ISBN-13: 978-0-13-273034-1
ISBN-10: 0-13-273034-0

Preface

To the Trainee

Welcome to your second year of training in power line work. If you are training under an NCCER Accredited Training Program Sponsor, you have successfully completed *Power Industry Fundamentals* and *Power Line Worker Level One*, and are well on your way to more advanced training.

Power Line Worker Level Two: Distribution builds on the knowledge you gained in *Power Line Worker Level One* and makes it more applicable to the area of power distribution. In this level you will learn about three-phase systems, aerial distribution equipment, underground residential distribution systems, and cable and conductor installation and removal.

When you successfully complete this skill-specific training, you will join the ranks of thousands of men and women whose primary responsibility is to provide and restore electricity to millions of businesses and residences across the nation.

You will be joining an occupation that is in high demand. As the nation works to modernize its aging energy infrastructure, it is expected that more than 20% of the current workforce will retire by the next decade. In power line work, variety and opportunity await those with the skills, interest, and willingness to learn.

We wish you success as you progress through this training program. Should you have any comments on how NCCER might improve upon this textbook, please complete the User Update form located at the back of each module and send it to us. We will always consider and respond to input from our customers.

We invite you to visit NCCER's website at **www.nccer.org** for information on the latest product releases and training, as well as online versions of the *Cornerstone* newsletter and Pearson's product catalog.

Your feedback is welcome. You may email your comments to **curriculum@nccer.org** or send general comments and inquiries to **info@nccer.org**.

NCCER Curricula

NCCER is a not-for-profit 501(c)(3) education foundation established in 1996 by the world's largest and most progressive construction companies and national construction associations. It was founded to address the severe workforce shortage facing the industry and to develop a standardized training process and curricula. Today, NCCER is supported by hundreds of leading construction and maintenance companies, manufacturers, and national associations. The NCCER Curricula was developed by NCCER in partnership with Pearson Education, Inc., the world's largest educational publisher.

Some features of the NCCER Curricula are as follows:

- An industry-proven record of success
- Curricula developed by the industry for the industry
- National standardization providing portability of learned job skills and educational credits
- Compliance with the Office of Apprenticeship requirements for related classroom training *(CFR 29:29)*
- Well-illustrated, up-to-date, and practical information

NCCER also maintains a National Registry that provides transcripts, certificates, and wallet cards to individuals who have successfully completed modules of the NCCER Curricula. *Training programs must be delivered by an NCCER Accredited Training Sponsor in order to receive these credentials.*

Special Features

In an effort to provide a comprehensive, user-friendly training resource, we have incorporated many different features for your use. Whether you are a visual or hands-on learner, this book will provide you with the proper tools to get started in the power line worker industry.

Introduction

This page is found at the beginning of each module and lists the Objectives, Performance Tasks, Trade Terms, and Required Trainee Materials for that module. The Objectives list the skills and knowledge you will need in order to complete the module successfully. The Performance Tasks give you the opportunity to apply your knowledge to the real world duties that power line workers perform. The list of Trade Terms identifies important terms you will need to know by the end of the module. Required Trainee Materials list the materials and supplies needed for the module.

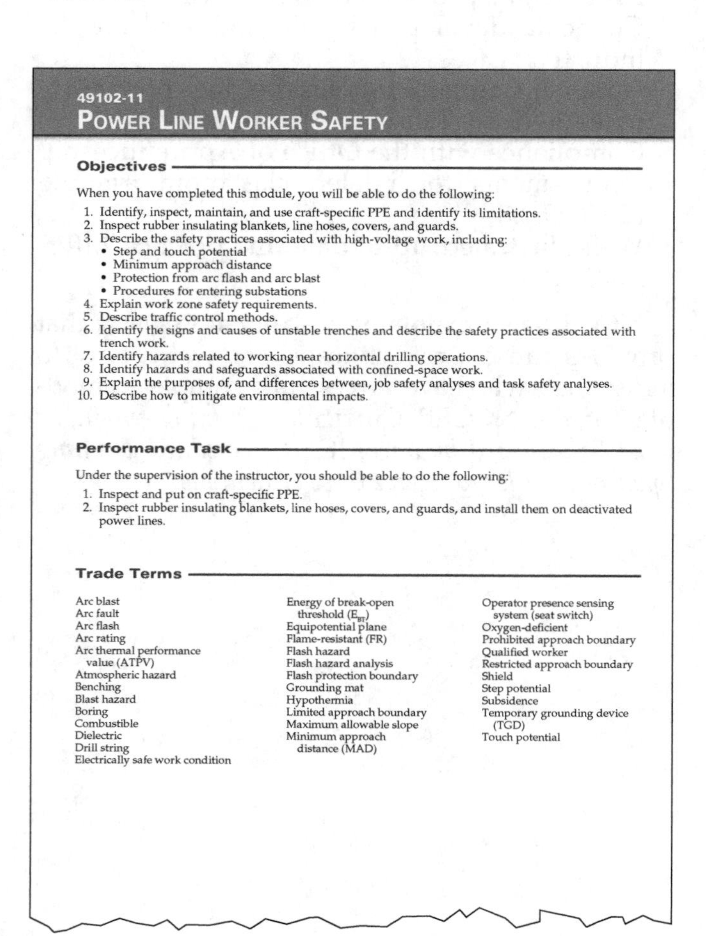

49102-11
POWER LINE WORKER SAFETY

Objectives

When you have completed this module, you will be able to do the following:

1. Identify, inspect, maintain, and use craft-specific PPE and identify its limitations.
2. Inspect rubber insulating blankets, line hoses, covers, and guards.
3. Describe the safety practices associated with high-voltage work, including:
   - Step and touch potential
   - Minimum approach distance
   - Protection from arc flash and arc blast
   - Procedures for entering substations
4. Explain work zone safety requirements.
5. Describe traffic control methods.
6. Identify the signs and causes of unstable trenches and describe the safety practices associated with trench work.
7. Identify hazards related to working near horizontal drilling operations.
8. Identify hazards and safeguards associated with confined-space work.
9. Explain the purposes of, and differences between, job safety analyses and task safety analyses.
10. Describe how to mitigate environmental impacts.

Performance Task

Under the supervision of the instructor, you should be able to do the following:

1. Inspect and put on craft-specific PPE.
2. Inspect rubber insulating blankets, line hoses, covers, and guards, and install them on deactivated power lines.

Trade Terms

Arc blast
Arc fault
Arc flash
Arc rating
Arc thermal performance value (ATPV)
Atmospheric hazard
Benching
Blast hazard
Boring
Combustible
Dielectric
Drill string
Electrically safe work condition
Energy of break-open threshold (E_{BT})
Equipotential plane
Flame-resistant (FR)
Flash hazard
Flash hazard analysis
Flash protection boundary
Grounding mat
Hypothermia
Limited approach boundary
Maximum allowable slope
Minimum approach distance (MAD)
Operator presence sensing system (seat switch)
Oxygen-deficient
Prohibited approach boundary
Qualified worker
Restricted approach boundary
Shield
Step potential
Subsidence
Temporary grounding device (TGD)
Touch potential

Color Illustrations and Photographs

Full-color illustrations and photographs are used throughout each module to provide vivid detail. These figures highlight important concepts from the text and provide clarity for complex instructions. Each figure reference is denoted in the text in *italic type* for easy reference.

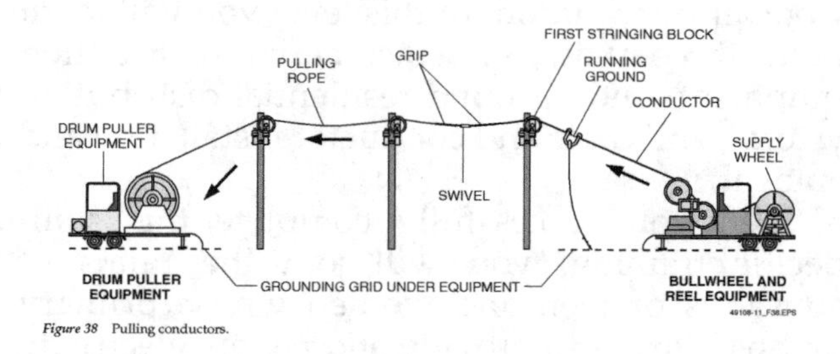

Figure 38 Pulling conductors.

Notes, Cautions, and Warnings

Safety features are set off from the main text in highlighted boxes and are organized into three categories based on the potential danger of the issue being addressed. Notes simply provide additional information on the topic area. Cautions alert you of a danger that does not present potential injury but may cause damage to equipment. Warnings stress a potentially dangerous situation that may cause injury to you or a co-worker.

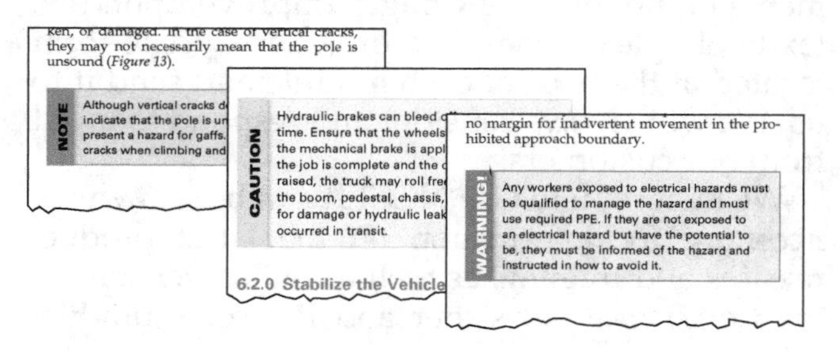

Did You Know?

The Did You Know? features offer hints, tips, and other helpful bits of information from the trade.

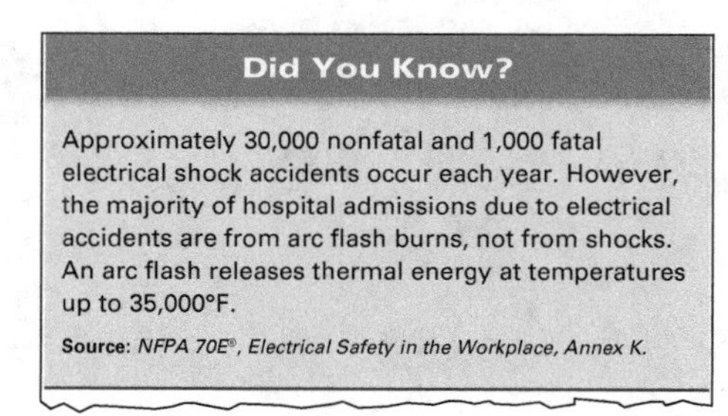

Did You Know?

Approximately 30,000 nonfatal and 1,000 fatal electrical shock accidents occur each year. However, the majority of hospital admissions due to electrical accidents are from arc flash burns, not from shocks. An arc flash releases thermal energy at temperatures up to 35,000°F.

Source: *NFPA 70E®, Electrical Safety in the Workplace, Annex K.*

On Site

On Site features provide a head start for those entering the electrical transmission and distribution fields by presenting technical tips and professional practices from power line workers on a variety of topics. On Site features often include real-life scenarios similar to those you might encounter on the job site.

On Site

Transformers

Large distribution transformers at power substations step down the power to the level required for local distribution. Pole transformers, such as the one shown here, step the power down further to the voltages needed for homes and businesses.

49102-11_SA02.EPS

Think About It

Think About It features use "What if?" questions to help you apply theory to real-world experiences and put your ideas into action.

Think About It

The Magic of Electricity

The effect of the flow of electrons occurs at a speed that is close to the speed of light, about 186,000 miles per second. How long does it take the light from the end of a flashlight to reach the floor? If you ran a light circuit from Maine to California and flipped the switch, how long would it take for the light to come on?

Step-by-Step Instructions

Step-by-step instructions are used throughout to guide you through technical procedures and tasks from start to finish. These steps show you not only how to perform a task but also how to do it safely and efficiently.

trically safe work condition. Most companies have detailed written procedures for performing this task.

Step 1 Determine whether any other crews are working on the circuit. All distribution lines are treated as if they are energized unless your team has performed a de-energizing procedure. If two or more crews are working on the same lines or equipment, each crew must independently perform a de-energizing procedure and apply their own lockout/tagout devices to energy controls.

Step 2 Designate one qualified member of the crew as the employee in charge of the electrical clearance.

Trade Terms

Each module presents a list of Trade Terms that are discussed within the text and defined in the Glossary at the end of the module. These terms are denoted in the text with **bold, blue type** upon their first occurrence. To make searches for key information easier, a comprehensive Glossary of Trade Terms from all modules is located at the back of this book.

You will hear the term *circuit* throughout your training. An electrical circuit contains, at minimum, a **voltage** source, a load, and **conductors** (wires) to carry the electrical **current** (*Figure 1*). The circuit should also have a means to stop and start the current, such as a switch.

Electricity is concerned with cause and effect. The presence of voltage (volts) in a closed circuit causes current (amps) to flow. The more voltage you apply, the more current will flow. However, the amount of current flow is also determined by how much **resistance**, in **ohms** (Ω), the load offers

Review Questions

Review Questions are provided to reinforce the knowledge you have gained. This makes them a useful tool for measuring what you have learned.

Review Questions

1. All of the following items are considered by OSHA to be acceptable fall prevention systems for working at heights above 6 feet *except* for _____.
   a. a guardrail system
   b. personal fall protection
   c. the two-climber buddy system
   d. a safety net system

2. Fall arrest systems _____.
   a. stop or take control of a fall in progress
   b. prevent falls from ever occurring
   c. ensure that the climber does not swing from side-to-side during a fall
   d. ensure that every fall is reported to the proper authorities

3. OSHA requires that a PFAS limits the maximum force imparted to the body during fall arrest to _____.
   a. 500 pounds
   b. 1,200 pounds
   c. 1,800 pounds
   d. 2,500 po

7. When an anchor point rated at 5,000 pounds is not available, the climber must then ensure that the chosen anchor position can at least withstand the maximum load that could be placed on it during a fall arrest, multiplied by a factor of _____.
   a. 1.5
   b. 2
   c. 3
   d. 5

8. Anchor points that are used as fall restraint anchors, and *not* as fall arrest anchors, must be load rated to handle _____.
   a. 2,500 pounds
   b. 3,000 pounds
   c. 5,000 pounds
   d. 7,500 pounds

9. The climber's personal fall arrest harness must be connected to the anchor point using _____.
   a. at least two lanyards

NCCER Curricula

NCCER's training programs comprise more than 80 construction, maintenance, pipeline, and utility areas and include skills assessments, safety training, and management education.

Boilermaking
Cabinetmaking
Carpentry
Concrete Finishing
Construction Craft Laborer
Construction Technology
Core Curriculum:
  Introductory Craft Skills
Drywall
Electrical
Electronic Systems Technician
Heating, Ventilating, and
  Air Conditioning
Heavy Equipment Operations
Highway/Heavy Construction
Hydroblasting
Industrial Coating and Lining
  Application Specialist
Industrial Maintenance
  Electrical and Instrumentation
  Technician
Industrial Maintenance
  Mechanic
Instrumentation
Insulating
Ironworking
Masonry
Millwright
Mobile Crane Operations
Painting
Painting, Industrial
Pipefitting
Pipelayer
Plumbing
Reinforcing Ironwork
Rigging
Scaffolding
Sheet Metal
Signal Person
Site Layout
Sprinkler Fitting
Tower Crane Operator
Welding

Green/Sustainable Construction

Building Auditor
Fundamentals of
  Weatherization
Introduction to Weatherization
Sustainable Construction
  Supervisor
Weatherization Crew Chief
Weatherization Technician
Your Role in the Green
  Environment

Energy

Alternative Energy
Introduction to the Power
  Industry
Introduction to Solar
  Photovoltaics
Introduction to Wind Energy
Power Industry Fundamentals
Power Generation Maintenance
  Electrician
Power Generation I&C
  Maintenance Technician
Power Generation Maintenance
  Mechanic
Power Line Worker: Distribution
Power Line Worker: Transmission
Solar Photovoltaic Systems
  Installer
Wind Turbine Maintenance
  Technician

Pipeline

Control Center Operations,
  Liquid
Corrosion Control
Electrical and Instrumentation
Field Operations, Liquid
Field Operations, Gas
Maintenance
Mechanical

Safety

Field Safety
Safety Orientation
Safety Technology

Management

Fundamentals of Crew
  Leadership
Project Management
Project Supervision

Supplemental Titles

Applied Construction Math
Careers in Construction
Tools for Success

Spanish Translations

Basic Rigging
  (Principios Básicos de
  Maniobras)
Carpentry Fundamentals
  (Introducción a la Carpintería,
  Nivel Uno)
Carpentry Forms
  (Formas para Carpintería,
  Nivel Trés)
Concrete Finishing, Level One
  (Acabado de Concreto,
  Nivel Uno)
Core Curriculum:
  Introductory Craft Skills
  (Currículo Básico:
  Habilidades Introductorias
  del Oficio)
Drywall, Level One
  (Paneles de Yeso, Nivel Uno)
Electrical, Level One
  (Electricidad, Nivel Uno)
Field Safety
  (Seguridad de Campo)
Insulating, Level One
  (Aislamiento, Nivel Uno)
Ironworking, Level One
  (Herrería, Nivel Uno)
Masonry, Level One
  (Albañilería, Nivel Uno)
Pipefitting, Level One
  (Instalación de Tubería
  Industrial, Nivel Uno)
Reinforcing Ironwork, Level One
  (Herreria de Refuerzo,
  Nivel Uno)
Safety Orientation
  (Orientación de Seguridad)
Scaffolding
  (Andamios)
Sprinkler Fitting, Level One
  (Instalación de Rociadores,
  Nivel Uno)

Acknowledgments

This curriculum was revised as a result of the farsightedness and leadership of the following sponsors:

Baltimore Gas & Electric
Cianbro Corporation
Gaylor, Inc.
Lakeland Electric
MasTec, Inc.
Quanta Services Inc.
Southeast Lineman Training Center
Vision Quest Academy

This curriculum would not exist were it not for the dedication and unselfish energy of those volunteers who served on the Authoring Team. A sincere thanks is extended to the following:

Don Daniel
David Champagne
Bruce Chesley
David Brzozowski
Robert Groner
Joe Holley
Gordon Johnson
Mark Lagasse
Paul Page
Michael A. Roedel
Jonathan Sacks
Antonio Vasquez
Randy Weeks
Terry Williams
Russell Zech

A final note: A special thanks is given to MasTec, Inc. and Baltimore Gas and Electric for the significant and generous support of their representatives.

NCCER Partners

American Fire Sprinkler Association
Associated Builders and Contractors, Inc.
Associated General Contractors of America
Association for Career and Technical Education
Association for Skilled and Technical Sciences
Carolinas AGC, Inc.
Carolinas Electrical Contractors Association
Center for the Improvement of Construction Management and Processes
Construction Industry Institute
Construction Users Roundtable
Construction Workforce Development Center
Design Build Institute of America
Merit Contractors Association of Canada
Metal Building Manufacturers Association
NACE International
National Association of Minority Contractors
National Association of Women in Construction
National Insulation Association
National Ready Mixed Concrete Association
National Technical Honor Society
National Utility Contractors Association
NAWIC Education Foundation
North American Technician Excellence
Painting & Decorating Contractors of America
Portland Cement Association
SkillsUSA
Steel Erectors Association of America
The Manufacturers Institute
U.S. Army Corps of Engineers
University of Florida, M.E. Rinker School of Building Construction
Women Construction Owners & Executives, USA

Contents

Module One

Alternating Current and Three-Phase Systems

Introduces the development of both single- and three-phase alternating current. Analyzes the relationship of AC phases and introduces key components used to refine AC power. Discusses the operation of transformers and introduces trainees to advanced AC concepts such as reactive power and the power factor. (Module ID 80201-11; 17.5 Hours)

Module Two

Aerial Distribution Equipment

Identifies the various equipment components found on overhead distribution system poles and describes the function of each, including transformers, reclosers, fuses, sectionalizers, capacitor banks, and voltage regulators. (Module ID 80202-11; 25 Hours)

Module Three

Cable and Conductor Installation and Removal

Describes the types of conductors and cables used in overhead and underground residential distribution systems and the equipment and procedures used to install and remove them. Includes methods used to splice conductors. (Module ID 80203-11; 20 Hours)

Module Four

Underground Residential Distribution (URD) Systems

Describes the methods used to distribute power in residential and commercial subdivisions, including the equipment used in the process, such as pad-mount transformers and switchgear. The module covers the components and methods used to connect primary and secondary power, as well as the protective devices used in URD systems and methods used to locate and repair buried cables. (Module ID 80204-11; 30 Hours)

Module Five

Overhead and URD Service Installations

Describes the methods and procedures used in terminating single-phase and three-phase aerial and URD systems at residential and commercial customer locations. Includes coverage of revenue meters and street light connections. (Module ID 80205-11; 15 Hours)

Module Six

Distribution Line Maintenance

Describes the inspection process and the methods and procedures used to inspect and maintain poles, conductors, and equipment used in aerial and URD systems. Includes coverage of transformer testing; location and correction of faults in URD systems; load management systems; and protective device coordination. (Module ID 80206-11; 50 Hours)

Glossary

Index

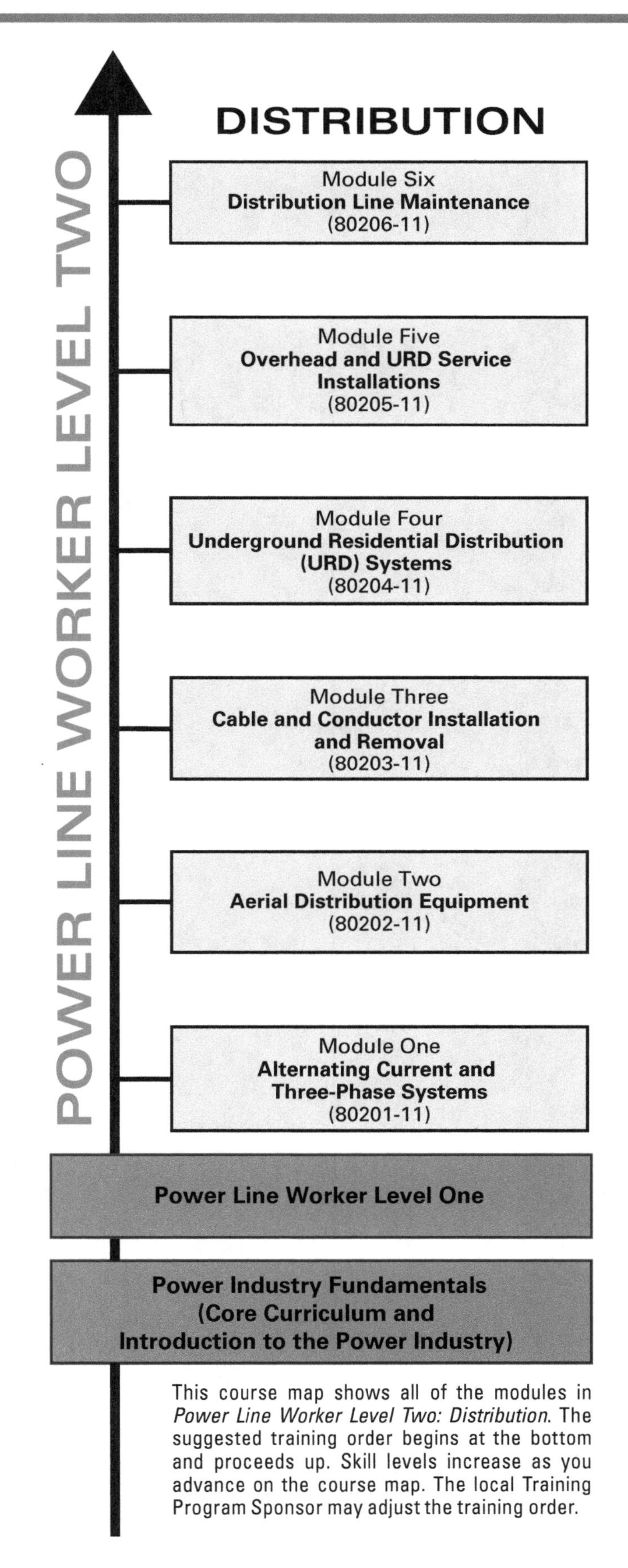

This course map shows all of the modules in *Power Line Worker Level Two: Distribution*. The suggested training order begins at the bottom and proceeds up. Skill levels increase as you advance on the course map. The local Training Program Sponsor may adjust the training order.

80201-11

Alternating Current and Three-Phase Systems

Module One

Trainees with successful module completions may be eligible for credentialing through NCCER's National Registry. To learn more, go to **www.nccer.org** or contact us at **1.888.622.3720**. Our website has information on the latest product releases and training, as well as online versions of our *Cornerstone* newsletter and Pearson's product catalog.

Your feedback is welcome. You may email your comments to **curriculum@nccer.org,** send general comments and inquiries to **info@nccer.org**, or fill out the User Update form at the back of this module.

V.1 12/11

80201-11
ALTERNATING CURRENT AND THREE-PHASE SYSTEMS

Objectives

When you have completed this module, you will be able to do the following:

1. Describe how single-phase and three-phase alternating current is developed.
2. Calculate the peak and effective voltage or current values for an AC waveform.
3. Describe phase relationships in AC circuits.
4. Describe impedance and explain how it affects AC circuits.
5. Describe the operating principles and functions of inductors.
6. Describe the operating principles and functions of capacitors.
7. Explain the principles and functions of transformers.
8. Explain the following terms as they relate to AC circuits:
   - True power
   - Apparent power
   - Reactive power
   - Power factor

Performance Tasks

This is a knowledge-based module; there are no performance tasks.

Trade Terms

Capacitance
Frequency
Hertz (Hz)
Impedance
Inductance
Micro
Peak voltage
Reactance
Resonance
Root-mean-square (rms)
Self-inductance

Required Trainee Materials

Scientific calculator

Industry Recognized Credentials

If you're training through an NCCER-accredited sponsor you may be eligible for credentials from NCCER's Registry. The ID number for this module is 80201-11. Note that this module may have been used in other NCCER curricula and may apply to other level completions. Contact NCCER's Registry at 888.622.3720 or go to nccer.org for more information.

Contents

Topics to be presented in this module include:

1.0.0 Introduction 1
2.0.0 Sine Wave Generation 1
3.0.0 Sine Wave Terminology 3
3.1.0 Frequency 3
3.2.0 Peak and Effective (RMS) Voltage Values 4
3.3.0 Average Value 5
4.0.0 AC Phase Relationships 6
4.1.0 Phase Angle 7
5.0.0 Resistance in AC Circuits 7
6.0.0 Inductance in AC Circuits 9
6.1.0 Factors Affecting Inductance 9
6.2.0 Voltage and Current in an Inductive AC Circuit 9
6.3.0 Inductive Reactance 11
7.0.0 Capacitance 12
7.1.0 Factors Affecting Capacitance 13
7.2.0 Calculating Equivalent Capacitance 14
7.3.0 Capacitor Specifications 14
7.3.1 Voltage Rating 14
7.3.2 Leak Resistance 15
7.4.0 Voltage and Current in a Capacitive AC Circuit 15
7.5.0 Capacitive Reactance 16
8.0.0 RL, RC, LC, and RLC Circuits 17
8.1.0 Resonance 19
9.0.0 Power In AC Circuits 19
9.1.0 True Power 19
9.2.0 Apparent Power 19
9.3.0 Reactive Power 20
9.4.0 Power Factor 20
9.5.0 Power Triangle 21
10.0.0 Transformers 22
10.1.0 Transformer Construction 22
10.1.1 Core Properties 22
10.1.2 Transformer Windings 23
10.2.0 Operating Properties 24
10.2.1 Energized with No Load 24
10.2.2 Phase Relationship 25
10.3.0 Turns and Voltage Ratios 25
10.4.0 Types of Transformers 26
10.4.1 Isolation Transformer 27
10.4.2 Autotransformer 29
10.4.3 Current Transformer 29
10.4.4 Potential Transformer 30
10.5.0 Transformer Selection 30
11.0.0 Three-Phase Power 31
11.1.0 Voltage and Current Imbalance in Three-Phase Systems 33

Figures and Tables

Figure 1 Conductor moving across a magnetic field 1
Figure 2 Angle versus rate of cutting lines of flux 2
Figure 3 One cycle of alternating voltage 3
Figure 4 Amplitude values for a sine wave 4
Figure 5 Frequency measurement 5
Figure 6 Root-mean-square (rms) amplitude 6
Figure 7 Voltage waveforms 90 degrees out of phase 7
Figure 8 Waves in phase 8
Figure 9 Resistive AC circuit 8
Figure 10 Voltage and current in a resistive AC circuit 9
Figure 11 Factors affecting the inductance of a coil 10
Figure 12 Inductor voltage and current relationship 11
Figure 13 Capacitors 12
Figure 14 Charging and discharging capacitor 13
Figure 15 Capacitors in parallel 14
Figure 16 Capacitors in series 14
Figure 17 Voltage and current in a capacitive AC circuit 16
Figure 18 Summary of AC circuit phase relationships 18
Figure 19 Power calculations in an AC circuit 20
Figure 20 RLC circuit calculation 21
Figure 21 Power triangle 23
Figure 22 Basic components of a transformer 23
Figure 23 Transformer action 24
Figure 24 Steel laminated core 24
Figure 25 Cutaway view of a transformer core 24
Figure 26 Transformer winding polarity 26
Figure 27 Transformer turns ratio 27
Figure 28 Tapped transformers 28
Figure 29 Importance of an isolation transformer 29
Figure 30 Autotransformer schematic diagram 29
Figure 31 Current transformer schematic diagram 29
Figure 32 Potential transformer 30
Figure 33 Three-phase voltage development 31
Figure 34 Wye-wye arrangement 32
Figure 35 Closed delta arrangement 32
Figure 36 Delta-wye arrangement 33
Figure 37 Three-phase transformer connections 34

1.0.0 INTRODUCTION

Alternating current (AC) reverses between positive and negative polarities and varies in amplitude with time. One complete waveform or cycle includes a complete set of variations, with two alternations in polarity. Many sources of voltage change direction with time and produce a waveform. The most common AC waveform is the sine wave.

This module covers the properties of the sine wave and other waveforms; the difference between AC and DC; the effects of capacitance and inductance on AC circuits; and the uses of transformers in AC circuits. Because of the alternating nature of AC, the components of an AC circuit react differently than those of a DC circuit. It is important to understand these differences, because they provide the basis for audio, video, and radio frequency systems.

2.0.0 SINE WAVE GENERATION

To understand how the AC sine wave is produced, some of the basic principles learned in magnetism should be reviewed. Two principles serve the basis of all electromagnetic reactions:

- An electric current in a conductor creates a magnetic field that surrounds the conductor.
- Relative motion between a conductor and a magnetic field creates a voltage in the conductor. This occurs when at least one component of that relative motion is in a direction that is perpendicular to the direction of the field.

Figure 1 shows how these principles are applied to generate an AC waveform in a simple one-loop rotary generator. The conductor loop rotates through the magnetic field to generate the

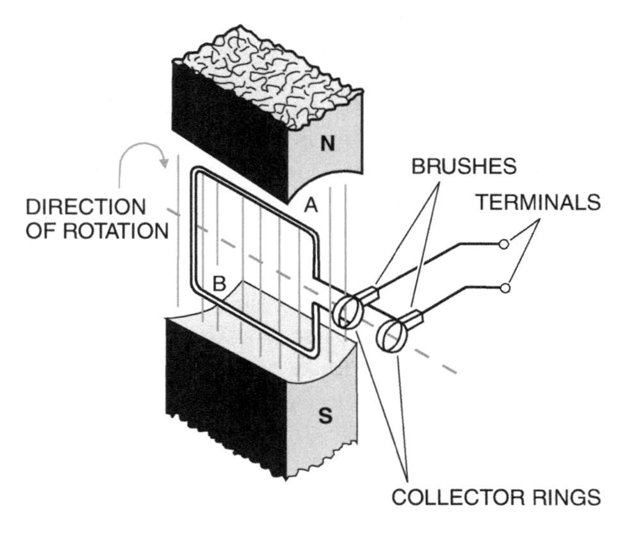

Figure 1 Conductor moving across a magnetic field.

On Site

Current Wars

After inventing the electric light bulb in 1879, Thomas Edison began work on a system for delivering electricity to homes and businesses. His system relied on direct current (DC)—electric current that always flows in one direction. However, DC transmission over long distances proved impractical. Transmitting direct current at the low voltages useful for lighting or motor operation required the use of thick, expensive copper wire. In fact, the service areas of Edison's DC generating stations were limited to about a square mile and mainly served the downtown areas of large cities. While Edison was pioneering DC distribution, electrical engineers in Europe were experimenting with alternating current (AC), which reverses direction at regular intervals.

The American businessman George Westinghouse saw the value in AC. High-voltage AC power could be distributed over longer distances using thinner, less expensive copper wires. Experts theorized that special devices (transformers) could step the voltage level up and down. Increasing the voltage level would allow it to be distributed across a wider region, while reducing the voltage level would enable the current to be used safely in homes and shops. Westinghouse hired a young electrical engineer named William Stanley, Jr., who developed the first effective transformer and demonstrated the first AC lighting system. Around the same time, the inventor Nikola Tesla filed patents for other devices run by AC.

After Westinghouse bought these patents, a full-scale industrial war known as the Current Wars erupted. At stake was whether Edison's direct current or Westinghouse's alternating current would electrify America. Edison claimed that alternating current was extremely dangerous and called for outlawing the high voltages transmitted by AC. Westinghouse countered that transformers safely reduced AC voltages before they entered buildings.

Within 10 years, the value of the alternating current system had been convincingly demonstrated. AC proved to be more practical and economical. Eventually, even Thomas Edison was forced to admit he had been wrong, and General Electric, the company he founded, began building and installing high-voltage AC transmission systems.

On Site

Why Do Power Companies Generate and Distribute AC Power Instead of DC Power?

The transformer is the key to AC power distribution. Power plants generate and distribute AC power because it permits the use of transformers, which makes power delivery more economical. Transformers used at generation plants step the AC voltage up, which decreases the current. Decreased current allows smaller-sized wires to be used for the power transmission lines. Smaller wire is less expensive and easier to support over the long distances that the power must travel from the generation plant to remotely located substations. At the substations, transformers are again used to step AC voltages back down to a level suitable for distribution to homes and businesses.

There is no such thing as a DC transformer. This means DC power would have to be transmitted at low voltages and high currents over very large wires, making the process very uneconomical. When DC is required for special applications, the AC voltage may be converted to DC voltage by using rectifiers, which make the change electrically, or by using AC motor-DC generator sets, which make the change mechanically.

induced AC voltage across its open terminals. The magnetic flux shown here is vertical.

There are several factors affecting the magnitude of voltage developed by a conductor through a magnetic field. They are the strength of the magnetic field, the length of the conductor, and the rate at which the conductor cuts directly across or at right angles to the magnetic field.

If the strength of the magnetic field and the length of the conductor making the loop are both constant, the voltage produced varies depending on the rate at which the loop cuts directly across the magnetic field.

The rate at which the conductor cuts the magnetic field depends on two things—the speed of the generator in revolutions per minute (rpm) and the angle at which the conductor travels through the field. If the generator is operated at a constant rpm, the voltage produced at any moment depends on the angle at which the conductor is cutting the field at that instant.

In *Figure 2*, the magnetic field is shown as parallel lines called lines of flux. These lines always go from the north to south poles in a generator. The motion of the conductor is shown by the large arrow.

Assume that the speed of the conductor is constant. As the angle between the flux and the conductor motion increases, the number of flux lines cut in a given time (the rate) increases. When the conductor is moving parallel to the lines of flux (angle of 0 degrees), it is not cutting any of them, and the voltage is zero.

The angle between the lines of flux and the motion of the conductor is called θ (theta). The magnitude of the voltage produced is proportional to the sine of the angle. Sine is a trigonometric function. Each angle has a sine value that never changes.

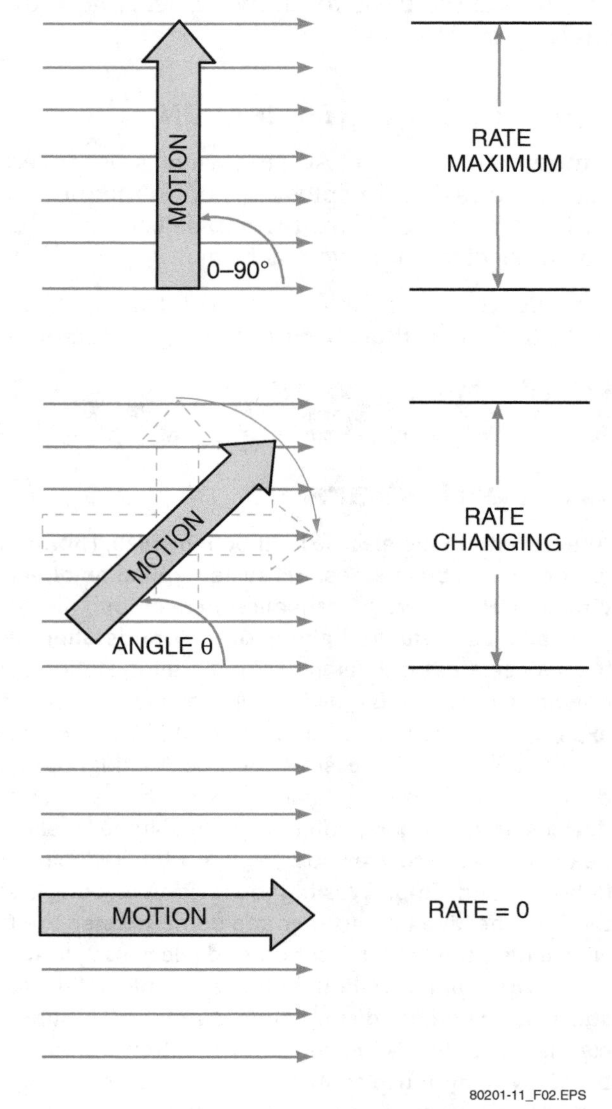

Figure 2 Angle versus rate of cutting lines of flux.

The sine of 0 degrees is 0. It increases to a maximum of 1 at 90 degrees. From 90 degrees to 180 degrees, the sine decreases back to 0. From 180 degrees to 270 degrees, the sine decreases to –1. Then from 270 degrees to 360 degrees (back to 0 degrees), the sine increases to its original 0.

Voltage is proportional to the sine of the angle. As the loop goes 360 degrees around the circle the voltage increases from 0 to its maximum at 90 degrees, back to 0 at 180 degrees, down to its maximum negative value at 270 degrees, and back up to 0 at 360 degrees, as shown in *Figure 3*.

Notice that at 180 degrees the polarity reverses. This is because the conductor has turned completely around and is now cutting the lines of flux in the opposite direction. The curve shown in *Figure 3* is called a sine wave because its shape is generated by the sine function. The value of voltage at any point along the sine wave can be calculated if the angle and the maximum obtainable voltage (E_{max}) are known. The formula used is as follows:

$$E = E_{max} \text{ sine } \theta$$

Where:

E = voltage induced
E_{max} = maximum induced voltage
θ = angle at which the voltage is induced

Using the above formula, the values of voltage anywhere along the sine wave in *Figure 4* can be calculated.

Sine values can be found using either a scientific calculator or trigonometric tables. With an E_{max} of 10 volts (V), the following values are calculated as examples:

$\theta = 0°$, sine = 0
$E = E_{max} \text{ sine } \theta$
$E = (10V)(0)$
$E = 0V$

$\theta = 45°$, sine = 0.707
$E = E_{max} \text{ sine } \theta$
$E = (10V)(0.707)$
$E = 7.07V$

$\theta = 90°$, sine = 1.0
$E = E_{max} \text{ sine } \theta$
$E = (10V)(1.0)$
$E = 10V$

$\theta = 135°$, sine = 0.707
$E = E_{max} \text{ sine } \theta$
$E = (10V)(0.707)$
$E = 7.07V$

$\theta = 180°$, sine = 0
$E = E_{max} \text{ sine } \theta$
$E = (10V)(0)$
$E = 0V$

$\theta = 225°$, sine = –0.707
$E = E_{max} \text{ sine } \theta$
$E = (10V)(-0.707)$
$E = -7.07V$

$\theta = 270°$, sine = –1.0
$E = E_{max} \text{ sine } \theta$
$E = (10V)(-1.0)$
$E = -10V$

$\theta = 315°$, sine = –0.707
$E = E_{max} \text{ sine } \theta$
$E = (10V)(-0.707)$
$E = -7.07V$

3.0.0 Sine Wave Terminology

In order to fully understand alternating current, you must understand the properties of the sine wave. The key properties are frequency, period, and voltage values.

3.1.0 Frequency

The frequency of a waveform is the number of times per second an identical pattern repeats itself. Each time the waveform changes from zero to a peak value and back to zero is called an al-

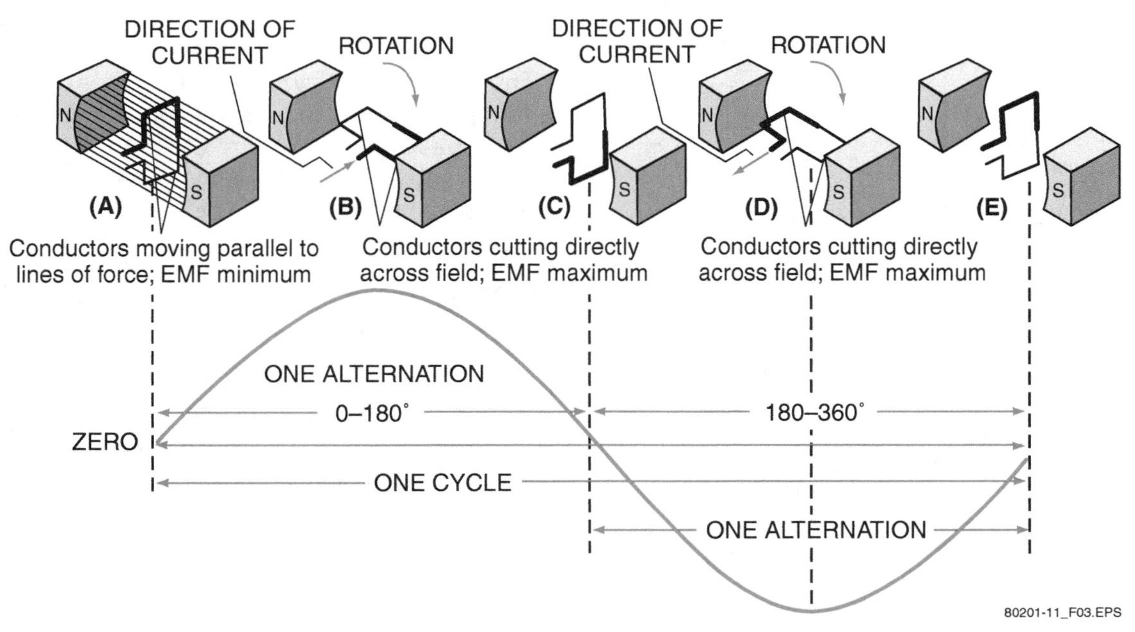

Figure 3 One cycle of alternating voltage.

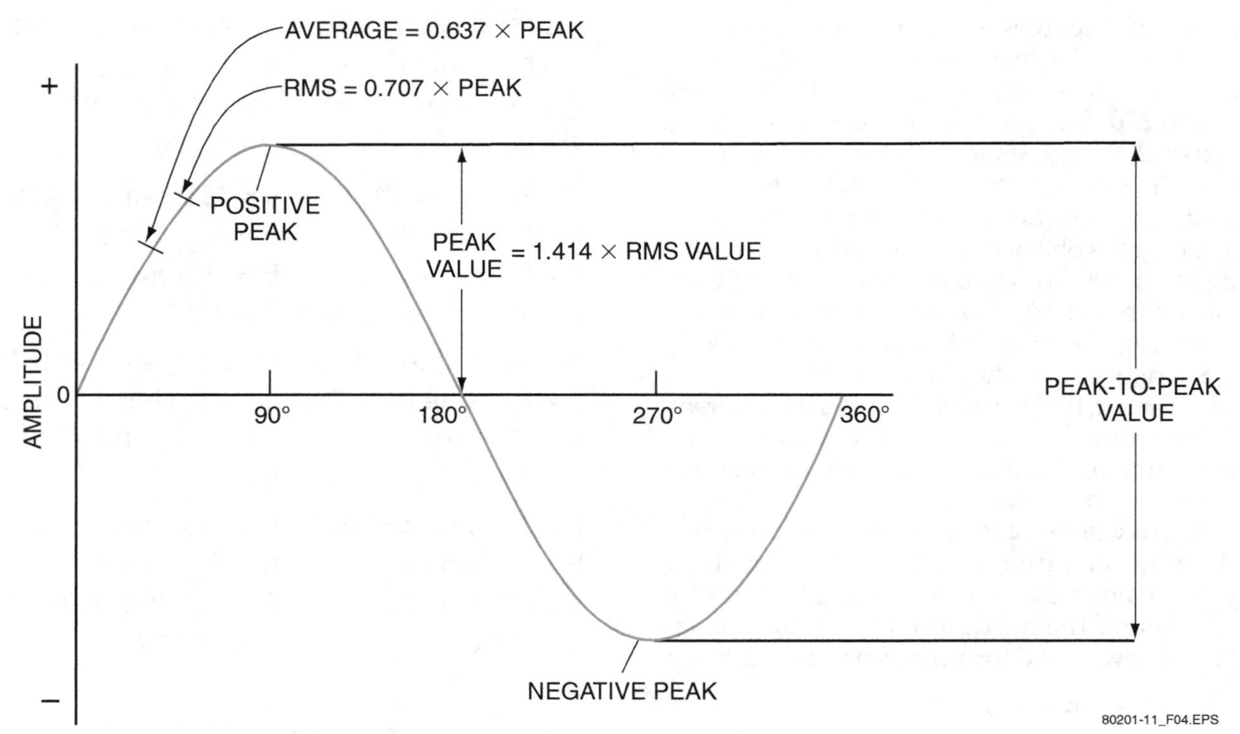

Figure 4 Amplitude values for a sine wave.

ternation. Two alternations form one cycle. The number of cycles per second is the frequency. The unit of frequency is **hertz (Hz)**. One hertz equals one cycle per second (cps).

For example, determine the frequency of the waveform shown in *Figure 5*. In one-half second, the basic sine wave is repeated five times. Therefore, the frequency (f) is:

$$f = \frac{5\ \text{cycles}}{0.5\ \text{second}} = 10\ \text{cycles per second (Hz)}$$

The period of a waveform is the time (t) required to complete one cycle. The period is the inverse of frequency:

$$t = \frac{1}{f}$$

Where:

t = period (seconds)
f = frequency (Hz or cps)

For example, determine the period of the waveform in *Figure 5*. If there are five cycles in one-half second, then the frequency for one cycle is 10 cps (5 ÷ 0.5 = 10). Therefore, the period is:

$$t = \frac{1}{\text{cps}}$$

$$t = \frac{1}{10} = 0.1\ \text{second}$$

3.2.0 Peak and Effective (RMS) Voltage Values

The peak value (*Figure 4*) is the maximum value of voltage (V_M) or current (I_M). Specifying that a sine wave has a **peak voltage** of 170V applies to either the positive or the negative peak. Meters used in AC circuits read a value called the effective value. The effective value is the value of the AC current or

On Site

High-Speed Travel

Electricity travels at the speed of light, which is 186,000 miles per second. The distance from Maine to California is about 3,000 miles, so it would take an electrical current about 0.016 second (3,000/186,000) to travel that distance.

Think About It

Frequency

The frequency of the utility power generated in the United States is normally 60Hz. In some European countries and elsewhere, utility power is often generated at a frequency of 50Hz. Which of these frequencies (60Hz or 50Hz) has the shortest period?

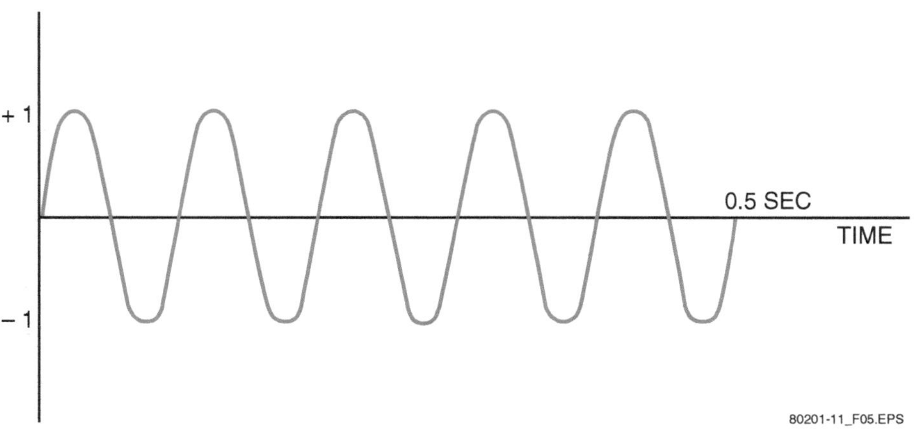

Figure 5 Frequency measurement.

voltage wave that indicates the same energy transfer as an equivalent direct current (DC) or voltage.

The direct comparison between DC and AC is in the heating effect of the two currents. Heat produced by current is a function of current amplitude only and is independent of current direction. Thus, heat is produced by both alternations of the AC wave, although the current changes direction during each alternation.

In a DC circuit, the current maintains a steady amplitude. Therefore, the heat produced is steady and is equal to I^2R. In an AC circuit, the current is continuously changing. To produce the same amount of heat from AC as from an equivalent amount of DC, the value of the AC must at times exceed the DC value.

By averaging the heating effects of all the instantaneous values during one cycle of alternating current, it is possible to find the average heat produced by the AC current during the cycle. The amount of DC required to produce that heat is equal to the effective value of the AC.

The most common method of specifying the amount of a sine wave of voltage or current is by stating its value at 45 degrees, which is 70.7 percent of the peak. This is its **root-mean-square (rms)** value. Therefore:

$$\text{Value of rms} = 0.707 \times \text{peak value}$$

For example, with a peak of 170V, the rms value is 0.707 × 170, or approximately 120V. This is the voltage of the common AC power source, which is always given in rms value.

To convert from rms voltage to peak voltage, multiply the rms value by 1.414.

Another way to view the concept of rms voltage is shown in *Figure 6*. The rms amplitude is the value assigned to an alternating voltage or current that results in the same power dissipation in a given resistance as DC voltage or current of the same amplitude. As shown, 120VAC (peak) does not produce the same light (350 lumens versus 500 lumens) as 120VDC from a 60W lamp. In order to produce the same light (500 lumens), 120V rms must be applied to the lamp. This requires that the applied sinusoidal AC waveform have a peak voltage of about 170V (170V × 0.707V = 120V).

3.3.0 Average Value

The average value is derived from all the values in a sine wave for one alternation or half cycle. The half cycle is used for the average because over

On Site

Effective Voltage

In service work, digital multimeters (DMM) like the one shown here are widely used by technicians to measure effective voltage. Unless stated otherwise, the voltages stamped on the nameplates of equipment refer to effective voltages. A true rms meter is a more sophisticated version of the DMM that reads actual rms voltage, instead of providing an estimation. Normally, a DMM estimates effective voltage because true rms is difficult to measure in an AC current that is not a perfect sine wave. The true rms meter can measure the voltage of imperfect waves.

80201-11_SA01.EPS

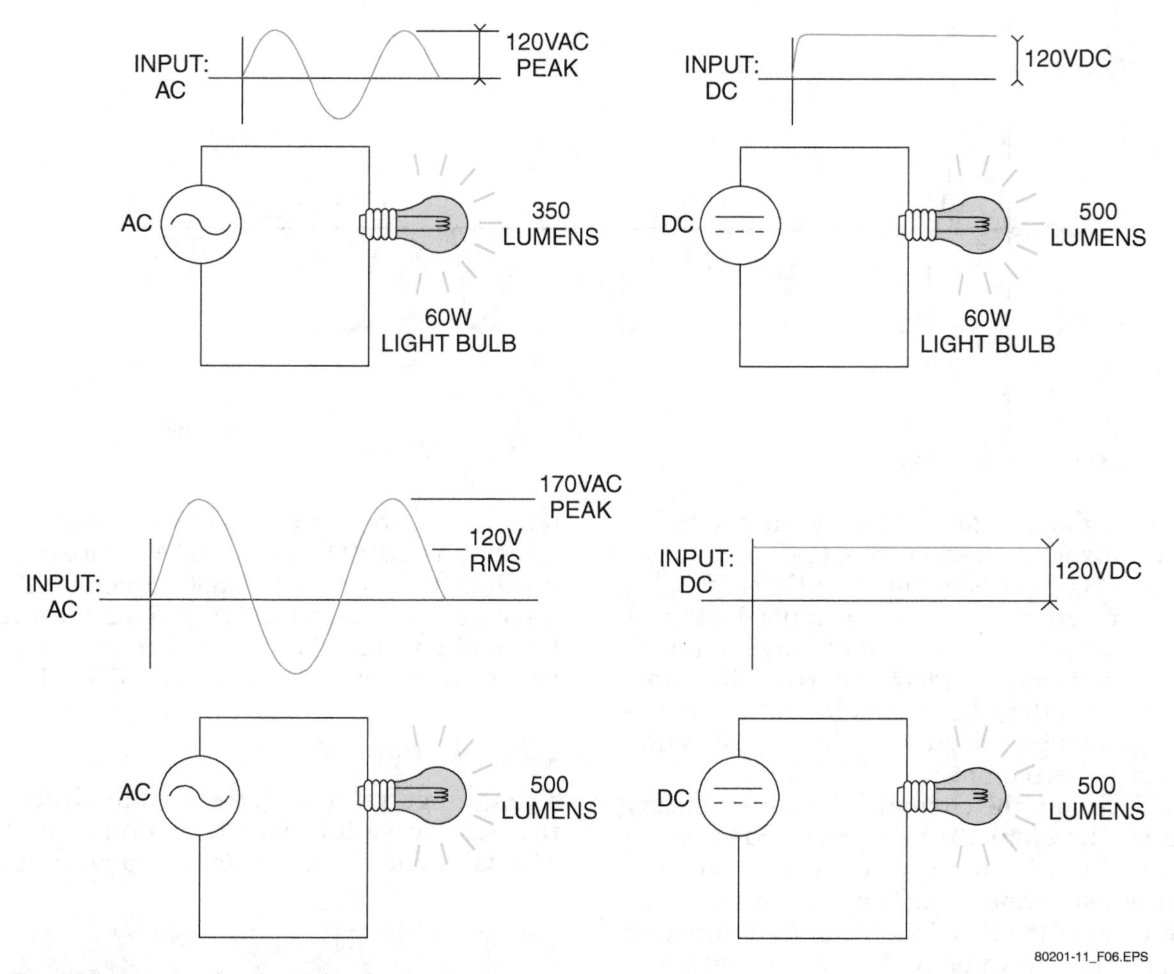

Figure 6 Root-mean-square (rms) amplitude.

a full cycle the average value is zero, which is useless for comparison purposes. If the sine values for all angles up to 180 degrees in one alternation are added and then divided by the number of values, this average equals 0.637.

Since the peak value of the sine is 1 and the average equals 0.637, the average value can be calculated as follows:

$$\text{Average value} = 0.637 \times \text{peak value}$$

For example, with a peak of 170V, the average value is 0.637 × 170V, which equals about 108V. *Figure 4* shows where the average value would fall on a sine wave.

4.0.0 AC Phase Relationships

In AC systems, phase is involved at the location of a point on a voltage or current wave with respect to the starting point of the wave or with respect to some corresponding point on the same wave. In the case of two waves of the same frequency, it is the time at which an event of one takes place with respect to a similar event of the other.

Often, the event is the starting of the waves at zero or the points at which the waves reach their maximum values. When two waves are compared in this manner, there is a phase lead or lag of one with respect to the other unless they are alternating in unison, in which case they are said to be in phase.

Suppose a generator started its cycle at 90 degrees where maximum voltage output is produced instead of starting at the point of zero output. The two output voltage waves are shown in *Figure 7*. Each is the same waveform of alternating voltage, but wave B starts at the maximum value while wave A starts at zero. The complete cycle of wave B through 360 degrees takes it back to the maximum value from which it started.

Wave A starts and finishes its cycle at zero. With respect to time, wave B is ahead of wave A in its values of generated voltage. The amount it leads in time equals one quarter revolution, which is 90 degrees. This angular difference is the phase angle between waves B and A. Wave B leads wave A by the phase angle of 90 degrees.

The 90-degree phase angle between waves B and A is maintained throughout the complete cycle and in all successive cycles as long as they both have the same frequency. At any instant in time, wave B has the value that A will have 90 degrees later. For instance, at 180 degrees, wave A is at zero, but B is already at its negative maximum value, the point where wave A will be later at 270 degrees.

To compare the phase angle between two waves, both waves must have the same frequency. Otherwise, the relative phase keeps changing. Both waves must also have sine wave variations, because this is the only kind of waveform that is measured in angular units of time. The amplitudes can be different for the two waves. The phases of two voltages, two currents, or a current with a voltage can be compared.

4.1.0 Phase Angle

To compare AC phases, it is much more convenient to use vector diagrams corresponding to the voltage waveforms, as shown in *Figure 7.* V_A and V_B represent the vector quantities corresponding to the generator voltages.

A vector is a quantity that has magnitude and direction. The length of the arrow indicates the magnitude of the alternating voltage in rms, peak, or any AC value as long as the same measure is used for all the vectors. The angle of the arrow with respect to the horizontal axis indicates the phase angle.

In *Figure 7,* the vector V_A represents the voltage wave A, with a phase angle of 0 degrees. This angle can be considered as the plane of the loop in the rotary generator where it starts with zero output voltage. The vector V_B is vertical to show the phase angle of 90 degrees for this voltage wave, corresponding to the vertical generator loop at the start of its cycle. The angle between the two vectors is the phase angle.

The symbol for a phase angle is θ (theta). In *Figure 7,* θ = 0 degrees. *Figure 8* shows the waveforms and phasor diagram of two waves that are in phase but have different amplitudes.

5.0.0 Resistance in AC Circuits

An AC circuit has an AC voltage source. Note the circular symbol with the sine wave inside it shown in *Figure 9.* It is used for any source of sine wave alternating voltage. This voltage connected across an external load resistance produces alternating current of the same waveform, frequency, and phase as the applied voltage.

According to Ohm's law, current (I) equals voltage (E) divided by resistance (R). When E is an rms value, I is also an rms value. For any instantaneous value of E during the cycle, the value of I is for the corresponding instant of time.

In an AC circuit with only resistance, the current variations are in phase with the applied voltage, as shown in *Figure 9.* This in-phase relationship between E and I means that such an AC circuit can be analyzed by the same methods used for DC circuits since there is not a phase angle to consider. Components that have only resistance include resistors, the filaments for incandescent light bulbs, and vacuum tube heaters.

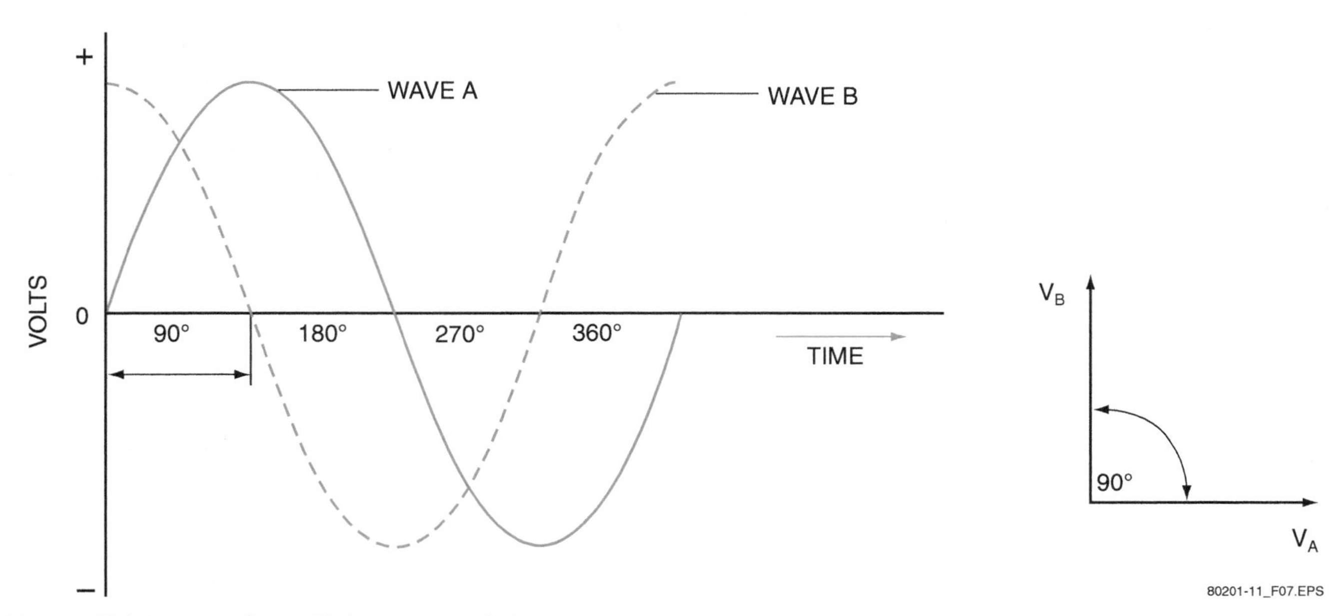

Figure 7 Voltage waveforms 90 degrees out of phase.

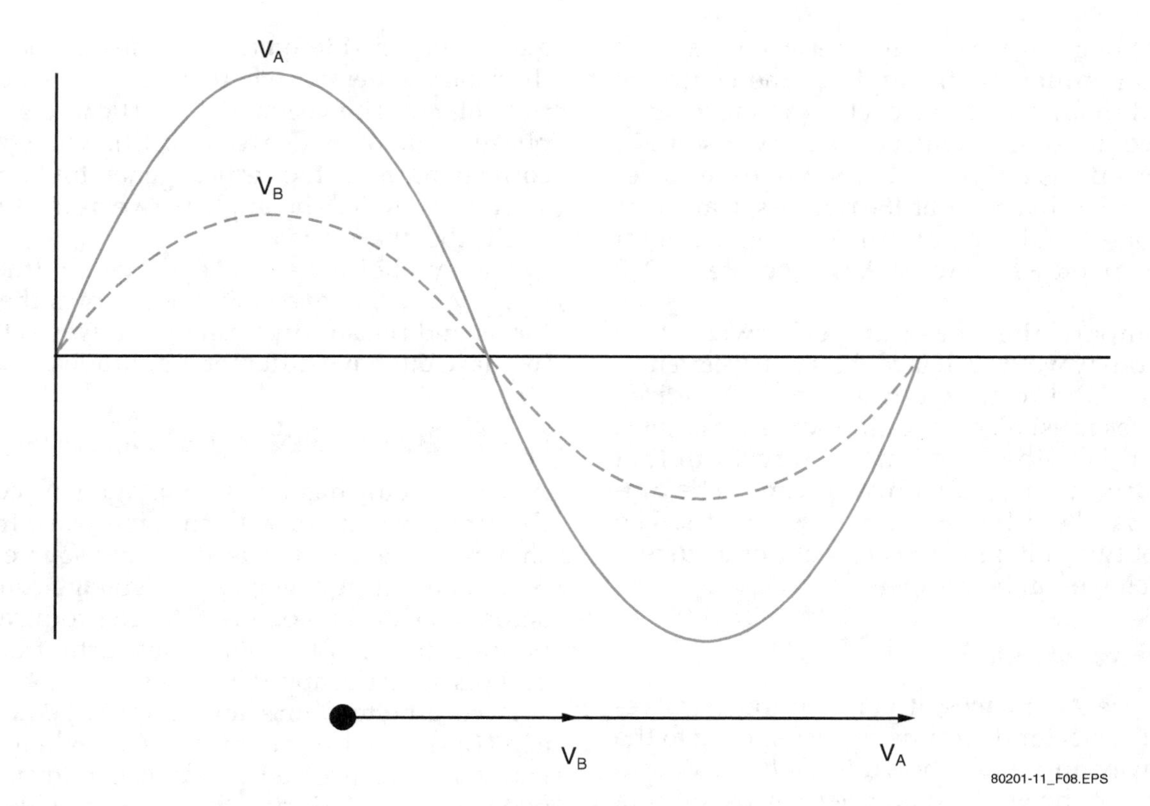

Figure 8 Waves in phase.

In purely resistive AC circuits, the voltage, current, and resistance are related by Ohm's law because the voltage and current are in phase:

$$I = \frac{E}{R}$$

Unless otherwise noted, the calculations in AC circuits are generally in rms values. For example, in *Figure 9*, the 120V applied across the 10V resistance R_L produces an rms current of 12A. This is determined as follows:

$$I = \frac{E}{R_L} = \frac{120V}{10\Omega} = 12A$$

Furthermore, the rms power (true power) dissipation is I^2R or:

$$P = (12A)^2 \times 10\Omega = 1{,}440W$$

Figure 10 shows the relationship between voltage and current in purely resistive AC circuits. The voltage and current are in phase, their cycles begin and end at the same time, and their peaks occur at the same time.

The value of the voltage shown in *Figure 10* depends on the applied voltage to the circuit. The value of the current depends on the applied voltage and the amount of resistance. If resistance is changed, it affects only the magnitude of the current.

The total resistance in any AC circuit, whether it is a series, parallel, or series-parallel circuit, is calculated using the same rules that you learned and applied to DC circuits with resistance.

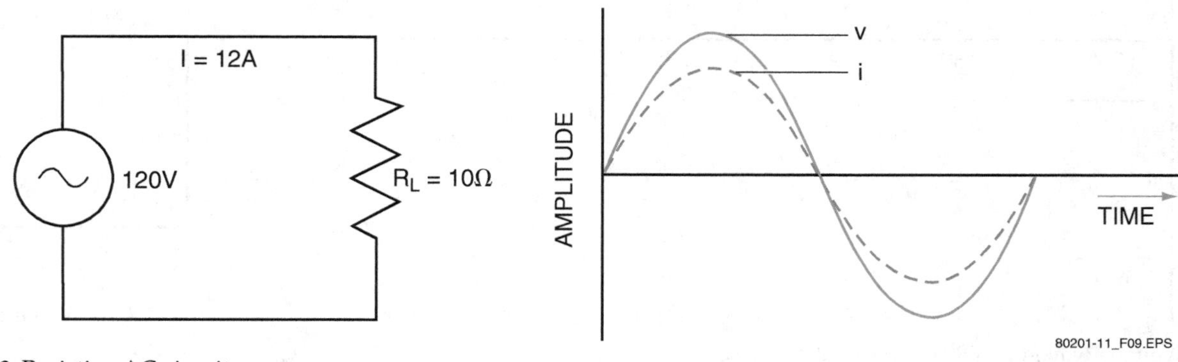

Figure 9 Resistive AC circuit.

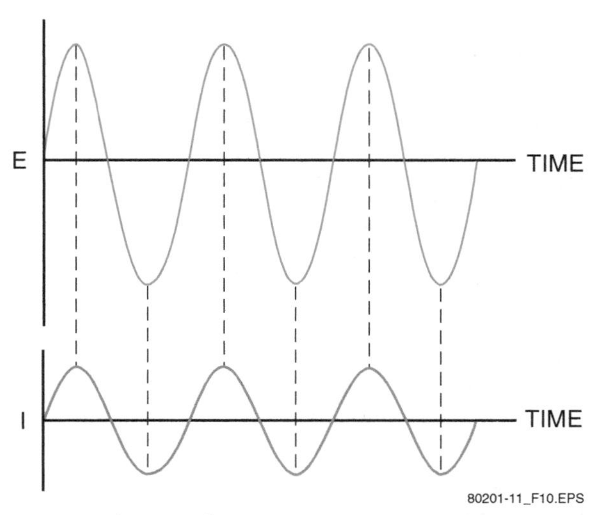

Figure 10 Voltage and current in a resistive AC circuit.

Power computation is discussed later in this module.

6.0.0 Inductance in AC Circuits

An inductor is a coil of wire wound around an iron core. As an alternating current flows through the coil, the coil becomes an electromagnet. Magnetic lines of force, called flux lines, radiate from the coil, expanding and collapsing as the AC waveform varies. The flux lines induce an opposing voltage into the coil. This induced voltage opposes the change in current flow and thereby causes the current to lag the voltage by 90 degrees. If the magnetic lines of force cut through any nearby conductor, a voltage will be induced into that conductor as well. This is the principle on which the transformer is based.

In DC circuits, a change must occur in the circuit to cause inductance. The current must change to provide motion of the flux. A steady DC of 10A cannot produce any induced voltage as long as the current value is constant. A current of 1A changing to 2A does induce voltage. Also, the faster the current changes, the higher the induced voltage becomes, because when the flux moves at a higher speed it can induce more voltage.

However, in an AC circuit the current is constantly changing and producing induced voltage. Lower frequencies of AC require more inductance to produce the same amount of induced voltage as a higher frequency current. The current can have any waveform as long as the amplitude is changing.

The ability of a conductor to induce voltage in itself when the current changes is referred to as self-inductance or simply inductance. The symbol for inductance is L and its unit is the henry (H). One henry is the amount of inductance that allows one volt to be induced when the current changes at the rate of one ampere per second.

6.1.0 Factors Affecting Inductance

An inductor is a coil of wire that may be wound on a core of metal or paper, or it may be self-supporting. It may consist of turns of wire placed side by side to form a layer of wire over the core or coil form. The inductance of a coil or inductor depends on its physical construction. The following factors affecting inductance are shown in *Figure 11*:

- *Number of turns* – The greater the number of turns, the greater the inductance. In addition, the spacing of the turns on a coil also affects inductance. A coil that has widely spaced turns has a lower inductance than one with the same number of more closely spaced turns. The reason for this higher inductance is that the closely wound turns produce a more concentrated magnetic field, causing the coil to show a greater inductance.
- *Coil diameter* – The inductance increases directly as the cross-sectional area of the coil increases.
- *Length of the coil* – When the length of the coil is decreased, the turn spacing is decreased, increasing the inductance of the coil.
- *Core material* – The core of the coil can be either a magnetic material (such as iron) or a non-magnetic material (such as paper or air). Coils wound on a magnetic core produce a stronger magnetic field than those with non-magnetic cores, giving them higher values of inductance. Air-core coils are used where small values of inductance are required.
- *Winding the coil in layers* – The more layers used to form a coil, the greater the effect the magnetic field has on the conductor. Layering a coil can increase the inductance.

6.2.0 Voltage and Current in an Inductive AC Circuit

The self-induced voltage across an inductance L is produced by a change in current with respect to time (DI/Dt) and can be stated as follows:

$$V_L = L\frac{\Delta I}{\Delta t}$$

Where:

Δ = change
V_L = volts
L = henrys
$\Delta I/\Delta t$ = amperes per second

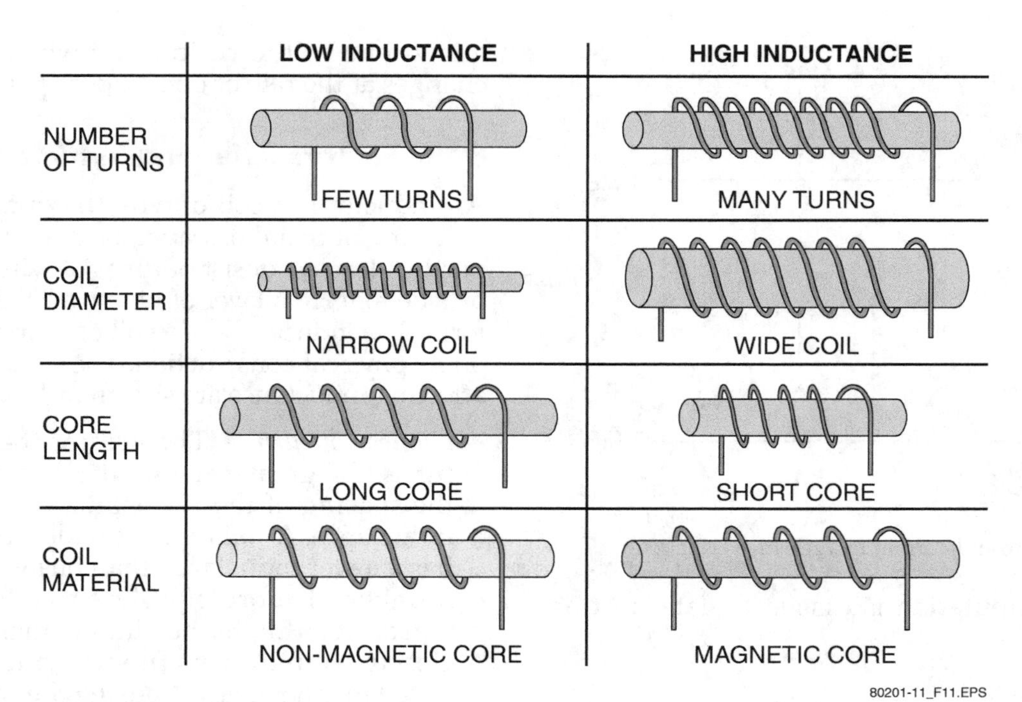

Figure 11 Factors affecting the inductance of a coil.

This gives the voltage in terms of how much magnetic flux is cut per second. When the magnetic flux associated with the current varies the same as I, this formula gives the same results for calculating induced voltage. Remember that the induced voltage across the coil is actually the result of inducing electrons to move in the conductor, so there is also an induced current.

For example, what is the self-induced voltage V_L across a 4h inductance produced by a current change of 12A per second?

$$V_L = L\frac{\Delta I}{\Delta t}$$

$$V_L = 4h\frac{12A}{1}$$

$$V_L = 4 \times 12$$

$$V_L = 48V$$

The current through a 200-microhenry (μh) inductor changes from 0 to 200 milliamps (mA) in 2 microseconds (μsec). The prefix **micro** means one-millionth. Determine the V_L:

$$V_L = L\frac{\Delta I}{\Delta t}$$

$$V_L = (200 \times 10^{-6})\frac{200 \times 10^{-3}}{2 \times 10^{-6}}$$

$$V_L = 20V$$

The induced voltage is an actual voltage that can be measured, although V_L is produced only while the current is changing. When DI/Dt is present for only a short time, V_L is in the form of a voltage pulse. With a sine wave current that is always changing, V_L is 90 degrees out of phase with I_L.

The current that flows in an inductor is induced by the changing magnetic field that surrounds the inductor. This changing magnetic field is produced by an AC voltage source that is applied to the inductor. The magnitude and polarity of the induced current depend on the field strength, direction, and rate at which the field cuts the inductor windings. The overall effect is that the current is out of phase and lags the applied voltage by 90 degrees.

At 270 degrees in *Figure 12*, the applied EMF is zero, but it is increasing in the positive direction at its greatest rate of change. Likewise, electron flow due to the applied EMF is also increasing at its greatest rate. As the electron flow increases, it produces a magnetic field that is building with it. The lines of flux cut the conductor as they move outward from it with the expanding field.

As the lines of flux cut the conductor, they induce a current into it. The induced current is at its maximum value because the lines of flux are expanding outward through the conductor at their greatest rate. The direction of the induced current opposes the force that generated it. Therefore, at 270 degrees the applied voltage is zero and is increasing to a positive value, while the current is at its maximum negative value.

At 0 degrees in *Figure 12*, the applied voltage is at its maximum positive value, but its rate of change is zero. Therefore, the field it produces

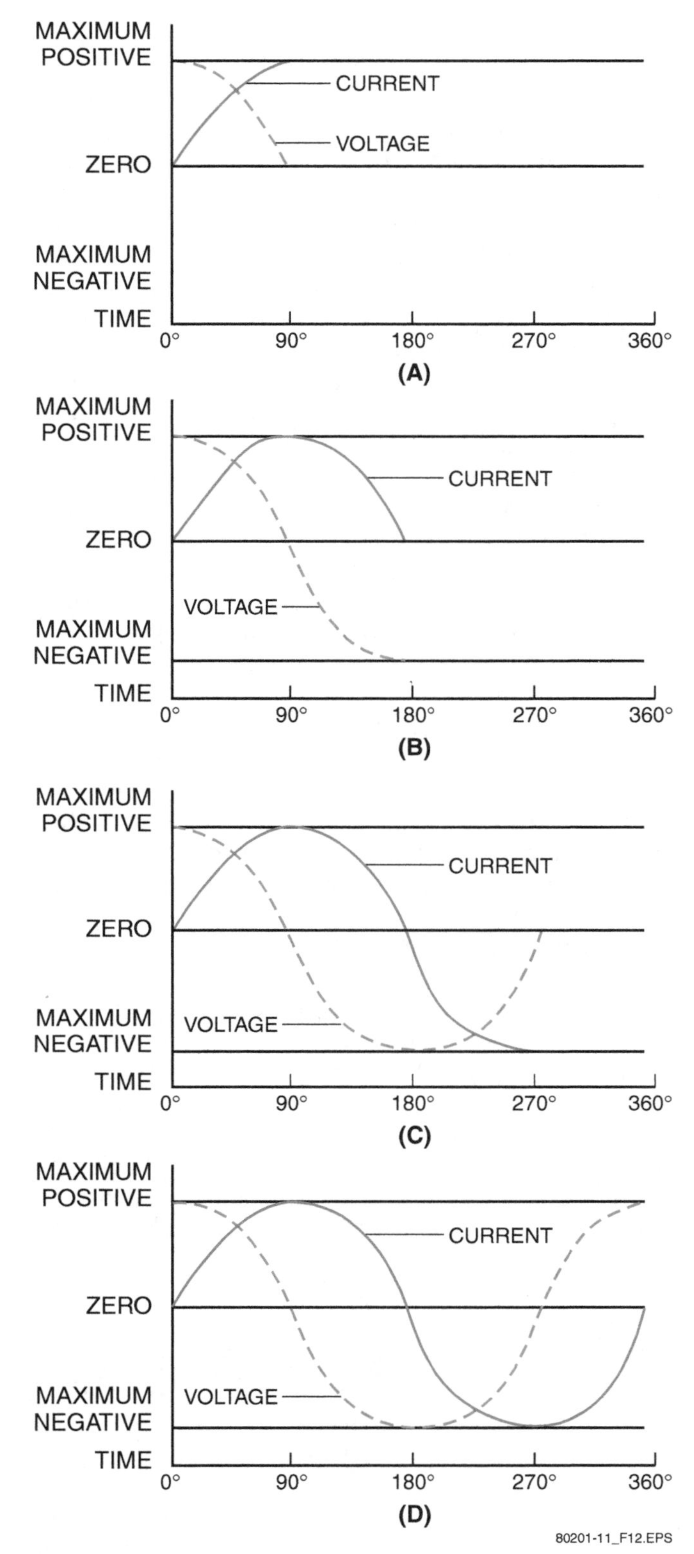

Figure 12 Inductor voltage and current relationship.

is no longer expanding and is not cutting the conductor. Because there is no relative motion between the field and conductor, no current is induced. Therefore, at 0 degrees voltage is at its maximum positive value, while current is zero.

At 90 degrees in *Figure 12*, voltage is once again zero, but this time it is decreasing toward negative at its greatest rate of change. Because the applied voltage is decreasing, the magnetic field is collapsing inward on the conductor. This has the effect of reversing the direction of motion between the field and conductor that existed at 0 degrees.

Therefore, the current will flow in a direction opposite of what it was at 0 degrees. Also, because the applied voltage is decreasing at its greatest rate, the field is collapsing at its greatest rate. This causes the flux to cut the conductor at the greatest rate, causing the induced current magnitude to be maximum. At 90 degrees, the applied voltage is zero decreasing toward negative, while the current is maximum positive.

At 180 degrees in *Figure 12*, the applied voltage is at its maximum negative value, but just as at 0 degrees, its rate of change is zero. At 180 degrees, therefore, current will be zero. This explanation shows that the voltage peaks positive first, then 90 degrees later the current peaks positive. Current thus lags the applied voltage in an inductor by 90 degrees. This can easily be remembered using this phrase: ELI the ICE man. ELI represents voltage (E), inductance (L), and current (I). In an inductor, the voltage leads the current just like the letter E leads or comes before the letter I. The word ICE will be explained in the section on capacitance.

6.3.0 Inductive Reactance

The opposing force that an inductor presents to the flow of alternating current cannot be called resistance since it is not the result of friction within a conductor. The name given to this force is inductive reactance because it is the reaction of the inductor to alternating current. Inductive reactance is measured in ohms and its symbol is X_L.

Remember that the induced voltage in a conductor is proportional to the rate at which magnetic lines of force cut the conductor. The greater the rate, or the higher the frequency, the greater the counter EMF (CEMF) will be. Also, the induced voltage increases with an increase in inductance; the more turns, the greater the CEMF will be. Reactance then increases with an increase of frequency and with an increase in inductance. The formula for inductive reactance is as follows:

$$X_L = 2\pi fL$$

Where:

X_L = inductive reactance in ohms
2π = a constant in which the Greek letter pi (π) represents 3.14 and 2 × pi = 6.28
f = frequency of the alternating current in hertz
L = inductance in henrys

For example, if f is equal to 60Hz and L is equal to 20h, find X_L:

$$X_L = 2\pi fL$$
$$X_L = 6.28 \times 60Hz \times 20h$$
$$X_L = 7{,}536\Omega$$

Once calculated, the value of X_L is used like resistance in a form of Ohm's law:

$$I = \frac{E}{X_L}$$

Where:

I = effective current (amps)
E = effective voltage (volts)
X_L = inductive reactance (ohms)

Unlike a resistor, there is no power dissipation in an ideal inductor. An inductor limits current, but it uses no net energy since the energy required to build up the field in the inductor is given back to the circuit when the field collapses.

7.0.0 Capacitance

Capacitance is the ability to store a charge. A capacitor is a device that stores an electric charge in a dielectric material. In storing a charge, a capacitor opposes a change in voltage. *Figure 13* shows a simple capacitor in a circuit as well as a schematic representation of two types of capacitors, and a photo of common capacitors.

Figure 14A shows a capacitor in a DC circuit. When voltage is applied, the capacitor begins to charge, as shown in *Figure 14B*. The charging continues until the potential difference across the capacitor is equal to the applied voltage. This charging current is transient or temporary since it flows only until the capacitor is charged to the applied voltage. Then there is no current in the circuit. *Figure 14C* shows this with the voltage across the capacitor equal to the battery voltage or 10V.

The capacitor can be discharged by connecting a conducting path across the dielectric. The stored charge across the dielectric provides the potential difference to produce a discharge current, as shown in *Figure 14D*. Once the capacitor is completely discharged, the voltage across it equals zero, and there is no discharge current.

In a capacitive circuit, the charge and discharge current must always be in opposite directions. Current flows in one direction to charge the capacitor and in the opposite direction when the capacitor is allowed to discharge.

Current flows in a capacitive circuit with AC voltage applied because of the capacitor charge and discharge current. There is no current through the dielectric, which is an insulator. While the capacitor is being charged by increasing applied voltage, the charging current flows in one direction to the plates. While the capacitor is discharging as the applied voltage decreases, the discharge current flows in the reverse direction. With alternating voltage applied, the capacitor alternately charges and discharges.

First, the capacitor is charged in one polarity, and then it discharges. Next, the capacitor is charged in the opposite polarity, and then it discharges again. The cycles of charge and discharge current provide alternating current in the circuit at the same frequency as the applied voltage. The

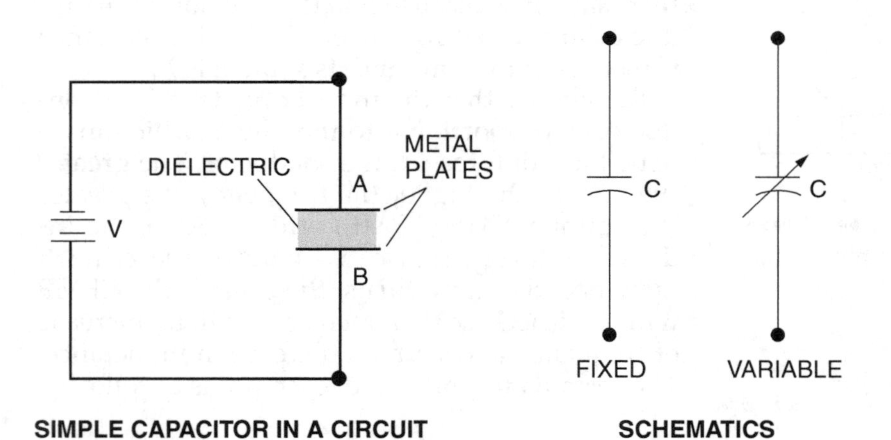

80201-11_F13.EPS

Figure 13 Capacitors.

On Site

ELI in ELI the ICE Man

Remembering the phrase ELI, as in ELI the ICE man, is an easy way to remember the phase relationships that always exist between voltage and current in an inductive circuit. An inductive circuit is a circuit where there is more inductive reactance than capacitive reactance. The L in ELI indicates inductance. The E (voltage) is stated before the I (current) in ELI, meaning that the voltage leads the current in an inductive circuit.

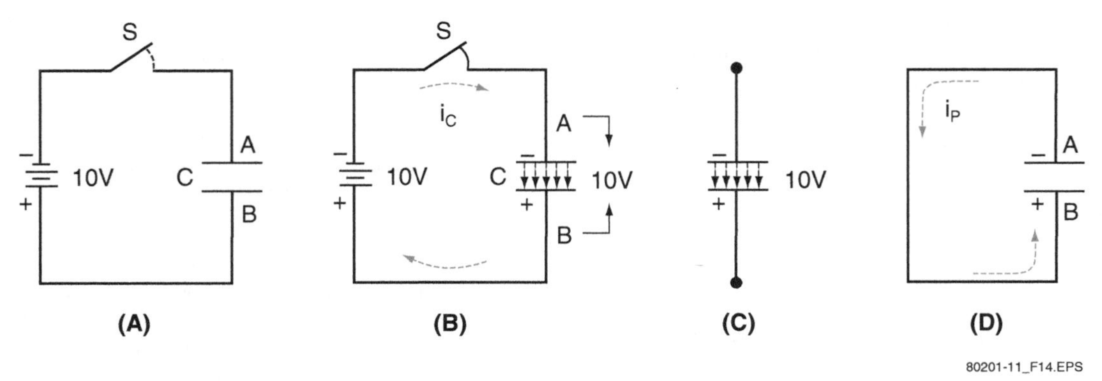

Figure 14 Charging and discharging capacitor.

amount of capacitance in the circuit determines how much current is allowed to flow.

Capacitance is measured in farads (F), where one farad is the capacitance when one coulomb is stored in the dielectric with a potential difference of one volt. Smaller values are measured in microfarads (μF). A small capacitance allows less charge and discharge current to flow than a larger capacitance. The smaller capacitor has more opposition to alternating current, because less current flows with the same applied voltage.

In summary, capacitance exhibits the following properties:

- DC is blocked by a capacitor. Once charged, no current will flow in the circuit.
- AC flows in a capacitive circuit with AC voltage applied.
- A smaller capacitance allows less current.

7.1.0 Factors Affecting Capacitance

A capacitor consists of two conductors separated by an insulating material called a dielectric. There are many types and sizes of capacitors with different dielectric materials. The capacitance is determined by three factors:

- *Area of the plates* – The initial charge displacement on a set of capacitor plates is related to the number of free electrons in each plate. Larger plates produce a greater capacitance than smaller ones. Therefore, the capacitance varies directly with the area of the plates. For example, if the area of the plates is doubled, the capacitance is doubled. If the size of the plates is reduced by 50 percent, the capacitance would also be reduced by 50 percent.
- *Distance between plates* – As two capacitor plates are brought closer together, more electrons will move away from the positively charged plate and into the negatively charged plate. This is because the mutual attraction between the opposite charges on the plates increases as the plates move closer together. This added movement of charge increases the capacitance of the capacitor. In a capacitor composed of two plates of equal area, the capacitance varies inversely with the distance between the plates. For ex-

On Site

Capacitance

The concept of capacitance, like many electrical quantities, is often hard to visualize or understand. A comparison with a balloon may help to make this concept clearer. Electrical capacitance has a charging effect similar to blowing up a balloon and holding it closed. The expansion capacity of the balloon can be changed by changing the thickness of the balloon walls. A balloon with thick walls will expand less (have less capacity) than one with thin walls. This is like a small 10μF capacitor that has less capacity and will charge less than a larger 100μF capacitor.

ample, if the distance between the plates is decreased by one-half, the capacitance is doubled. If the distance between the plates is doubled, the capacitance would be one-half as great.

- *Dielectric permittivity* – Another factor that determines the value of capacitance is the permittivity of the dielectric. The dielectric is the material between the capacitor plates in which the electric field appears. Relative permittivity expresses the ratio of the electric field strength in a dielectric to that in a vacuum. Permittivity has nothing to do with the dielectric strength of the medium or the breakdown voltage. An insulating material that withstands a higher applied voltage than some other substance does not always have a higher dielectric permittivity. Many insulating materials have a greater dielectric permittivity than air. For a given applied voltage, a greater attraction exists between the opposite charges on the capacitor plates, and an electric field can be set up more easily than when the dielectric is air. The capacitance of the capacitor is increased when the permittivity of the dielectric is increased if all the other factors remain unchanged.

7.2.0 Calculating Equivalent Capacitance

Connecting capacitors in parallel is equivalent to adding the plate areas. Therefore, the total capacitance is the sum of the individual capacitances, as shown in *Figure 15*.

A 10μF capacitor in parallel with a 5μF capacitor, for example, provides a 15μF capacitance for the parallel combination. The voltage is the same across the parallel capacitors. Note that adding parallel capacitance is opposite to the case of inductances in parallel and resistances in parallel.

Connecting capacitances in series is equivalent to increasing the thickness of the dielectric. Therefore, the combined capacitance is less than the smallest individual value. The combined equivalent capacitance is calculated by the reciprocal formula, as shown in *Figure 16*.

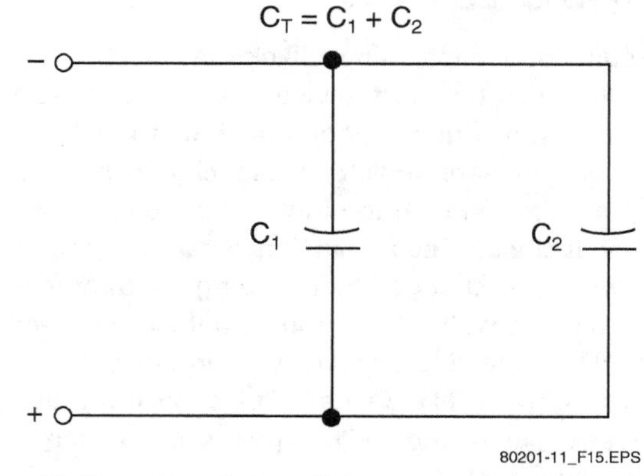

Figure 15 Capacitors in parallel.

> **Think About It**
>
> **Capacitor Substitution**
>
> Suppose you had a motor with a bad 30μF starting capacitor, and no 30μF direct replacement capacitor was available. As a temporary measure, you are authorized to substitute two equal-value capacitors in its place. What size capacitors (μF) should be used if you are connecting them in parallel?

Capacitors connected in series are combined like resistors in parallel. Any of the shortcut calculations for the reciprocal formula apply. For example, the combined capacitance of two equal capacitances of 10μF in series is 5μF. Capacitors are used in series to provide a higher voltage breakdown rating for the combination. For instance, each of three equal capacitances in series has one-third the applied voltage.

In series, the voltage across each capacitor is inversely proportional to its capacitance. The smaller capacitance has the larger proportion of the applied voltage. The reason is that the series capacitances all have the same charge because they are in one current path. With equal charge, a smaller capacitance has a greater potential difference.

7.3.0 Capacitor Specifications

In addition to its capacitance rating, the capacitor is rated by its operating voltage and leakage resistance. These factors must be considered in the selection of a capacitor.

7.3.1 Voltage Rating

This rating specifies the maximum potential difference that can be applied across the plates without puncturing the dielectric. Usually, the voltage rating is for temperatures up to about 60°C. High temperatures result in a lower voltage rating. Voltage ratings for general-purpose paper, mica,

$$C_T = \frac{1}{\frac{1}{C_1} + \frac{1}{C_2}}$$

C_1 C_2

80201-11_F16.EPS

Figure 16 Capacitors in series.

and ceramic capacitors are typically 200V to 500V. Ceramic capacitors with ratings of 1 to 5kV are also available.

Electrolytic capacitors are commonly used in 25V, 150V, and 450V ratings. In addition, 6V and 10V electrolytic capacitors are often used in transistor circuits. For applications where a lower voltage rating is permissible, more capacitance can be obtained in a smaller physical size.

The potential difference across the capacitor depends on the applied voltage and is not necessarily equal to the voltage rating. A voltage rating higher than the potential difference applied across the capacitor provides a safety factor for long life in service. With electrolytic capacitors, however, the actual capacitor voltage should be close to the rated voltage to produce the oxide film that provides the specified capacitance.

The voltage ratings are for applied DC voltage. The breakdown rating is lower for AC voltage because of the internal heat produced by continuous charge and discharge.

7.3.2 Leak Resistance

Consider a capacitor charged by a DC voltage source. After the charging voltage is removed, a perfect capacitor would keep its charge indefinitely. After a long period of time, however, the charge will be neutralized by a small leakage current through the dielectric and across the insulated case between terminals, because there is no perfect insulator. For paper, ceramic, and mica capacitors, the leakage current is very slight, or inversely, the leakage resistance is very high. For these types of capacitors, leakage resistance is 100MV or more. However, electrolytic capacitors may have a leakage resistance of 0.5MV or less.

WARNING!

The residual charge on a capacitor can be dangerous even after the power is turned off. The charge must be bled off before the capacitor terminals can be touched.

7.4.0 Voltage and Current in a Capacitive AC Circuit

In a capacitive circuit driven by an AC voltage source, the voltage is continuously changing. Thus, the charge on the capacitor is also continuously changing. The four parts of *Figure 17* show the variation of the alternating voltage and current in a capacitive circuit for each quarter of one cycle.

The solid line represents the voltage across the capacitor, and the dotted line represents the current. The line running through the center is the zero or reference point for both the voltage and the current. The bottom line marks off the time of the cycle in terms of electric degrees. Assume that the AC voltage has been acting on the capacitor for some time before the time represented by the starting point of the sine wave.

At the beginning of the first quarter-cycle (0 to 90 degrees), the voltage has just passed through zero and is increasing in the positive direction. Since the zero point is the steepest part of the sine wave, the voltage is changing at its greatest rate.

The charge on a capacitor varies directly with the voltage; therefore, the charge on the capacitor is also changing at its greatest rate at the beginning of the first quarter-cycle. In other words, the greatest number of electrons are moving off one plate and onto the other plate. Thus, the capacitor current is at its maximum value.

As the voltage proceeds toward maximum at 90 degrees, its rate of change becomes lower and lower, making the current decrease toward zero. At 90 degrees, the voltage across the capacitor is maximum, the capacitor is fully charged, and there is no further movement of electrons from plate to plate. That is why the current at 90 degrees is zero.

At the end of the first quarter-cycle, the alternating voltage stops increasing in the positive direction and starts to decrease. It is still a positive voltage; but to the capacitor, the decrease in voltage means that the plate that has an excess of electrons, and must lose some of them. The current flow must reverse its direction. The second part of the figure shows the current curve to be below the zero line (negative current direction) during the second quarter-cycle (90 to 180 degrees).

At 180 degrees, the voltage has dropped to zero. This means that, for a brief instant, the electrons are equally distributed between the two plates; the current is maximum because the rate of change of voltage is maximum.

Just after 180 degrees, the voltage has reversed polarity and starts building to its maximum negative peak, which is reached at the end of the third quarter-cycle (180 to 270 degrees). During the third quarter-cycle, the rate of voltage change gradually decreases as the charge builds to a maximum at 270 degrees. At this point, the capacitor is fully charged and carries the full impressed voltage. Because the capacitor is fully charged, there is no further exchange of electrons and the current flow is zero at this point. The conditions are exactly the same as at the end of the first quarter-cycle (90 degrees), but the polarity is reversed.

Just after 270 degrees, the impressed voltage once again starts to decrease, and the capacitor must lose electrons from the negative plate. It must discharge, starting at a minimum rate of flow and rising to a maximum. This discharging action continues through the last quarter-cycle (270 to 360 degrees) until the impressed voltage has reached zero. The beginning of the entire cycle is 360 degrees, and everything starts over again.

In *Figure 17*, note that the current always arrives at a certain point in the cycle 90 degrees ahead of the voltage because of the charging and discharging action. This voltage-current phase relationship in a capacitive circuit is exactly opposite to that in an inductive circuit. The current through a capacitor leads the voltage across the capacitor by 90 degrees. A convenient way to remember this is the phrase ELI the ICE man (ELI refers to inductors, as previously explained). ICE pertains to capacitors as follows:

I = current
C = capacitor
E = voltage

In capacitors (C), current (I) leads voltage (E) by 90 degrees.

It is important to realize that the current and voltage are both going through their individual cycles at the same time during the period the AC voltage is impressed. The current does not go through part of its cycle (charging or discharging) and then stop and wait for the voltage to catch up. The amplitude and polarity of the voltage and the amplitude and direction of the current are continually changing.

Their positions, with respect to each other and to the zero line at any electrical instant or any degree between 0 and 360 degrees, can be seen by reading upward from the time-degree line. The current swing from the positive peak at 0 degrees to the negative peak at 180 degrees is not a measure of the number of electrons or the charge on the plates. It is a picture of the direction and strength of the current in relation to the polarity and strength of the voltage appearing across the plates.

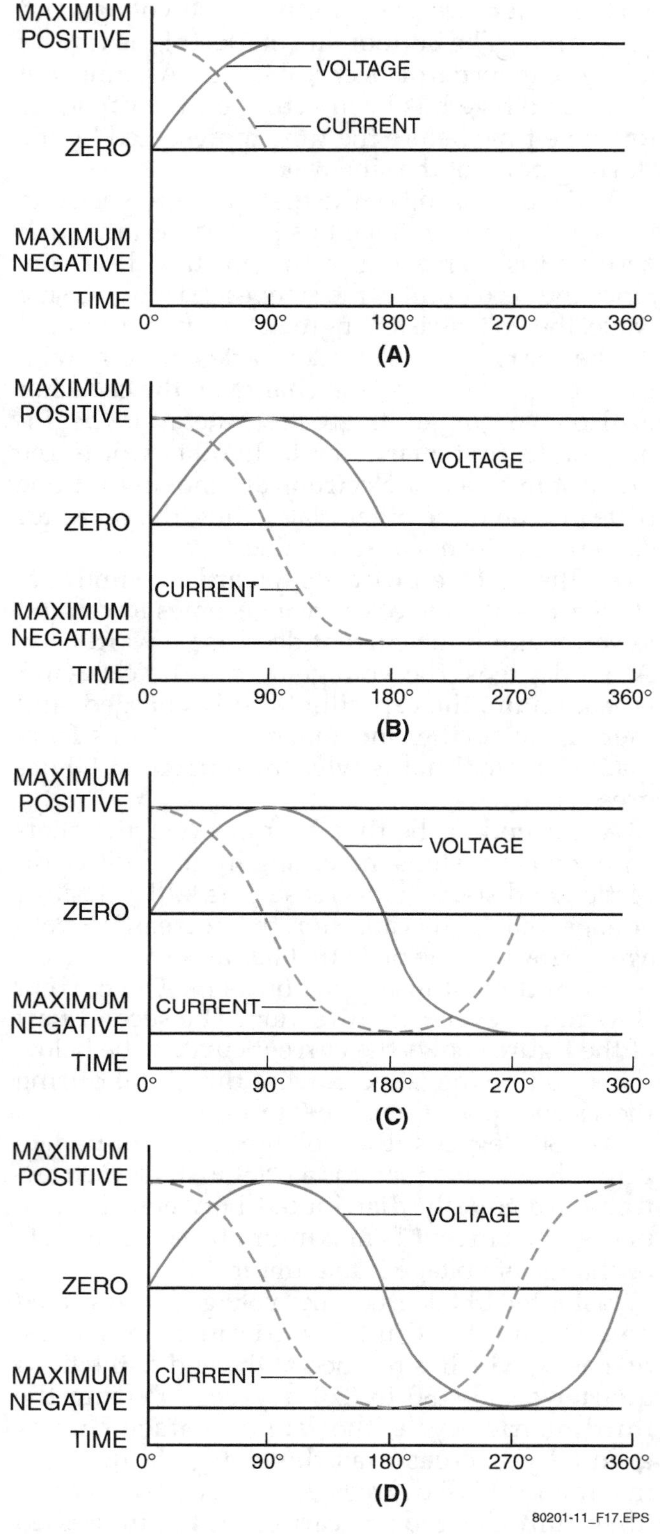

Figure 17 Voltage and current in a capacitive AC circuit.

7.5.0 Capacitive Reactance

Capacitors offer a very real opposition to current flow. This opposition arises from the fact that, at a given voltage and frequency, the number of electrons that go back and forth from plate to plate is limited by the storage ability or the capacitance

of the capacitor. As the capacitance is increased, a greater number of electrons changes plates every cycle. Since current is a measure of the number of electrons passing a given point in a given time, the current is increased.

Increasing the frequency also decreases the opposition offered by a capacitor. This occurs because the number of electrons that the capacitor is capable of handling at a given voltage changes plates more often. As a result, more electrons pass a given point in a given time (greater current flow). The opposition that a capacitor offers to alternating current is therefore inversely proportional to frequency and capacitance. This opposition is called capacitive reactance. Capacitive reactance decreases with increasing frequency or, for a given frequency, the capacitive reactance decreases with increasing capacitance. The symbol for capacitive reactance is X_C. The formula is:

$$X_c = \frac{1}{2\pi fC}$$

Where:

X_C = capacitive reactance in ohms
f = frequency in hertz
C = capacitance in farads
$2\pi = 6.28\ (2 \times 3.14)$

For example, what is the capacitive reactance of a 0.05μF capacitor in a circuit whose frequency is 1 megahertz?

$$X_c = \frac{1}{2\pi fC}$$

$$X_c = \frac{1}{6.28(10^6 \text{ hertz})(5 \times 10^{-8} \text{ farads})}$$

$$X_c = \frac{1}{6.28(5 \times 10^{-2})}$$

$$X_c = \frac{1}{31.4 \times 10^{-2}}$$

$$X_c = \frac{1}{0.314} = 3.18 \text{ ohms}$$

The capacitive reactance of a 0.05μF capacitor operated at a frequency of 1 megahertz is 3.18 ohms. Suppose this same capacitor is operated at a lower frequency of 1,500 hertz instead of 1 megahertz. What is the capacitive reactance now? Substituting where 1,500 = 1.5×10^3 hertz:

$$X_c = \frac{1}{2\pi fC}$$

$$X_c = \frac{1}{6.28(1.5 \times 10^3 \text{ hertz})(5 \times 10^{-8} \text{ farads})}$$

$$X_c = \frac{1}{6.28(7.5 \times 10^{-5})}$$

$$X_c = \frac{1}{47.1 \times 10^{-5}}$$

$$X_c = \frac{1}{0.000471} = 2{,}123 \text{ ohms}$$

Note a very interesting point from these two examples. As frequency is decreased from 1 megahertz to 1,500 hertz, the capacitive reactance increases from 3.18 ohms to 2,123 ohms. Capacitive reactance increases as the frequency decreases.

8.0.0 RL, RC, LC, AND RLC CIRCUITS

AC circuits often contain inductors, capacitors, and/or resistors connected in series or parallel combinations. When this is done, it is important to determine the phase relationship between the applied voltage and the current in the circuit. The simplest method of combining factors that have different phase relationships is vector addition. Each quantity is represented as a vector, and the resultant vector and phase angle are then calculated.

In purely resistive circuits, the voltage and current are in phase. In inductive circuits, the voltage leads the current by 90 degrees. In capacitive circuits, the current leads the voltage by 90 degrees. *Figure 18* shows the phase relationships of these components used in AC circuits. Recall that these properties are summarized by the phrase ELI the ICE man.

ELI = E Leads I (inductive)
ICE = I Capacitive (leads) E

The **impedance** Z of a circuit is defined as the total opposition to current flow. The magnitude of the impedance Z is given by the following equation in a series circuit:

$$Z = \sqrt{R^2 + X^2}$$

Where:

Z = impedance (ohms)
R = resistance (ohms)
X = net reactance (ohms)

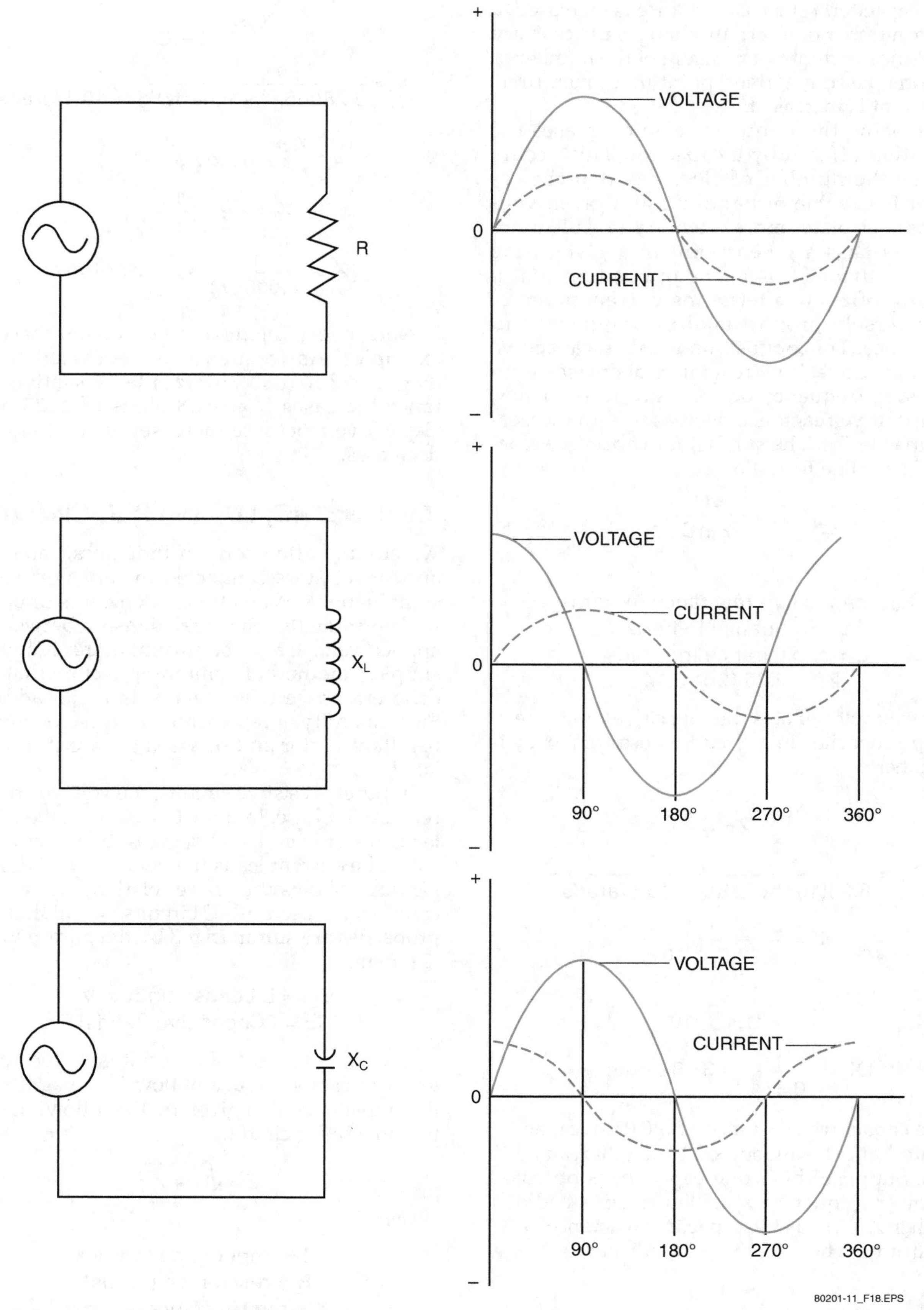

Figure 18 Summary of AC circuit phase relationships.

Think About It

Frequency and Capacitive Reactance

A variable capacitor is used in the tuner of an AM radio to tune the radio to the desired station. Will its capacitive reactance value be higher or lower when it is tuned to the low end of the frequency band (550kHz) than it would be when tuned to the high end of the band (1,440kHz)?

The current through a resistance is always in phase with the voltage applied to it; thus resistance is shown along the 0-degree axis. The voltage across an inductor leads the current by 90 degrees; thus inductive reactance is shown along the 90-degree axis. The voltage across a capacitor lags the current by 90 degrees; thus capacitive reactance is shown along the 90-degree axis. The net reactance is the algebraic difference between the inductive reactance and the capacitive reactance:

$$X = \text{net reactance (ohms)}$$
$$X_L = \text{inductive reactance (ohms)}$$
$$X_C = \text{capacitive reactance (ohms)}$$

The impedance Z is the vector sum of the resistance R and the net reactance X. The angle, called the phase angle, gives the phase relationship between the applied voltage and current.

8.1.0 Resonance

Every circuit containing inductors and capacitors has a frequency at which resonance occurs. This is the frequency at which the X_L and X_C are equal. In a series circuit, the impedances oppose each other and therefore cancel each other out at the resonant frequency. Because of this cancelling effect, the voltage output of the circuit is maximized. A parallel resonant circuit works the opposite: impedance is maximum at the resonant frequency, and therefore the voltage is minimized. Practical applications of this principle can be seen in filters. In a series LC filter, signals at the resonant frequency or a band of frequencies pass readily, while others are rejected. In a parallel LC filter, signals at the resonant frequency are rejected while others are passed.

9.0.0 Power In AC Circuits

In DC circuits, the power consumed is the sum of all the I^2R heating in the resistors. It is also equal to the power produced by the source, which is the product of the source voltage and current. In AC circuits containing only resistors, the same relationship also holds true.

9.1.0 True Power

The power consumed by resistance is called true power and is measured in watts. True power is the product of the resistor current squared and the resistance:

$$P_T = I^2R$$

This formula applies because current and voltage have the same phase across a resistance.

To find the corresponding value of power as a product of voltage and current, this product must be multiplied by the cosine of the phase angle θ:

$$P_T = I^2R \text{ or } P_T = EI\cos\theta$$

Where E and I are in rms values to calculate the true power in watts, multiplying I by the cosine of the phase angle provides the resistive component for true power equal to I^2R.

For example, a series RL circuit has 2A through a 100V resistor in series with the X_L of 173V. Therefore:

$$P_T = I^2R$$
$$P_T = 4 \times 100$$
$$P_T = 400W$$

Furthermore, in this circuit the phase angle is 60 degrees with a cosine of 0.5. The applied voltage is 400V. Therefore:

$$P_T = EI \times \cos\theta$$
$$P_T = 400 \times 2 \times 0.5$$
$$P_T = 400W$$

In both cases, the true power is the same (400W) because this is the amount of power supplied by the generator and dissipated in the resistance. Either formula can be used for calculating the true power.

9.2.0 Apparent Power

In ideal AC circuits containing resistors, capacitors, and inductors, the only mechanism for power consumption is $I^2_{eff}R$ heating in the resistors. Inductors and capacitors consume no power. The only function of inductors and capacitors is to store and release energy. However, because of the phase shifts that are introduced by these elements, the power consumed by the resistors is not equal to the product of the source voltage and current. The product of the source voltage and current is called apparent power and has units of volt-amperes (VA).

The apparent power is the product of the source voltage and the total current. Therefore, apparent power is actual power delivered by the source. The formula for apparent power is:

$$P_A = (E_A)(I)$$

Figure 19 shows a series RL circuit and its associated vector diagram.

This formula is used to calculate the apparent power and compare it to the circuit's true power:

$$P_A = (E_A)(I)$$
$$P_A = (400V)(2A)$$
$$P_A = 800VA$$
$$P_T = EI \times \cos\theta$$
$$\theta = \frac{R}{X_L} = \frac{173}{100} = 60°$$

$$P_T = (400V)(2A)(\cos 60°)$$
$$P_T = (400V)(2A)(0.5)$$
$$P_T = 400W$$

Note that the apparent power formula is the product of EI alone without considering the cosine of the phase angle.

9.3.0 Reactive Power

Reactive power is that portion of the apparent power that is caused by inductors and capacitors in the circuit. Inductance and capacitance are always present in real AC circuits. No work is performed by reactive power; the power is stored in the inductors and capacitors, then returned to the circuit. Therefore, reactive power is always 90 degrees out of phase with true power. The units for reactive power are volt-amperes-reactive (VARs).

In general, for any phase angle θ between E and I, multiplying EI by sine θ gives the vertical component at 90 degrees for the value of the VARs. In *Figure 19*, the value of sine 60 degrees is 800 × 0.866 = 692.8 VARs.

Note that the factor sine θ for the volt-amperes-reactive (VARs) gives the vertical or reactive component of the apparent power EI. However, multiplying EI by cosine θ as the power factor gives the horizontal or resistive component for the real power.

9.4.0 Power Factor

Because it indicates the resistive component, cosine θ is the power factor (pf) of the circuit, converting the EI product to real power. For series circuits, use the following formula:

$$pf = \cos\theta = \frac{R}{Z}$$

For parallel circuits, use this formula:

$$pf = \cos\theta = \frac{I_R}{I_T}$$

In *Figure 19* as an example of a series circuit, R and Z are used for the calculations:

$$pf = \cos\theta = \frac{R}{Z}$$

$$pf = \frac{100\Omega}{200\Omega} = 0.5$$

The power factor is not an angular measure but a numerical ratio with a value between 0 and 1, equal to the cosine of the phase angle. With all re-

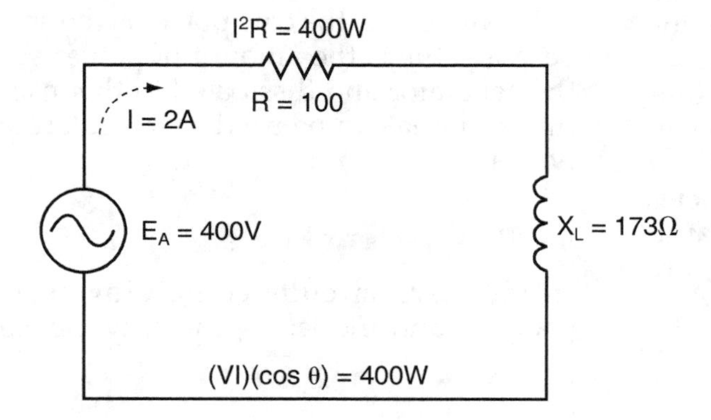

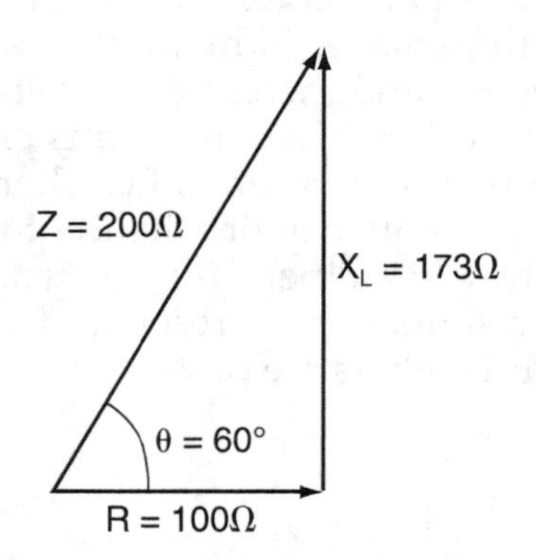

Figure 19 Power calculations in an AC circuit.

sistance and zero reactance, R and Z are the same for a series circuit of I_R and I_T and are the same for a parallel circuit. The ratio is 1. Therefore, unity power factor means a resistive circuit. At the opposite extreme, all reactance with zero resistance makes the power factor zero, meaning that the circuit is all reactive.

The power factor gives the relationship between apparent power and true power. The power factor can thus be defined as the ratio of true power to apparent power:

$$pf = \frac{P_T}{P_A}$$

For example, calculate the power factor of the circuit shown in *Figure 20*.

The true power is the product of the resistor current squared and the resistance:

$$P_T = I^2R$$
$$P_T = 10A^2 \times 10\Omega$$
$$P_T = 1{,}000W$$

The apparent power is the product of the source voltage and total current:

$$P_A = (I_T)(E)$$
$$P_A = 10.2A \times 100V$$
$$P_A = 1{,}020VA$$

Calculating total current:

$$I_T = \sqrt{I_R^2 + (I_C - I_L)^2} = \sqrt{10A^2 + (4A - 2A)^2}$$
$$I_T = 10.2A$$

The power factor is the ratio of true power to apparent power:

$$pf = \frac{P_T}{P_A}$$

$$pf = \frac{1{,}000}{1{,}020}$$

$$pf = 0.98$$

As illustrated in the previous example, the power factor is determined by the system load. If the load contained only resistance, the apparent power would equal the true power and the power factor would be at its maximum value of one. Purely resistive circuits have a power factor of unity or one. If the load is more inductive than capacitive, the apparent power will lag the true power and the power factor lags. If the load is more capacitive than inductive, the apparent power leads the true power and the power factor leads. If there is any reactive load on the system, the apparent power is greater than the true power and the power factor is less than one.

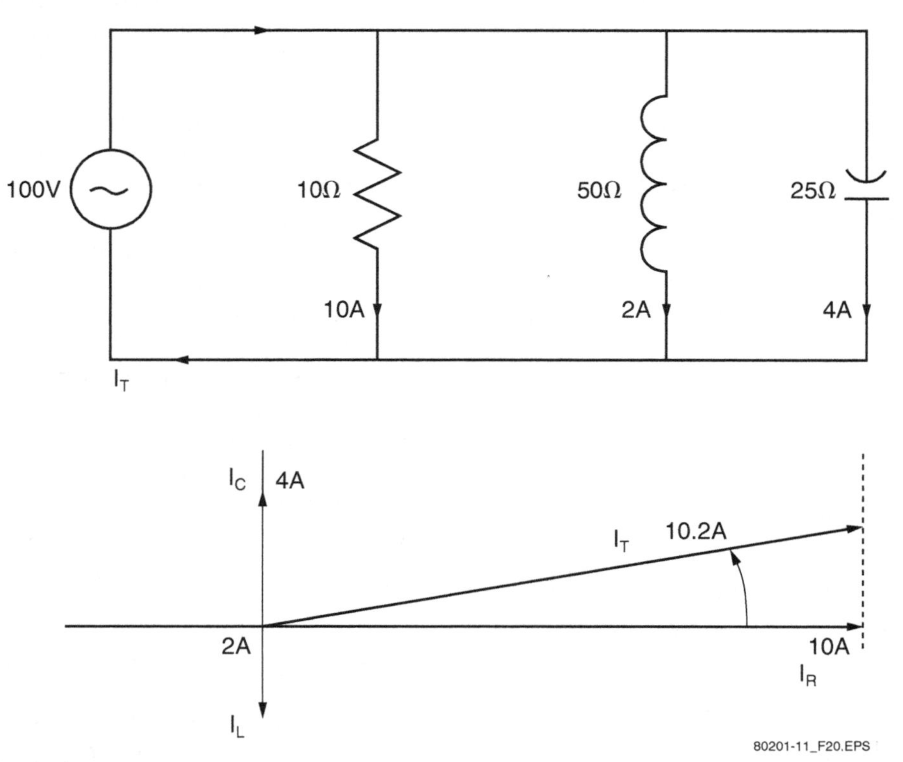

Figure 20 RLC circuit calculation.

On Site

Managing Power Factor

In hot weather, everyone turns on air conditioners. Air conditioners have compressors and fans that are driven by motors. These motors are inductive loads. The addition of so many inductive loads in the system can result in a poor system power factor and excessive power losses.

Banks of capacitors are automatically switched into the system so that their capacitive reactance cancels out as near as possible the inductive reactance presented by the motors. Remember that inductors shift the current and voltage away from each other in one direction (ELI), and capacitors shift them away from each other in the opposite direction (ICE). This maintains the system power factor at an acceptable value and keeps system losses to a minimum.

9.5.0 Power Triangle

The phase relationships among the three types of AC power are easily visualized on the power triangle shown in *Figure 21*. The true power (W) is the horizontal leg, the apparent power (VA) is the hypotenuse, and the cosine of the phase angle between them is the power factor. The vertical leg of the triangle is the reactive power and has units of volt-amperes-reactive (VAR).z

As illustrated on the power triangle (*Figure 21*), the apparent power is always greater than the true power or reactive power. Also, the apparent power is the result of the vector addition of true and reactive power. The power magnitude relationships shown in *Figure 21* can be derived from the Pythagorean theorem for right triangles:

$$c^2 = a^2 + b^2$$

Therefore, c also equals the square root of $a^2 + b^2$, as shown:

$$c = \sqrt{a^2 + b^2}$$

10.0.0 Transformers

A transformer is a device that transfers electrical energy from one circuit to another by electromagnetic induction (transformer action). The electrical energy is always transferred without a change in frequency, but the transfer may involve changes in the magnitudes of voltage and current. Because a transformer works on the principle of electromagnetic induction, it must be used with an input source voltage that varies in amplitude.

10.1.0 Transformer Construction

Figure 22 shows the basic components of a transformer. In its most basic form, a transformer consists of the following:

- Primary coil or winding
- Secondary coil or winding
- Core that supports the coils or windings

A simple transformer action is shown in *Figure 23*. The primary winding is connected to a 60Hz AC voltage source. The magnetic field or flux builds up (expands) and collapses (contracts) around the primary winding. The expanding and contracting magnetic field around the primary winding cuts the secondary winding and induces an alternating voltage into the winding. This voltage causes AC to flow through the load. The voltage may be stepped up or down depending on the design of the primary and secondary windings.

10.1.1 Core Properties

Commonly used core materials are air, soft iron, and steel. Each of these materials is suitable for particular applications and unsuitable for others. Generally, air-core transformers are used when the voltage source has a high frequency (above 20kHz). Iron-core transformers are generally used when the source frequency is low (below 20kHz). A soft-iron transformer is very useful where the transformer must be physically small yet efficient. The iron-core transformer provides better power transfer than the air-core transformer. Laminated sheets of steel are often used in a transformer to reduce one type of power loss known as eddy currents. These are undesirable currents, induced into the core, which circulate around the core. Laminating the core reduces these currents to smaller levels. These steel laminations are insulated with a nonconducting material, such as varnish, and then formed into a core as shown in *Figure 24*. It takes about 50 such laminations to make a core one-inch thick. The most efficient transformer core is one that offers the best path for the most lines of flux, with the least loss in magnetic and electrical energy.

10.1.2 Transformer Windings

A transformer consists of two coils called windings, which are wrapped around a core. The transformer operates when a source of AC voltage

GOING GREEN

Power Factor

In power distribution circuits, it is desirable to achieve a power factor approaching a value of 1 in order to obtain the most efficient transfer of power. In AC circuits where there are large inductive loads such as in motors and transformers, the power factor can be considerably less than 1. For example, in a highly inductive motor circuit, if the voltage is 120V, the current is 12A, and the current lags the voltage by 60 degrees, the power factor is 0.5 or 50 percent (cosine of 60 degrees = 0.5). The apparent power is 1,440VA (120V × 12A), but the true power is only 720W [120V × (0.5 × 12A) = 720W]. This is a very inefficient circuit. What would you do to this circuit in order to achieve a circuit having a power factor as close to 1 as possible?

$$P_A = \sqrt{(P_T^2) + (P_{RX}^2)}$$

$$P_T = \sqrt{(P_A^2) - (P_{RX}^2)}$$

$$P_{RX} = \sqrt{(P_A^2) + (P_T^2)}$$

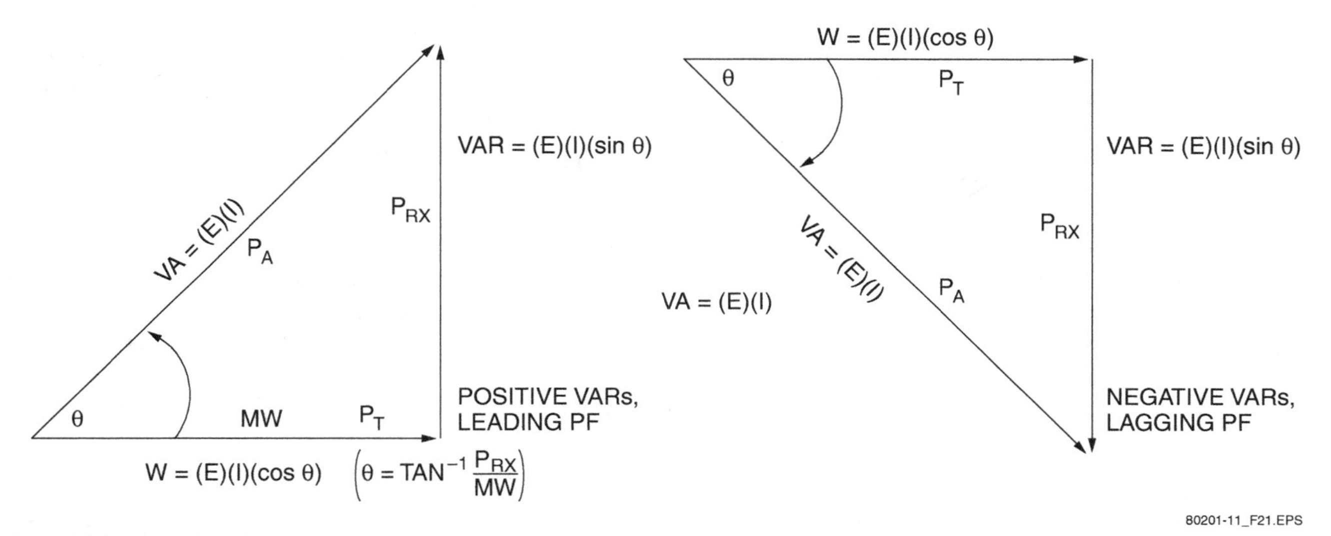

Figure 21 Power triangle.

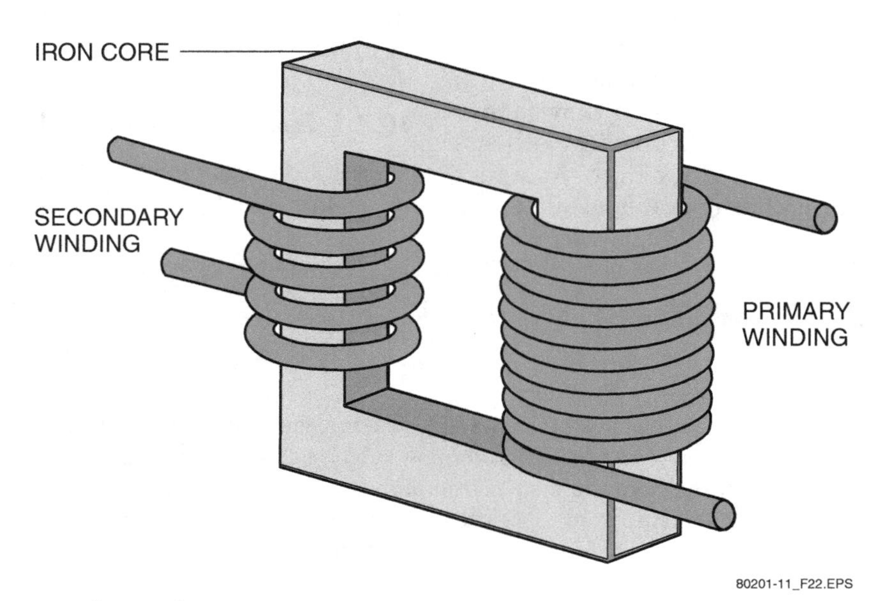

Figure 22 Basic components of a transformer.

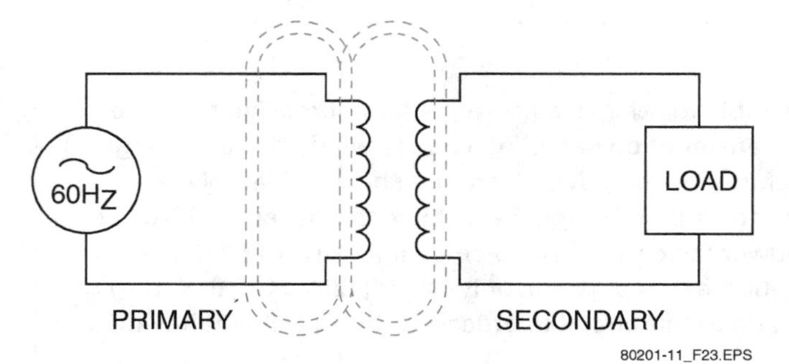

Figure 23 Transformer action.

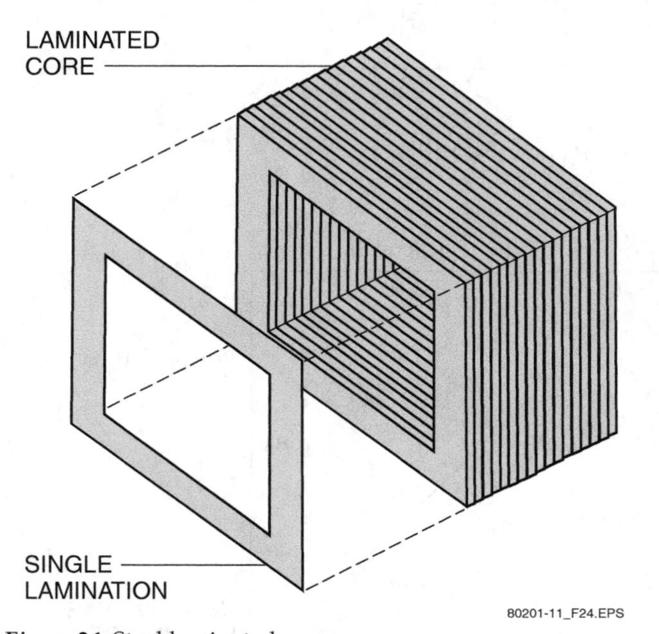

Figure 24 Steel laminated core.

is connected to one of the windings and a load device is connected to the other. The winding that is connected to the source is called the primary winding. The winding that is connected to the load is called the secondary winding. *Figure 25* shows a cutaway view of a typical transformer.

The wire is coated with varnish so that each turn of the winding is insulated from every other turn. In a transformer designed for high-voltage applications, sheets of insulating material such as

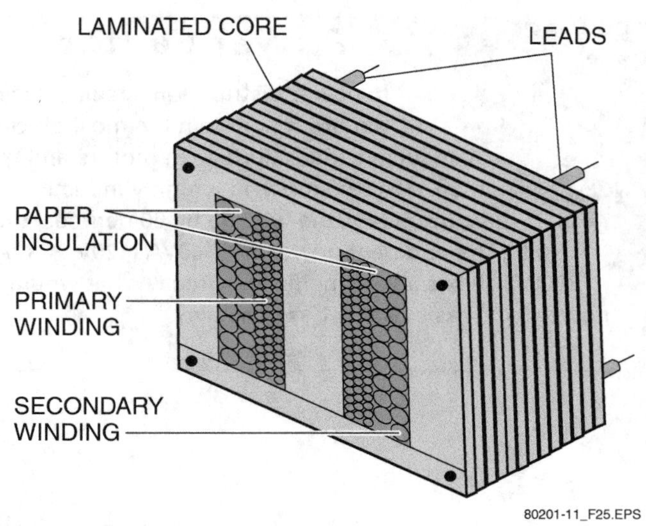

Figure 25 Cutaway view of a transformer core.

paper are placed between the layers of windings to provide additional insulation.

When the primary winding is completely wound, it is wrapped in insulating paper or cloth. The secondary winding is then wound on top of the primary winding. After the secondary winding is complete, it too is covered with insulating paper. Next, the core is inserted into and around the windings as shown.

Sometimes terminals may be provided on the enclosure for connections to the windings. *Figure 25* shows four leads, two from the primary and two from the secondary. These leads must be connected to the source and load, respectively.

10.2.0 Operating Properties

Regardless of type, most transformers operate in the same way. This section covers the no-load operation and phase relationships in a simple transformer.

10.2.1 Energized with No Load

A no-load condition is said to exist when a voltage is applied to the primary, but no load is con-

On Site

Transformers

Transformers are essential to all electrical systems and all types of electronic equipment. They are especially crucial to the operation of AC high-voltage power distribution systems. Transformers are used to both step up voltage and step down voltage throughout the distribution process. For example, a typical power generation plant might generate AC power at 13,800V, step it up to 230,000V for distribution over long transmission lines, step it down to 13,800V again at substations located at different points for local distribution, and finally step it down again to 240V and 120V for lighting and local power use.

nected to the secondary. Assume the output of the secondary is connected to a load by an open switch. Because of the open switch, there is no current flowing in the secondary winding. With the switch open and an AC voltage applied to the primary, there is, however, a very small amount of current, called exciting current, flowing in the primary. This current excites the coil of the primary to create a magnetic field. The amount of exciting current is determined by three factors: the amount of voltage applied (E_A); the resistance (R) of the primary coil's wire and core losses; and the X_L, which depends on the frequency of the exciting current. These factors are all controlled by transformer design.

This very small amount of exciting current serves two functions:

- Most of the exciting energy is used to support the magnetic field of the primary.
- A small amount of energy is used to overcome the resistance of the wire and core. This is dissipated in the form of heat (power loss).

Exciting current flows in the primary winding at all times to maintain this magnetic field, but no transfer of energy takes place as long as the secondary circuit is open.

10.2.2 Phase Relationship

The secondary voltage of a simple transformer may be either in phase or out of phase with the primary voltage. This depends on the direction in which the windings are wound and the arrangement of the connection to the external circuit (load). Simply, this means that the two voltages may rise and fall together, or one may rise while the other is falling. Transformers in which the secondary voltage is in phase with the primary are referred to as like-wound transformers, while those in which the voltages are 180 degrees out of phase are called unlike-wound transformers.

Dots are used to indicate points on a transformer schematic symbol that have the same polarity (points that are in phase). The use of phase-indicating dots is illustrated in *Figure 26.* In the first part of the figure, both the primary and secondary windings are wound from top to bottom in a clockwise direction, as viewed from above the windings. When built in this manner, the top lead of the primary and the top lead of the secondary have the same polarity. This is indicated by the dots on the transformer symbol.

The second part of the figure shows a transformer in which the primary and secondary are wound in opposite directions. As viewed from above the windings, the primary is wound in a clockwise direction from top to bottom, while the secondary is wound in a counterclockwise direction. Notice that the top leads of the primary and secondary have opposite polarities. This is indicated by the dots being placed on opposite ends of the transformer symbol. Thus, the polarity of voltage at the terminals of the transformer secondary depends on the direction in which the secondary is wound with respect to the primary.

10.3.0 Turns and Voltage Ratios

To understand how a transformer can be used to step up or step down voltage, the concept of turns ratio must be understood. The total voltage induced into the secondary winding of a transformer is determined mainly by the ratio of the number of turns in the primary to the number of turns in the secondary, and by the amount of voltage applied to the primary. Therefore, to set up a formula:

$$\text{Turns ratio} = \frac{\text{number of turns in the primary}}{\text{number of turns in the secondary}}$$

The first transformer in *Figure 27* shows a transformer whose primary consists of ten turns of wire, and whose secondary consists of a single turn of wire. As lines of flux generated by the primary expand and collapse, they cut both the ten turns of the primary and the single turn of the secondary. Since the length of the wire in the secondary is roughly the same as the length of the wire in each turn of the primary, the EMF induced into the secondary will be the same as the EMF induced into each turn of the primary.

This means that if the voltage applied to the primary winding is 10 volts, the EMF in the primary is almost 10 volts. Thus, each turn in the primary has an induced EMF of one-tenth of the total applied voltage, or one volt. Since the same flux lines cut the turns in both the secondary and the primary, each turn has an EMF of one volt induced into it. The first transformer in *Figure 27* has only one turn in the secondary, thus, the EMF across the secondary is one volt.

The second transformer in *Figure 27* has a ten-turn primary and a two-turn secondary. Since the flux induces one volt per turn, the total voltage across the secondary is two volts. Notice that the volts per turn are the same for both primary and secondary windings. Since the EMF in the primary is equal (or almost) to the applied voltage, a proportion may be set up to express the value of the voltage induced in terms of the voltage applied to the primary and the number of turns in each winding. This proportion also shows the relationship between the number of turns in each

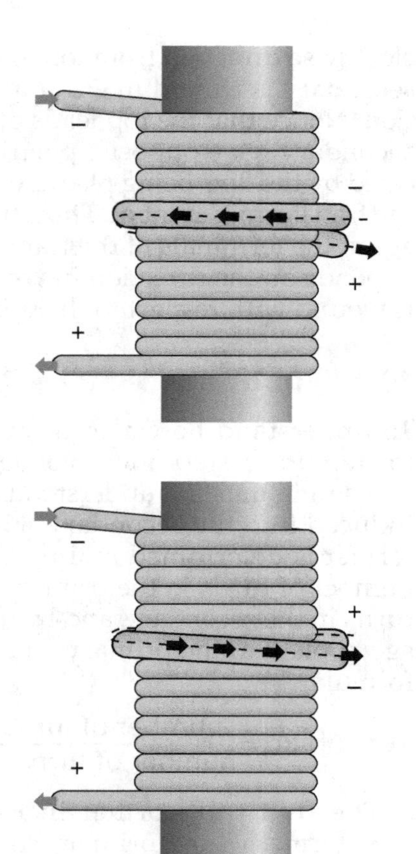
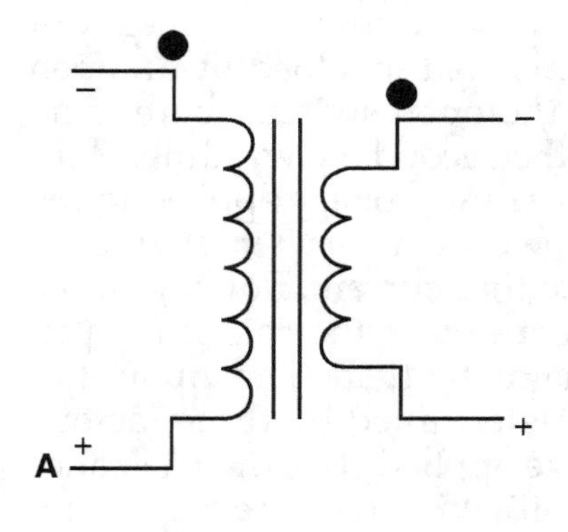

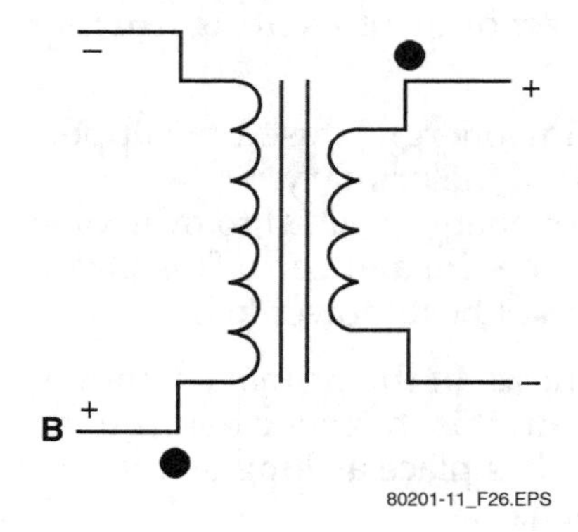

Figure 26 Transformer winding polarity.

winding and the voltage across each winding, and is expressed by the equation:

$$\frac{E_S}{E_P} = \frac{N_S}{N_P}$$

Where:

N_P = number of turns in the primary
E_P = voltage applied to the primary
E_S = voltage induced in the secondary
N_S = number of turns in the secondary

The equation shows that the ratio of secondary voltage to primary voltage is equal to the ratio of secondary turns to primary turns. The equation can be written as:

$$E_P N_S = E_S N_P$$

For example, a transformer has 100 turns in the primary, 50 turns in the secondary, and 120VAC applied to the primary (E_P). What is the voltage across the secondary (E_S)?

N_P = 100 turns
N_S = 50 turns
E_P = 120VAC

$$\frac{E_S}{E_P} = \frac{N_S}{N_P} \quad or \quad E_S = \frac{E_P N_S}{N_P}$$

$$E_S = \frac{120V \times 50 \text{ turns}}{100 \text{ turns}} = 60VAC$$

The transformers in *Figure 27* have fewer turns in the secondary than in the primary. As a result, there is less voltage across the secondary than across the primary. A transformer in which the voltage across the secondary is less than the voltage across the primary is called a step-down transformer. The ratio of a 10-to-1 step-down transformer is written as 10:1.

A transformer that has fewer turns in the primary than in the secondary produces a greater voltage across the secondary than the voltage applied to the primary. A transformer in which the voltage across the secondary is greater than the voltage applied to the primary is called a step-up transformer. The ratio of a 1-to-4 step-up transformer should be written 1:4. Notice in the two ratios that the value of the primary winding is always stated first.

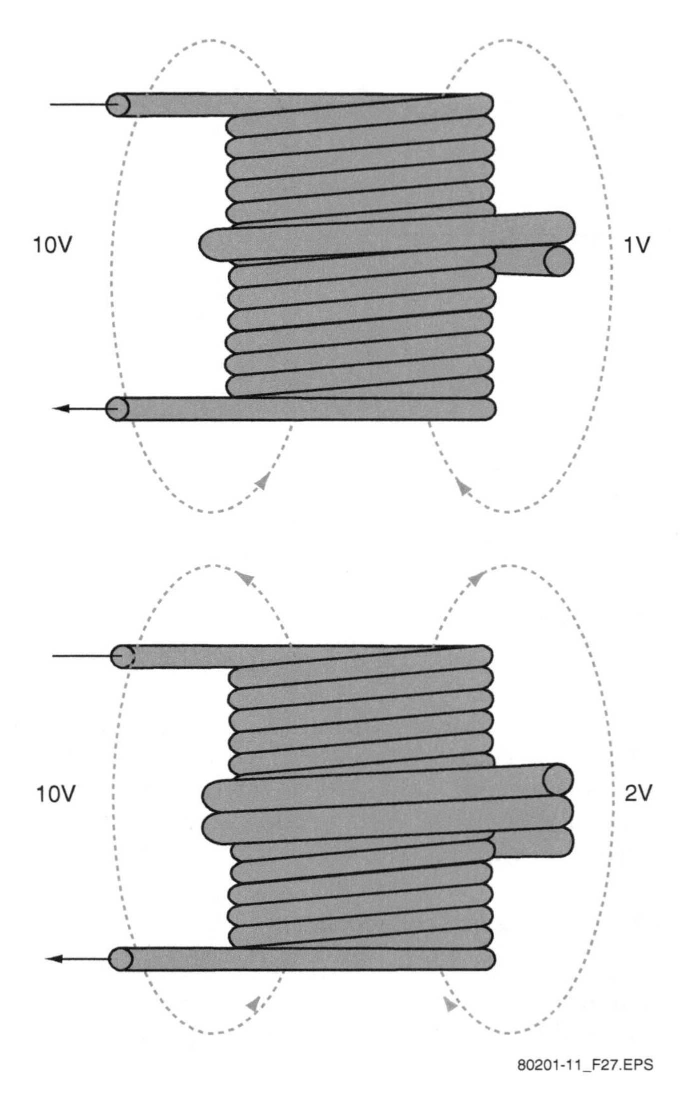

Figure 27 Transformer turns ratio.

10.4.0 Types of Transformers

Transformers are widely used to permit the use of trip coils and instruments of moderate current and voltage capacities and to measure the properties of high-voltage and high-current circuits. Since secondary voltage and current are directly related to primary voltage and current, measurements can be made under the low-voltage or low-current conditions of the secondary circuit and still determine primary properties. Tripping transformers and instrument transformers are examples of this use of transformers.

The primary or secondary coils of a transformer can be tapped to permit multiple input and output voltages. *Figure 28* shows examples of tapped transformers. The center-tapped transformer is particularly important because it can be used in conjunction with other components to convert an AC input to a DC output. There are many types of transformers, but the basic principles are the same for all transformers.

10.4.1 Isolation Transformer

Isolation transformers are wound so that their primary and secondary voltages are equal. Their purpose is to electrically isolate a piece of electrical equipment from the power distribution system.

Many pieces of electronic equipment use the metal chassis on which the components are mounted as part of the circuit (*Figure 29*). Personnel working with this equipment may accidentally come in contact with the chassis, completing the circuit to ground, and receive a shock, as shown

Think About It

Turns and Voltage Ratios

What is the magnitude of the voltage and current supplied by the secondary of the transformer in the circuit shown here?

12VDC
10:1
10Ω 60MH 80μF 30Ω
100μF 60Ω 40MH
80201-11_SA02.EPS

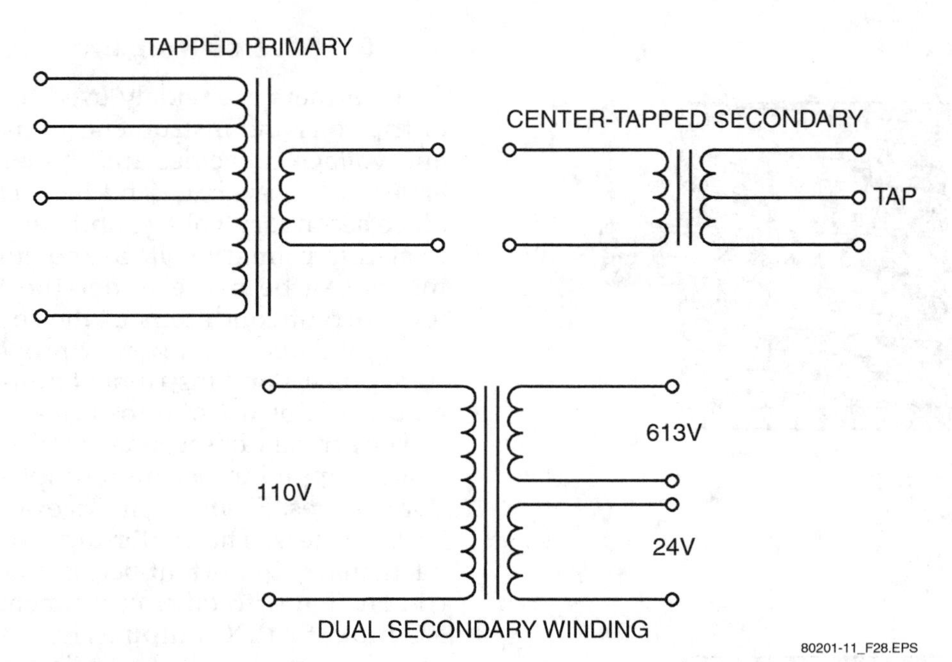

Figure 28 Tapped transformers.

in *Figure 29A*. If the resistances of a worker's body and the ground path are low, the shock can be fatal. Placing an isolation transformer in the circuit, as shown in *Figure 29B*, breaks the ground current path that includes the worker. Current can no longer flow from the power supply through the chassis and worker to ground; however, the equipment is still supplied with the normal operating voltage and current.

10.4.2 Autotransformer

In a transformer, the primary and secondary do not have to be separate and distinct windings. *Figure 30* is a schematic diagram of what is known as an autotransformer. Note that a single coil of wire is tapped to produce what is electrically both a primary and a secondary winding.

The voltage across the secondary winding has the same relationship to the voltage across the primary that it would have if they were two distinct windings. The movable tap in the secondary is used to select a value of output voltage either higher or lower than E_P, within the range of the transformer. When the tap is at Point A, E_S is less than E_P; when the tap is at Point B, E_S is greater than E_P.

Autotransformers rely on self-induction to induce their secondary voltage. The term *autotransformer* can be broken down into two words: *auto*, meaning self; and *transformer*, meaning to change potential. The autotransformer is made of one winding that acts as both a primary and a secondary winding. It may be used as either a step-up or step-down transformer. Some common uses of autotransformers are as variable AC voltage supplies and fluorescent light ballast transformers, and to reduce the line voltage for various types of low-voltage motor starters.

10.4.3 Current Transformer

A current transformer differs from other transformers in that the primary is a conductor to the load and the secondary is a coil wrapped around the wire to the load. Just as any ammeter is connected in line with a circuit, the current transformer is connected in series with the current to be measured. *Figure 31* is a diagram of a current transformer.

WARNING!

Do not open a current transformer under load. Although the use of a current transformer completely isolates the secondary and the related ammeter from the high-voltage lines, the secondary of a current transformer should never be left open circuited. To do so may result in dangerously high voltage being induced in the secondary.

Since current transformers are series transformers, the usual voltage and current relationships do not apply. Current transformers vary in rated primary current, but are usually designed with ampere-turn ratios such that the secondary delivers five amperes at full primary load.

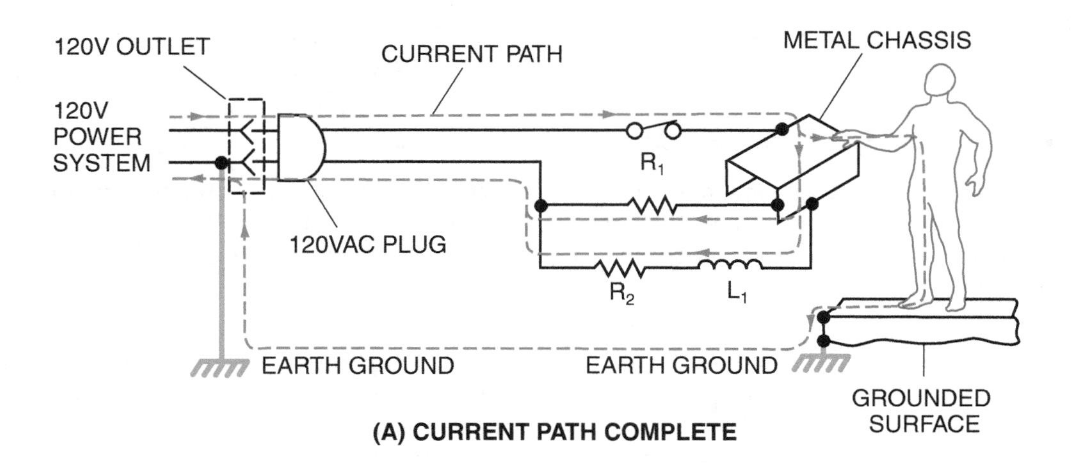

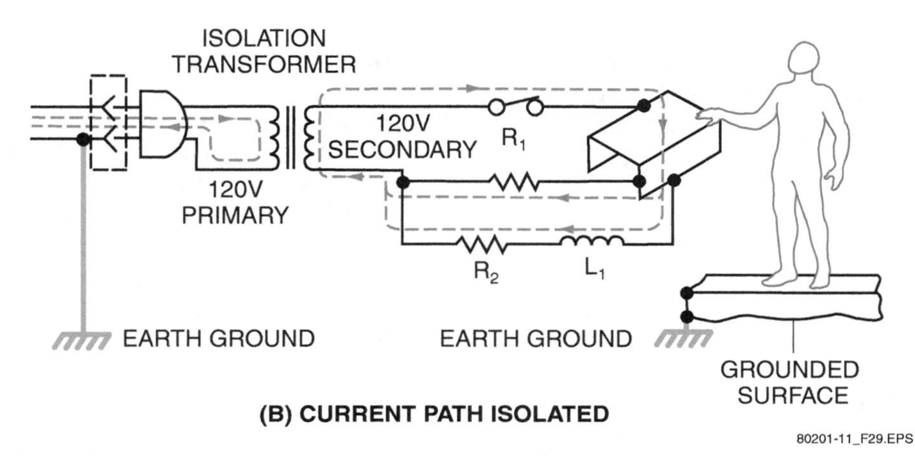

Figure 29 Importance of an isolation transformer.

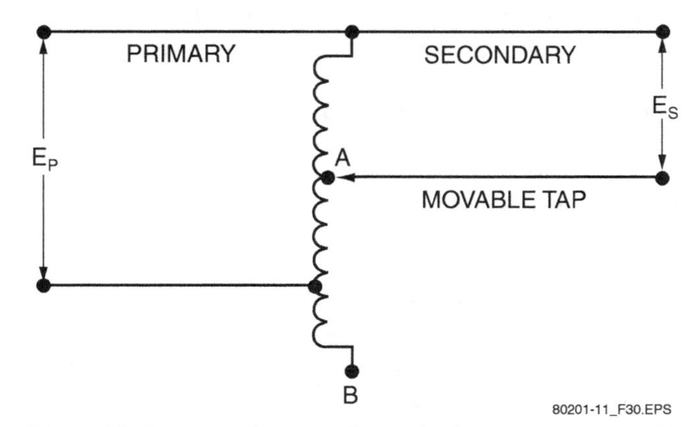

Figure 30 Autotransformer schematic diagram.

Current transformers are generally built with only a few turns or no turns in the primary. The voltage in the secondary is induced by the changing magnetic field that exists around a single conductor. The secondary is wound on a circular core, and the large conductor that makes up the primary passes through the hole in its center. Because the primary has few or no turns, the secondary must have many turns (providing a high turns ratio) in order to produce a usable

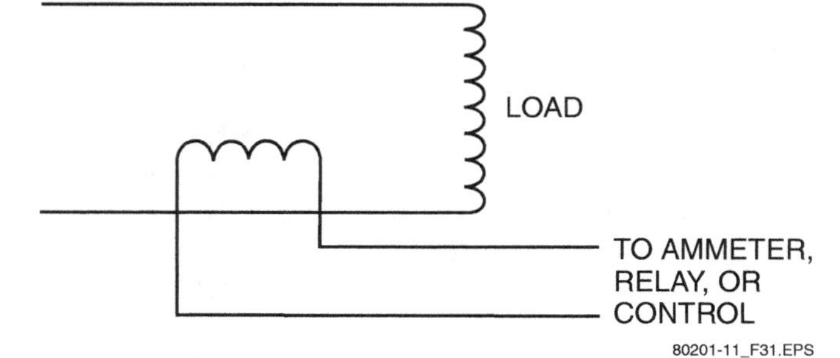

Figure 31 Current transformer schematic diagram.

voltage. The advantage is that the output off the secondary is proportional to the current flowing through the primary, without much voltage drop across the primary. This is because the primary voltage equals the current times the impedance. The impedance is kept near zero by using no or very few primary turns. The disadvantage is that the secondary circuit cannot be opened with the primary energized. To do so would cause the secondary current to drop rapidly to zero. This would cause the magnetic field generated by the secondary current to collapse rapidly. The rapid collapse of the secondary field through the many

On Site

Isolation Transformers

In addition to being used to protect personnel from electrical shocks, shielded isolation transformers are widely used to prevent electrical disturbances on power lines from being transmitted into related load circuits. The shielded isolation transformer has a grounded shield between the primary and secondary windings that acts to direct unwanted signals to ground.

turns of the secondary winding would induce a high voltage in the secondary, creating an equipment and personnel hazard.

Because the output of current transformers is proportional to the current in the primary, they are most often used to power current-sensing meters and relays. This allows the instruments to respond to primary current without having to handle extreme magnitudes of current.

10.4.4 Potential Transformer

The primary of a potential transformer is connected across or in parallel with the voltage to be measured, just as a voltmeter is connected across a circuit. *Figure 32* shows the schematic diagram for a potential transformer.

Potential transformers are basically the same as any other single-phase transformer. Although primary voltage ratings vary widely according to the specific application, secondary voltage ratings are usually 120V, a convenient voltage for meters and relays.

The output of potential transformers is proportional to the phase-to-phase voltage of the primary. For that reason, they are often used to power voltage-sensing meters and relays. This allows the instruments to respond to primary voltage while having to handle only 120V. Also, potential transformers are essentially single-phase, step-down transformers. Therefore, power to operate low-voltage auxiliary equipment associated with high-voltage switchgear can be supplied off the high-voltage lines that the equipment serves via potential transformers.

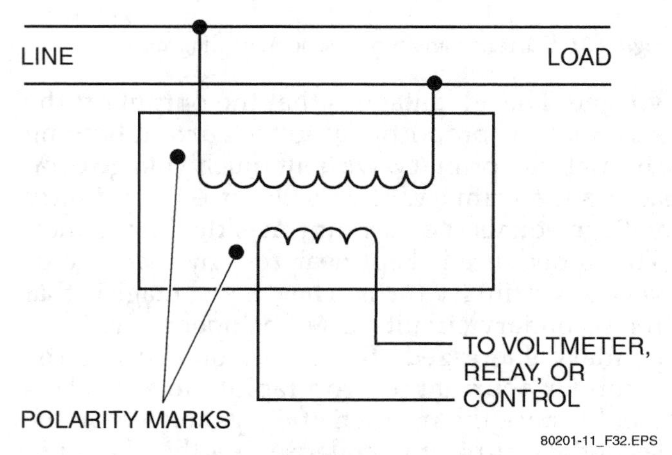

Figure 32 Potential transformer.

10.5.0 Transformer Selection

When replacing a defective transformer that is known to be original equipment, try to replace it with an exact duplicate. If it cannot be determined whether the transformer is original equipment or not, make sure that the replacement transformer safely and efficiently handles the primary voltage and current, and supports the voltage and current demands of the load.

Depending on its use, greater load demand properties may have to be considered when choosing a replacement control transformer. Three important properties are total steady-state VA (sometimes referred to as sealed VA), total inrush VA, and inrush load power factor. These three factors usually apply to transformers used with inductive circuits such as motor control circuitry, and transformers are used with electromagnetic control devices such as relays.

The total steady-state VA requirement is the rating that provides sufficient volt-amperes to the load in order to hold the load in a continuous energized state.

Electromagnetic control devices such as magnetic motor starter or relay coils may take anywhere from 20 to 60 milliseconds to become totally energized. The coil may draw up to 10 times the normal operating current during this time. This is referred to as the inrush VA and the transformer must be rated to handle these inrush current levels.

The true inrush load power factor cannot be determined without applying complex vector analysis to control transformer load components. Therefore, in determining the inrush load power factor, it is safe to apply a power factor of 40 percent. This means that from the transformer's available inrush power, only 40 percent is usable inrush power for the load. In other words,

On Site

Dangers of Working with AC Power

Working with AC power can be dangerous unless proper safety methods and procedures are followed. The National Institute for Occupational Safety and Health (NIOSH) investigated 224 incidents of electrocutions that resulted in occupational fatalities. One hundred twenty-one of the victims were employed in the construction industry, while 221 of the incidents (99 percent) involved AC. Of the 221 AC electrocutions, 74 (33 percent) involved AC voltages less than 600V and 147 (66 percent) involved 600V or more. Of the 74 lower-voltage electrocutions, 40 accidents involved 120/240V.

Factors relating to the causes of these electrocutions included the lack of enforcement of existing employer policies, including the use of personal protective equipment and the lack of supervisory intervention when existing policies were being violated. Of the 224 victims, 194 (80 percent) had some type of electrical safety training, while 39 victims had no training at all. It is notable that 100 of the victims had been on the job less than one year.

Never assume that you are safe when working at lower voltages. All voltage levels must be considered potentially lethal. Also, safety training is useless unless put it into practice every day. Always put safety first.

the transformer's total inrush VA rating must be about 160 percent of the inrush requirements of the loads if the total load consists of electromagnetic devices. Most replacement transformers have their inrush VA rating calculated for a power factor of 40 percent.

If an isolation transformer must be replaced, make sure to select a transformer designed for isolation. Its source and load impedance must match those of the original transformer. The replacement should also be designed to handle the signal levels and frequency range.

NOTE

Although an isolation transformer may have a 1:1 turns ratio, it should not be reversed in the circuit because a change in operating performance could result.

11.0.0 Three-Phase Power

Single-phase power is adequate to supply residences and small commercial businesses. In commercial and industrial uses, the power demand is greater, especially where large electric motors are used. Motors larger than one horsepower are usually three-phase motors.

The discussion of power generation has focused on a single conductor rotating in a magnetic field. If three rotating conductors are placed 120 degrees apart, three equal voltages are generated. As shown in *Figure 33*, they occur 120 degrees apart in time; in other words, they are 120 degrees out-of-phase with one another.

There are several ways in which three-phase power sources can be connected, depending on the voltage(s) and the amount of current required. Three-phase transformers selection is based on the voltage levels required by the customer(s). The transformers are rated for certain voltage ranges, typically 120V or 240V. The actual voltages produced depend on the way the transformers are connected. The four-wire wye-wye arrangement (*Figure 34*) is by far the most common in power distribution systems. It typically produces 120/208 volts or 277/480 volts. The 277 volts is used for fluorescent lighting. The delta-delta (closed delta) arrangement (*Figure 35*) can provide 120/240 volts or 240/480 volts. A delta-wye configuration (*Figure 36*) can produce 120/208V or 277/480 volts.

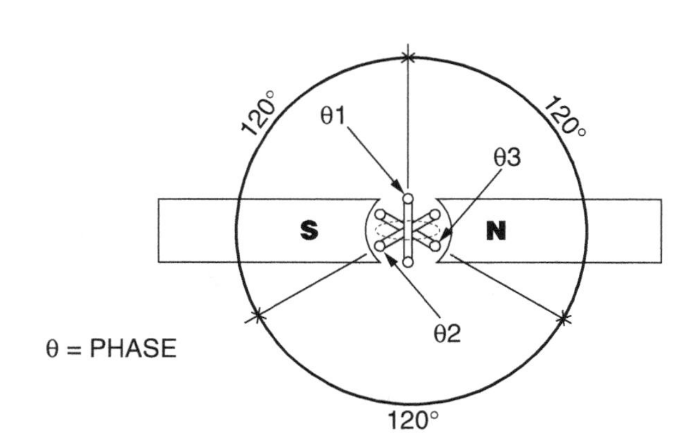

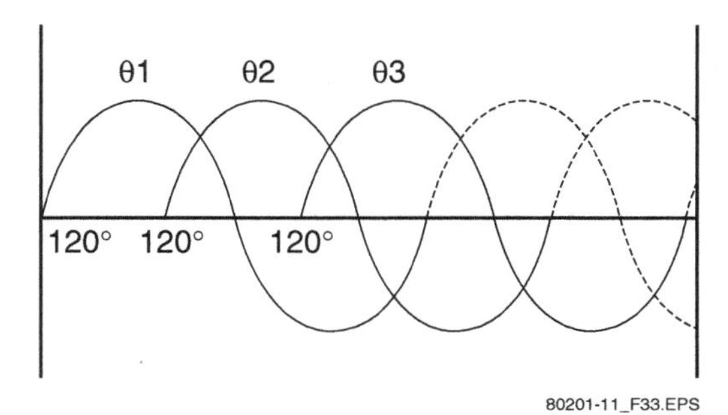

Figure 33 Three-phase voltage development.

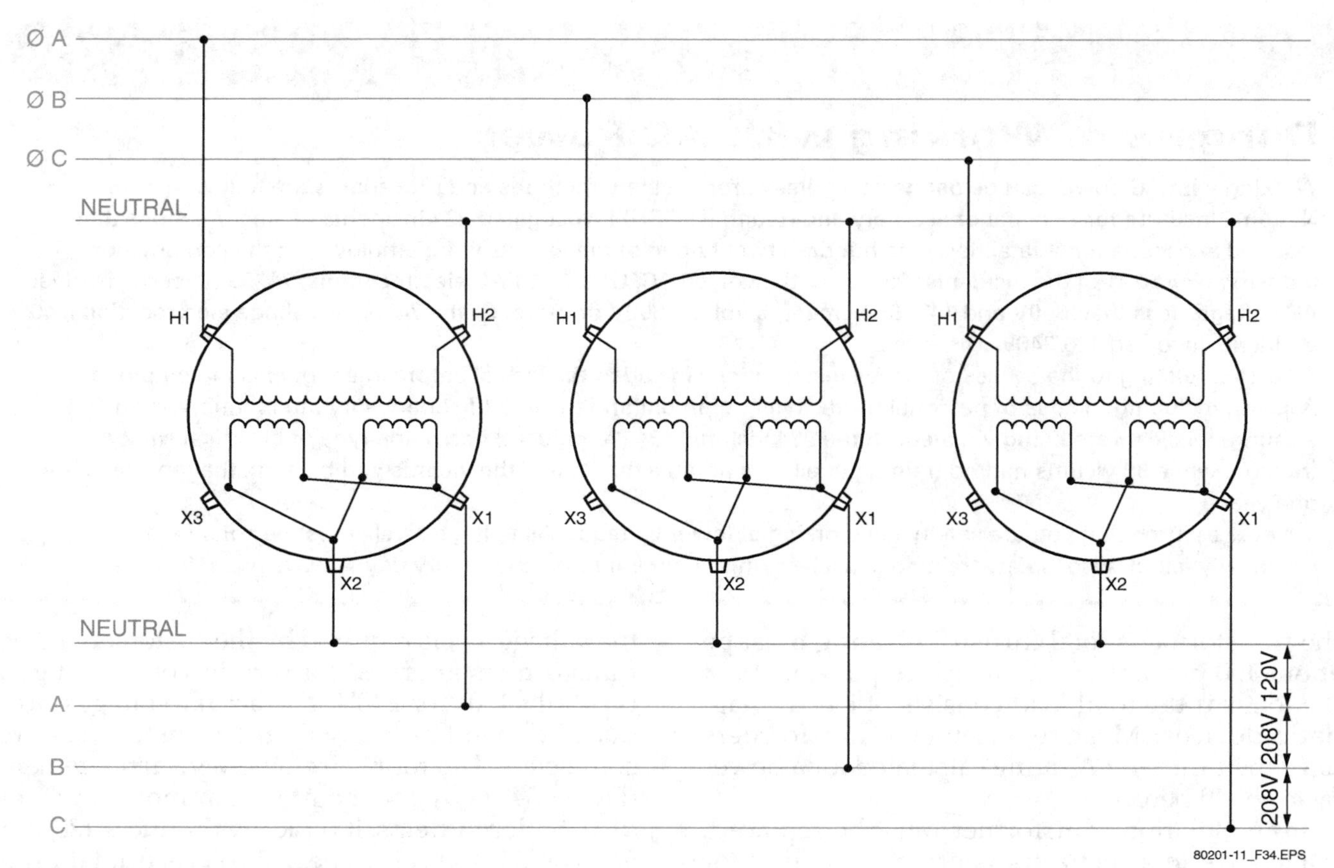

Figure 34 Wye-wye arrangement.

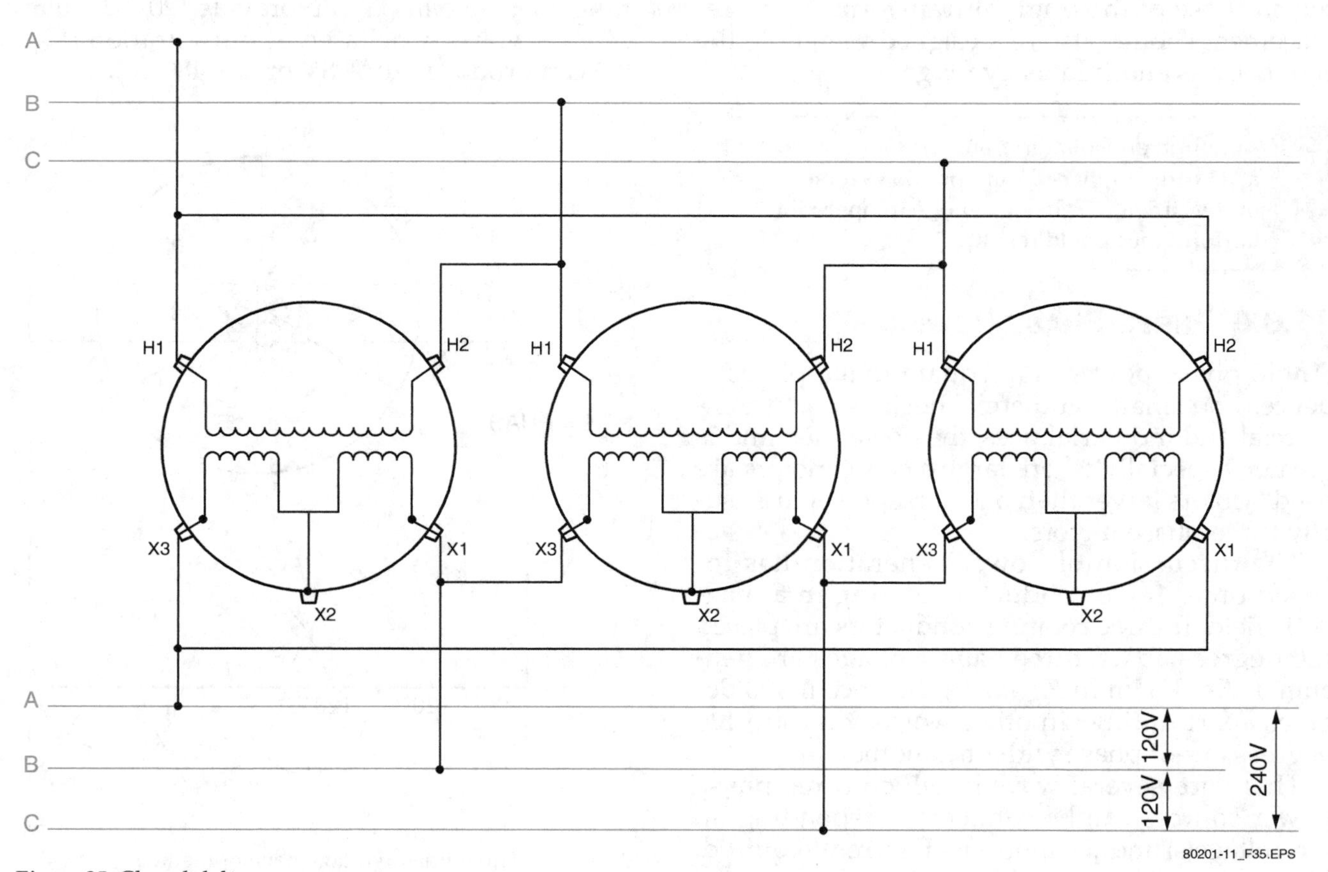

Figure 35 Closed delta arrangement.

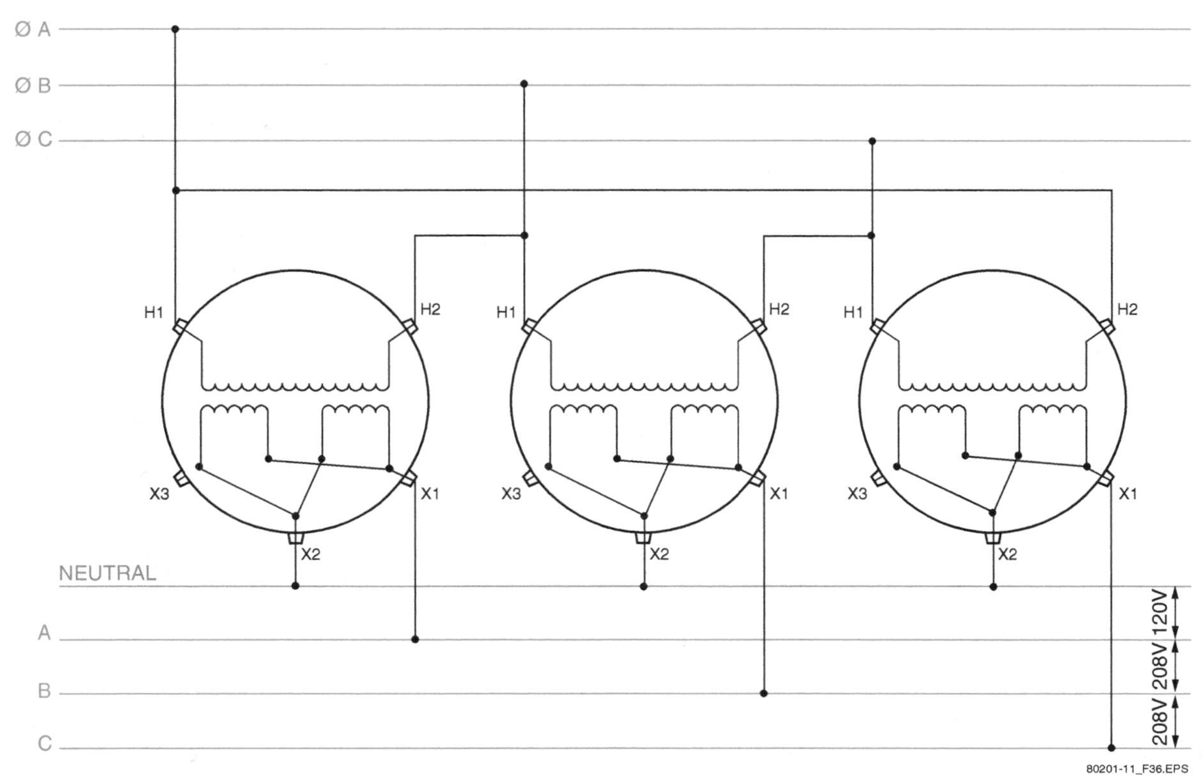

Figure 36 Delta-wye arrangement.

Primary voltages in power distribution systems typically carry three-phase power ranging from 2,400V to 34,000V. *Figure 37* shows an example of the transformer connections for a typical three-phase distribution system.

11.1.0 Voltage and Current Imbalance in Three-Phase Systems

In a three-phase system, a current imbalance in one leg can cause overheating in the other two legs. Three-phase systems therefore need to be checked periodically to make sure they are balanced. If the current is out of balance by more than 10 percent or the voltage is out of balance by more than 2 percent, the imbalance must be corrected. Sometimes the problem is at the source and must be corrected by the power company. This is a common occurrence following a power outage.

A small imbalance in the phase-to-phase voltages can result in a much greater current imbalance. With a current imbalance, the heat generated in motor windings and other inductive loads is increased. Both current and heat can cause nuisance overload trips and may cause motor/equipment failures. For this reason, the voltage imbalance between any two legs of the voltage applied to a three-phase motor or system should not exceed

On Site

Power Distribution

Many commercial and industrial operations need the higher voltages provided by three-phase power systems. Three-phase transformers provide that power, while single-phase transformers provide power to homes and small businesses. In modern residential subdivisions and business developments, power is routed underground and distributed from pad-mount transformers.

80201-11_SA03.EPS

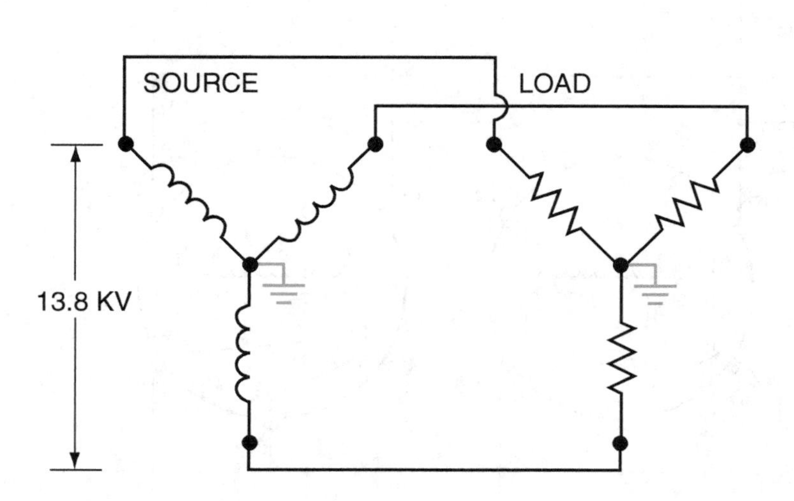

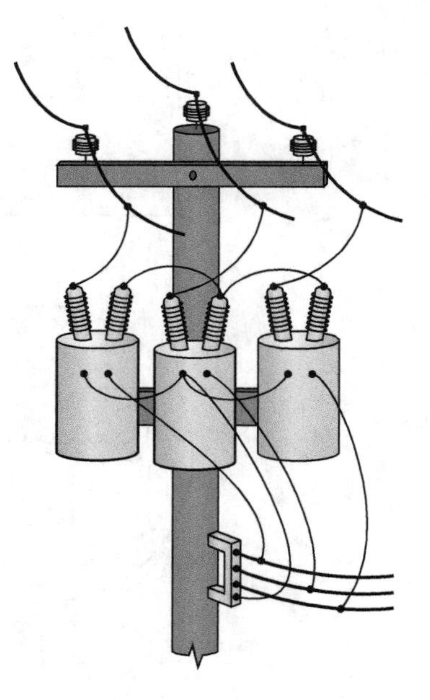

80201-11_F37.EPS

Figure 37 Three-phase transformer connections.

2 percent. If a voltage imbalance of more than 2 percent exists at the input to the equipment, the problem in the building or utility power distribution system should be corrected before operating the equipment.

Current imbalance between any two legs of a three-phase system should not exceed 10 percent. A current imbalance may occur without a voltage imbalance. This can happen when an electrical terminal, contact, etc., becomes loose or corroded, causing a high resistance in the leg. Since current follows the path of least resistance, the current in the other two legs will increase, causing more heat to be generated in the devices supplied by those legs.

SUMMARY

The process by which current is produced electromagnetically is called induction. As the conductor moves across the magnetic field, it cuts the lines of force, and electrons within the conductor flow, creating an electromotive force (EMF).

Inductive and capacitance components offer impedance to current flow that is called reactance, rather than resistance. Inductive and capacitance reactances are out of phase with each other and with any DC resistive components in the circuit. Therefore, the total impedance to current flow must take the phase shift into account. In addition, reactive components do not consume power. Thus, the true power consumed by an AC circuit cannot be determined in the same way it is in a DC circuit.

Transformers are key components in AC circuits. They are used to provide circuit isolation and to increase (step up) or decrease (step down) AC voltage. The ratio of wire turns between the transformer primary and secondary determines the amount the voltage that will be stepped up or down.

Power is the amount of energy consumed by a circuit. In an AC circuit, a decrease in voltage occurs as the conductor intersects the magnetic field at an angle less than 90 degrees. The greatest current is produced when the conductor intersects the magnetic field at right angles to the flux lines.

Review Questions

1. An electric current always produces _____.
   a. mutual inductance
   b. a magnetic field
   c. capacitive reactance
   d. high voltage

2. The number of cycles that an alternating electric current undergoes per second is known as _____.
   a. amperage
   b. frequency
   c. voltage
   d. resistance

3. If the frequency of an AC current is 60 cycles per second, the hertz rating of the circuit is _____.
   a. 15 hertz
   b. 30 hertz
   c. 60 hertz
   d. 90 hertz

4. The period of a waveform is the _____.
   a. inverse of frequency
   b. square root of frequency
   c. sine of frequency
   d. square of frequency

5. The peak voltage in a 120VAC circuit is _____.
   a. 117 volts
   b. 120 volts
   c. 150 volts
   d. 170 volts

6. Which of the following conditions exists in a circuit containing pure resistance?
   a. The voltage and current are in phase.
   b. The voltage and current are 90 degrees out of phase.
   c. The voltage and current are 120 degrees out of phase.
   d. The voltage and current are 180 degrees out of phase.

7. In a purely resistive circuit where 240V is applied across a 10-ohm resistor, the amperage is _____.
   a. 10A
   b. 24A
   c. 60A
   d. 120A

8. Which of the following will cause the inductance of a coil to decrease?
   a. Adding turns of wire
   b. Decreasing coil diameter
   c. Decreasing turn spacing
   d. Using a magnetic core

9. The inductive reactance of a 15-henry inductor in a 60Hz circuit is _____.
   a. 2,132V
   b. 4,712V
   c. 5,652V
   d. 6,937V

10. Reactance increases with an increase in inductance.
    a. True
    b. False

11. Capacitance is measured in _____.
    a. farads
    b. joules
    c. henrys
    d. amps

12. The total capacitance of a series circuit containing a 60μF capacitor and a 40μF capacitor is approximately _____.
    a. 24μF
    b. 32μF
    c. 40μF
    d. 62μF

13. The total capacitance of a 30-microfarad capacitor in parallel with a 20-microfarad capacitor is _____.
    a. 20 microfarads
    b. 30 microfarads
    c. 50 microfarads
    d. 600 microfarads

14. The opposition to current flow offered by the capacitance of a circuit is known as _____.
    a. mutual inductance
    b. pure resistance
    c. inductive reactance
    d. capacitive reactance

15. Current leads the voltage by 90 degrees in a(n) _____ .
    a. capacitive circuit
    b. resistive circuit
    c. inductive circuit
    d. RC circuit

16. Which of the following conditions exist in a circuit of pure inductance?
    a. The voltage and current are in phase.
    b. The voltage and current are 90 degrees out of phase.
    c. The voltage and current are 120 degrees out of phase.
    d. The voltage and current are 180 degrees out of phase.

17. The total opposition to current flow in an AC circuit is known as _____.
    a. resistance
    b. capacitive reactance
    c. inductive reactance
    d. impedance

18. The formula $\sqrt{R^2 + X^2}$ is used to solve for the _____.
    a. total impedance of an AC circuit
    b. frequency of a parallel RC circuit
    c. total current in an AC circuit
    d. power factor of an AC circuit

19. True power is best described as the _____.
    a. power consumed by the combination of the resistive and reactive components of a circuit
    b. power consumed by the resistive components of a circuit
    c. power consumed by the inductors and capacitors in a circuit
    d. product of source voltage and current in an AC circuit

20. A power factor is not an angular measure, but a numerical ratio with a value between 0 and 1, equal to the _____.
    a. sine of the phase angle
    b. tangent of the phase angle
    c. cosine of the phase angle
    d. cotangent of the phase angle

21. The two windings of a conventional transformer are known as the _____.
    a. mutual and inductive windings
    b. high- and low-voltage windings
    c. primary and secondary windings
    d. step-up and step-down windings

22. A transformer has 300 turns on the primary, 100 turns on the secondary and 120VAC applied to the primary. The voltage across the secondary is _____.
    a. 33V
    b. 40V
    c. 90V
    d. 120V

23. A transformer with a primary of 200 turns and a secondary of 400 turns has a _____.
    a. step-up ratio of 4:2
    b. step-down ratio of 2:4
    c. step-up ratio of 1:2
    d. step-down ratio of 2:1

24. Inrush current is a factor in selecting transformers for _____.
    a. RC circuits
    b. resistive circuits
    c. capacitive circuits
    d. inductive circuits

25. In a three-phase system, problems are likely to occur if the voltage imbalance between any two legs is greater than _____.
    a. 2 percent
    b. 5 percent
    c. 8 percent
    d. 10 percent

Trade Terms Introduced in This Module

Capacitance: The storage of electricity in a capacitor; capacitance produces an opposition to voltage change. The unit of measurement for capacitance is the farad (F) or microfarad (μF).

Frequency: The number of cycles an alternating electric current, sound wave, or vibrating object undergoes per second.

Hertz (Hz): A unit of frequency; one hertz equals one cycle per second.

Impedance: The opposition to current flow in an AC circuit; impedance includes resistance (R), capacitive reactance (X_C), and inductive reactance (X_L). Impedance is measured in ohms.

Inductance: The creation of a voltage due to a time-varying current; also, the opposition to current change, causing current changes to lag behind voltage changes. The unit of measure for inductance is the henry (H).

Micro: Prefix designating one-millionth of a unit. For example, one microfarad is one-millionth of a farad.

Peak voltage: The peak value of a sinusoidally varying (cyclical) voltage or current is equal to the root-mean-square (rms) value multiplied by the square root of two (1.414). AC voltages are usually expressed as rms values; that is, 120 volts, 208 volts, 240 volts, 277 volts, 480 volts, etc., are all rms values. The peak voltage, however, differs. For example, the peak value of 120 volts (rms) is actually $120 \times 1.414 = 169.68$ volts.

Reactance: The imaginary part of impedance. Also, the opposition to alternating current (AC) due to capacitance (X_C) and/or inductance (X_L).

Resonance: A condition reached in an electrical circuit when the inductive reactance neutralizes the capacitance reactance, leaving ohmic resistance as the only opposition to the flow of current.

Root-mean-square (rms): The square root of the average of the square of the function taken throughout the period. The rms value of a sinusoidally varying voltage or current is the effective value of the voltage or current.

Self-inductance: A magnetic field induced in the conductor carrying the current.

Additional Resources

This module presents thorough resources for task training. The following resource material is suggested for further study.

Principles of Electric Circuits: Conventional Current Version, 2009. Thomas L. Floyd. New York: Prentice Hall.

Figure Credits

Topaz Publications, Inc., Module opener, Figure 13 (photo), and SA03
Fluke Corporation reproduced with permission, SA01

NCCER CURRICULA — USER UPDATE

NCCER makes every effort to keep its textbooks up-to-date and free of technical errors. We appreciate your help in this process. If you find an error, a typographical mistake, or an inaccuracy in NCCER's curricula, please fill out this form (or a photocopy), or complete the online form at **www.nccer.org/olf**. Be sure to include the exact module ID number, page number, a detailed description, and your recommended correction. Your input will be brought to the attention of the Authoring Team. Thank you for your assistance.

Instructors – If you have an idea for improving this textbook, or have found that additional materials were necessary to teach this module effectively, please let us know so that we may present your suggestions to the Authoring Team.

NCCER Product Development and Revision

13614 Progress Blvd., Alachua, FL 32615

Email: curriculum@nccer.org
Online: www.nccer.org/olf

❑ Trainee Guide ❑ AIG ❑ Exam ❑ PowerPoints Other ____________________

Craft / Level: Copyright Date:

Module ID Number / Title:

Section Number(s):

Description:

Recommended Correction:

Your Name:

Address:

Email: Phone:

80202-11

Aerial Distribution Equipment

Module Two

Trainees with successful module completions may be eligible for credentialing through NCCER's National Registry. To learn more, go to **www.nccer.org** or contact us at **1.888.622.3720**. Our website has information on the latest product releases and training, as well as online versions of our *Cornerstone* newsletter and Pearson's product catalog.

Your feedback is welcome. You may email your comments to **curriculum@nccer.org,** send general comments and inquiries to **info@nccer.org**, or fill in the User Update form at the back of this module.

V.1 12/11

80202-11
Aerial Distribution Equipment

Objectives

When you have completed this module, you will be able to do the following:

1. Describe the types of transformers and how they are used in aerial distribution systems.
2. Explain the construction of an aerial distribution transformer.
3. Describe the functions of aerial load management devices including:
   - Regulators
   - Reclosers
   - Capacitors
   - Fault indicators
   - Fuses and cutouts
   - Switches
4. Assemble overhead street lights.
5. Energize or de-energize a single-phase transformer using a proper hot stick.
6. Open a disconnect switch using a load break tool.

Performance Tasks

Under the supervision of your instructor, you should be able to do the following:

1. Energize or de-energize a single-phase transformer using a hot stick.
2. Open a disconnect switch using a load break tool.
3. Assemble an overhead street light.
4. Hook up a three-phase transformer per diagrams and instructions provided by the instructor.

Trade Terms

Additive polarity
Air break switch
Amorphous metal
Basic impulse level (BIL)
Bushing
Completely self-protected (CSP) transformer
Disconnect switch
Faulted circuit indicator (FCI)
Feeder voltage regulator
Fused cutout
Fusing ratio
Lightning arrestor
Luminaire
Mast arm
Metal-oxide varistor (MOV)
Oil switch
Recloser
Sectionalizer
Self-protected (SP) transformer
Shell
Shell-type construction
Subtractive polarity
Transformer taps

Industry Recognized Credentials

If you're training through an NCCER-accredited sponsor you may be eligible for credentials from NCCER's Registry. The ID number for this module is 80202-11. Note that this module may have been used in other NCCER curricula and may apply to other level completions. Contact NCCER's Registry at 888.622.3720 or go to nccer.org for more information.

Contents

Topics to be presented in this module include:

1.0.0 Introduction 1
2.0.0 Safety 1
3.0.0 Aerial Distribution Transformers 2
3.1.0 Aerial Transformer Characteristics 2
3.2.0 Aerial Distribution Transformer Types 2
3.3.0 Other Distribution Transformers 4
4.0.0 Transformer Construction 5
4.1.0 Transformer Terminals 6
4.2.0 Common Transformer Connections 7
4.3.0 Single-Phase Light and Power Systems 8
4.4.0 Three-Phase Power Systems 8
4.5.0 Connecting Transformers 9
4.5.1 Transformer Primary Terminals 9
4.5.2 Transformer Secondary Terminals 9
5.0.0 Load Management and Protective Devices 11
5.1.0 Current Surge Protection Devices 11
5.1.1 Lightning Arrestors 11
5.1.2 Fused Cutouts 12
5.1.3 Enclosed Cutout 13
5.1.4 Open Cutout 13
5.1.5 Open-Link Cutout 13
5.1.6 Cutout Selection 13
5.1.7 Fuse Selection 14
5.2.0 Voltage Regulators 15
5.2.1 Transformer Taps 15
5.2.2 Feeder Voltage Regulators 16
5.2.3 Capacitors 16
5.3.0 Isolating Devices 17
5.3.1 Air Break Switches 17
5.3.2 Air Break Switch Installation 18
5.3.3 Oil Switches 18
5.3.4 Reclosers 18
5.3.5 Sectionalizers 18
5.3.6 Fault Indicators 20
6.0.0 Insulated Tools 21
6.1.0 Hot Sticks 21
6.2.0 Clamp Stick (Shotgun Stick) 22
6.3.0 Extendo Stick 22
6.4.0 Load Break Tool 22
7.0.0 Overhead Street Lights 23
7.1.0 Street Light Power Supply 24
7.2.0 Pole Requirements 24
7.3.0 Luminaire Installation 24
7.4.0 Luminaire Control 25

Figures and Tables

Figure 1 Pole-mounted transformer 1
Figure 2 Transformer with heat exchanger 3
Figure 3 Three-phase transformer 4
Figure 4 Conventional single-phase transformer 4
Figure 5 Completely self-protected single-phase transformer 4
Figure 6 Transformers supported by two poles 5
Figure 7 Pad-mounted distribution transformer 5
Figure 8 Transformer construction 5
Figure 9 Completely self-protected transformer 6
Figure 10 Single-phase transformer terminal designations 7
Figure 11 Additive polarity 7
Figure 12 Subtractive polarity 8
Figure 13 Single-phase transformer connection 9
Figure 14 Wye-wye transformer connections 10
Figure 15 Primary clamp-type terminal 10
Figure 16 Secondary clamp-type terminals 10
Figure 17 Secondary spade terminals 11
Figure 18 MOV lightning arrestor 11
Figure 19 Cutout with fuse installed 13
Figure 20 Enclosed cutout indicating blown fuse 13
Figure 21 Open cutout with blown fuse 14
Figure 22 Open-link cutout attached to transformer high-voltage bushing 14
Figure 23 Fuse protecting a wye-connected single-phase transformer 14
Figure 24 Fuses protecting three wye-connected transformers 15
Figure 25 Tapped transformer primary 15
Figure 26 Tap-changing switch 16
Figure 27 Feeder voltage regulator 16
Figure 28 Capacitor current and voltage characteristics 16
Figure 29 Capacitor bank 17
Figure 30 Three-pole air break switch 17
Figure 31 Disconnect switch 18
Figure 32 Conductor connected to switch 18
Figure 33 Vertical-mounted disconnect switch 19
Figure 34 Oil switch 19
Figure 35 Oil-type recloser. 19
Figure 36 Vacuum recloser 20
Figure 37 Sectionalizer 20
Figure 38 Clip-on fault indicator 20
Figure 39 Line worker installing fault indicators 20
Figure 40 Fault indicator attached to conductor 21
Figure 41 Hot stick 21
Figure 42 Hot stick with universal head 22
Figure 43 Shotgun stick 22
Figure 44 Shotgun stick with universal head 22

Figures and Tables (*continued*)

Figure 45 Telescoping extendo stick 23
Figure 46 Disconnect attachments 23
Figure 47 Load break tool 23
Figure 48 Load break tool operation 24
Figure 49 Street light on dedicated pole 24
Figure 50 Clamp-on mast arm base 25
Figure 51 Power supplied to a street light 26
Figure 52 Neutral conductor supporting cable 26
Figure 53 Photocell control 26

Table 1 Characteristics of Single-Phase Aerial Transformers 3
Table 2 Fuse Link Type and Response Time 12
Table 3 Recommended Transformer Fuse Sizes 14

1.0.0 INTRODUCTION

The power produced at a power station must be carried over a great distance with minimal losses. To do this, output transformers at the power plant increase the generated voltage (24,000V maximum) to a voltage as high as 765kV. At a higher voltage, there is less current flowing through the system's conductors. Lower current keeps transmission losses low. Substations located near customer loads use transformers to further lower the voltage for distribution to customer drops. Transformers are used yet again at or near the customer's location to lower the voltage to a level that is usable. Depending on the customer (commercial or residential), these voltages may be 480V or 575V three-phase or 240V single-phase. Pole-mounted transformers (*Figure 1*) are used with overhead (aerial) distribution systems. This module covers aerial distribution systems with a focus on the transformers and devices used with them.

2.0.0 SAFETY

All power line workers are required to wear personal protective equipment (PPE) to prevent on-the-job injuries.

- *Clothing* – Clothing should be made of natural fibers, such as cotton or wool, and treated to be arc-resistant and fire-retardant. Avoid clothing made of certain synthetic fibers, such as rayon, nylon, and polyester, that could burn or melt and cling to skin. Pant legs should extend over the top of shoes or boots. Never wear shorts on the job. Shirts should be long-sleeved to protect the skin from the sun, to minimize injuries caused by wood splinters, and to provide protection from chemicals used to treat the wood against rot. On most job sites, workers are required to wear a bright-colored shirt or vest to increase their visibility.
- *Footwear* – Only safety-toe leather shoes or boots should be worn on the job. High-top boots have the added advantage of greater support for the ankles. Many companies require dielectrically rated boots or overshoes. Never wear sandals or canvas-type shoes on the job.
- *Eye protection* – Safety glasses with side shields in accordance with *ANSI Standard Z87.1* must be worn at all times while on the job. If a power tool such as a chain saw is used on the job, a full face-shield may be necessary in addition to the safety glasses.
- *Hearing protection* – If power tools or noisy machinery, such as a chain saw, trencher, or air compressor, are used on the job, approved earmuffs or earplugs are generally required.

80202-11_F01.EPS

Figure 1 Pole-mounted transformer.

- *Dust mask* – If a wood pole is cut or drilled, a dust mask must be worn. The preservatives in the resulting saw dust may be harmful if inhaled.
- *Gloves* – Working on wood poles used in aerial distribution systems can be rough on hands. Leather work gloves provide protection from rope burns, wood splinters, and other hazards.
- *Rubber gloves and sleeves* – Power line workers often work near energized lines or equipment. Rubber gloves and sleeves protect against electrical shock. Use an air and water test on the gloves and roll the sleeves in each direction to look for damage. Inspect all other PPE, and tools according to the manufacturer's recommendations to ensure they are safe to use.
- *Safety helmet/hard hat/hard cap* – An approved *ANSI Standard Z89.1 Class E* safety helmet made of a non-conducting plastic material, such as polypropylene or polycarbonate, must be worn at all times while on the job site. The helmet protects the head from injuries caused by bumps and falling objects, and it provides shade from the sun.

In addition to wearing the proper PPE, follow these safety precautions when working with transformers and their related devices:

- Avoid installing or removing aerial equipment in wet or bad weather except in emergencies.
- Use protective covers, line hose, rubber blankets, and shields around conductors and other components when working near energized lines.

- Before starting any job, inspect and test rubber gloves and sleeves using the water and/or air method. Inspect all PPE and tools to ensure they are safe to use.
- Only use tools and equipment that are approved for power line use.
- Use fall protection equipment when working at heights above six feet (four feet for unqualified climbers).
- If an insulated lift bucket is used for bare hand work, test the insulating arm for excessive leakage current each day before it is used. A non-barehand bucket requires an annual boom leakage test.
- Have a qualified observer at all job sites to ensure that minimum approach distances to energized equipment are observed, safe work practices are being followed, and correct PPE and tools are being used.

NOTE

A qualified observer is a member of the work crew with the primary duty of ensuring that clearances to energized equipment are maintained, safe work practices are followed, and appropriate PPE and protective shields are being used. That person is familiar with all aspects of installation and construction of distribution systems and is trained in first aid. While performing the duties of a qualified observer, that person cannot perform any other tasks.

3.0.0 Aerial Distribution Transformers

Pole-mounted transformers must be easy to install and able to withstand extreme weather conditions. They must also be able to handle the electrical load placed on them.

3.1.0 Aerial Transformer Characteristics

In the United States, distribution system voltages can vary based on local conditions. For example, in thinly populated rural areas, 7,200V systems are common. In more densely populated areas, 4,800V systems are found. In other areas, system voltages can be as low as 2,400V or as high as 24,940V. The transformer must have a power rating that enables it to meet the power demands of the customers without overloading. For example, single-phase transformers are typically rated up to 167kVA. Aerial transformers must be protected from current surges caused by lightning strikes. The **basic impulse level (BIL)** measures the transformer's ability to withstand these current surges. *Table 1* contains dimensions, power ratings, basic impulse levels, and weights of single-phase aerial transformers.

Aerial transformers have a temperature-rise rating. The rating is the average temperature permitted in the windings when the transformer is operated in a maximum 40°C (104°F) ambient temperature. For example, if a transformer has a temperature-rise rating of 65°C (149°F), the average winding temperature cannot exceed 105°C (221°F) (65°C + 40°C = 105°C).

Aerial transformers may be equipped with heat exchanger fins on the **shell** (*Figure 2*) that cool the insulating oil. Cooling the insulating oil cools the transformer windings.

3.2.0 Aerial Distribution Transformer Types

Pole-mounted transformers are available in single-phase and three-phase types. It is common to use three single-phase transformers to supply three-phase power. Dedicated three-phase units

GOING GREEN

Wildlife Protectors

Squirrels and birds can perch on overhead lines and transformers. This can result in a short circuit that can kill them and cause an outage. To prevent this, insulating devices called wildlife protectors are placed on overhead conductors and other places where birds are likely to perch.

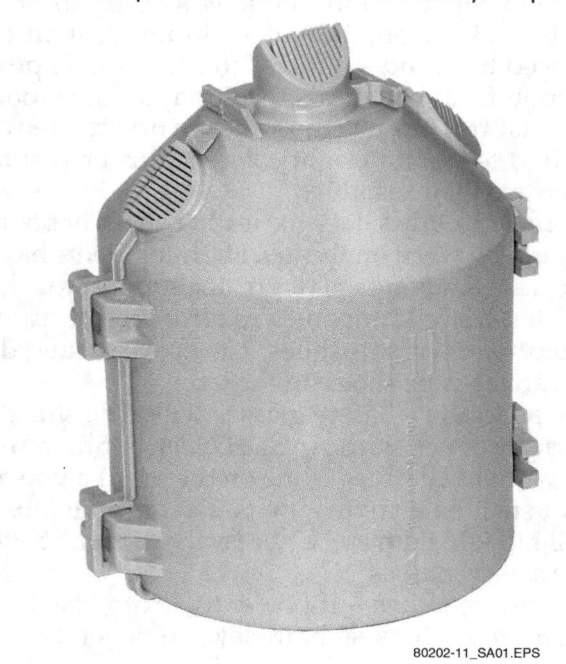

80202-11_SA01.EPS

Table 1 Characteristics of Single-Phase Aerial Transformers

Product Scope:
kVA: 5-167
Primary Voltage: 2400-19,920 V
Secondary Voltage: 120-600 V

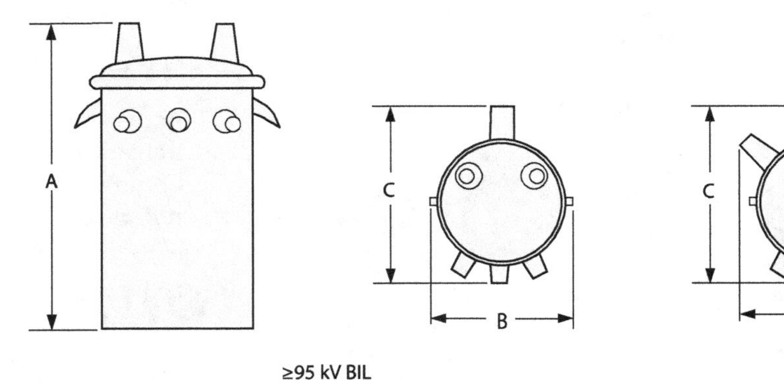

TABLE 1
Typical Dimensions and Weights[2,3]

kVA	Dimensions (in.) "A" ≤75 kV BIL	"A" 95 kV BIL	"A" 125 kV BIL	"A" 150 kV BIL	"B" ≤75 kV BIL	"B" ≥95 kV BIL	"C" [1]	Approx. Weight (lbs.)
5	26	32	42	45	28[1]	17	20	220
10	26	32	42	45	28[1]	17	20	220
15	30	35	46	49	28[1]	17	20	280
25	31	38	48	51	30[1]	20	22	350
37.5	33	40	52	55	31[1]	20	24	450
50	36	44	52	55	33[1]	22	25	600
75	39	51	54	57	33[1]	24	28	820
100	40	55	58	61	33[1]	27	31	1100
167	47	55	58	61	35[1]	35	37	1400

[1] Includes sidewall mount H.V. bushings.
[2] Includes radiators.
[3] Weights, gallons of fluid and dimensions are for reference only, and not for construction. Please contact Cooper Power Systems for exact dimensions.

80202-11 T01.EPS

(*Figure 3*) are also available. They consist of three single-phase transformers inside a single shell. Their major advantages are easier and less costly installation, along with the ability to handle large motor loads.

Common methods for connecting transformers in a power distribution system are covered later in this module. Single-phase transformers are available in three general types:

- Conventional transformer
- Self-protected (SP) transformer
- Completely self-protected (CSP) transformer

The conventional type (*Figure 4*) requires a fused cutout and lightning arrestor on the primary conductor feeding the transformer. The self-protected type has a built-in lightning arrestor. The completely self-protected type (*Figure 5*) has a current overload device built in and has more standard protection than the other two types.

80202-11_F02.EPS

Figure 2 Transformer with heat exchanger.

NOTE

The amount and type of protective devices available on the three different types of single-phase transformers may vary among manufacturers.

Transformers are usually mounted on one wood pole as a single unit or in groups of three. In some cases, more than one transformer may be mounted between a pair of poles (*Figure 6*) or on a platform between two poles.

80202-11_F03.EPS

Figure 3 Three-phase transformer.

80202-11_F04.EPS

Figure 4 Conventional single-phase transformer.

On Site

Transformer Names

Across the United States, pole-mounted transformers are known by a variety of slang terms. Common slang terms include *can, kettle, pot, and tub.*

NOTE

All pole-mounted transformers have the same basic features. However, some manufacturers may include other features that make installation or service easier.

3.3.0 Other Distribution Transformers

There are conditions where pole-mounted transformers cannot be used. Examples include transformers that are too heavy for pole mounting or areas where the distribution system is underground. In those cases, transformers may be installed on a pad or in a vault. The transformers may be oil-cooled or air-cooled and are often square or rectangular shaped (*Figure 7*).

80202-11_F05.EPS

Figure 5 Completely self-protected single-phase transformer.

80202-11_F06.EPS

Figure 6 Transformers supported by two poles.

4.0.0 Transformer Construction

A pole-mounted transformer is contained in a welded steel shell or tank that sheds water and is treated to resist rust. The top of the shell can be removed to take out or service the transformer. The transformer within the tank can be of core-type or shell-type construction (*Figure 8*). The laminations in the core can be made of silicon steel or more efficient amorphous metal. The transformer is immersed in fire-resistant insulating oil. Mineral oil is widely used. The completely self-protected single-phase transformer shown in *Figure 9* is equipped with these features:

- Wildlife protectors
- High-voltage bushing
- Low-voltage bushings
- Overload signal light
- Ground strap
- Secondary circuit breaker handle

Other features that are not shown but may be provided include the following:

- External tap changer switch
- Oil drain fitting
- Oil level indicator
- Oil/winding temperature indicator
- Pressure-relief valve
- Mounting bracket

A single-phase pole-mounted transformer may have one or two high-voltage bushings. A transformer with one high-voltage bushing can only be used with a well-grounded distribution system with a wye primary connection. A transformer with two high-voltage bushings is suitable for use in a wye or delta distribution system.

80202-11_F07.EPS

Figure 7 Pad-mounted distribution transformer.

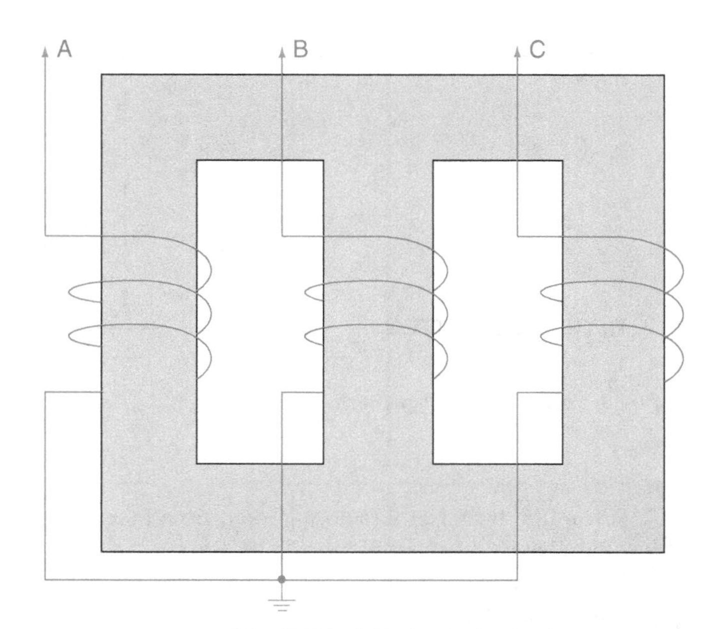

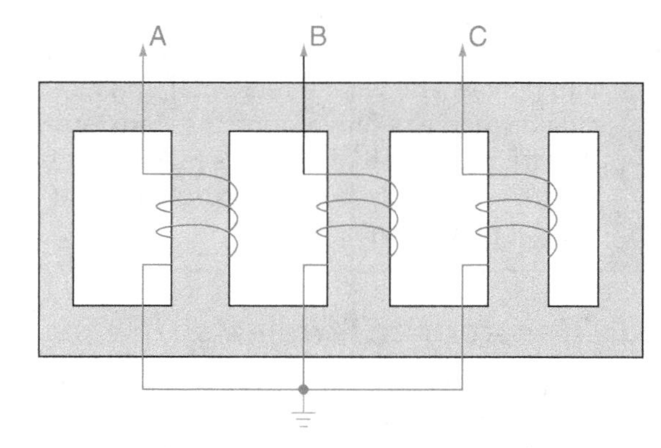

80202-11_F08.EPS

Figure 8 Transformer construction.

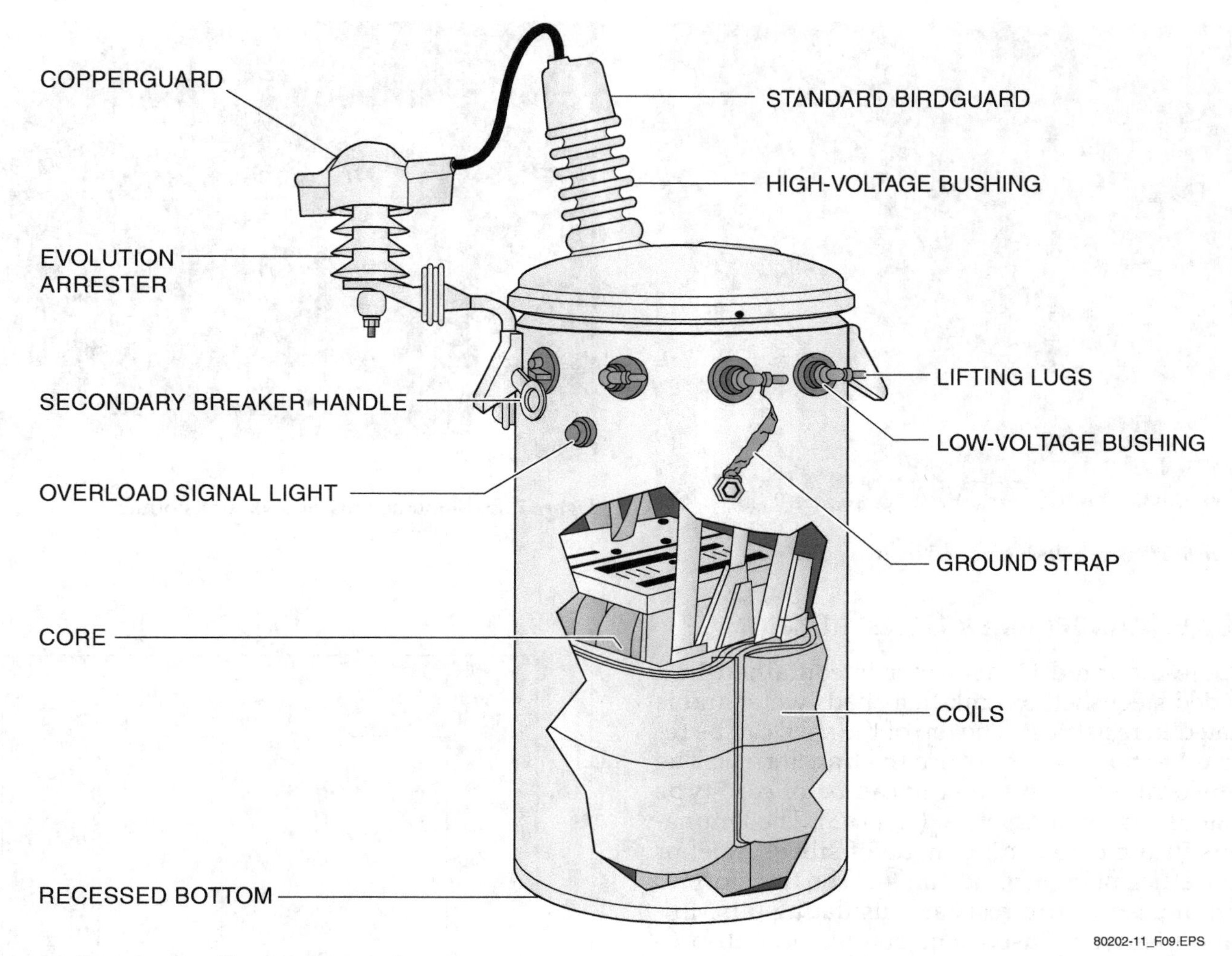

Figure 9 Completely self-protected transformer.

WARNING!

The first transformers contained pure mineral oil. Polychlorinated biphenyls (PCBs) were later added to make the oil fire resistant. Older transformers (made before 1978) are likely to have insulating oil that contains a small percentage of toxic PCBs. If you find that a leaking transformer contains PCBs, you must wear disposable PCB-resistant clothing, boots, and gloves, and wear a respirator and face shield when working around the transformer. Any transformer that contains PCBs must be stored and disposed of in accordance with federal regulations.

4.1.0 Transformer Terminals

Transformers are equipped with high- and low-voltage terminals called bushings. When facing the secondary bushings, the high-voltage terminal on the left is always designated H1. The other high-voltage terminal is designated H2 (*Figure 10*). This is an industry standard. Low-voltage terminals are designated X1, X2, and X3. The positions of X1 and X3 depend on the transformer polarity.

Transformers have either **additive polarity** or **subtractive polarity**, depending on the direction in which the coils are wound. This in turn determines the direction of current flow. In an additive transformer, the X1 terminal is diagonally opposite the H1 terminal. In a subtractive transformer, the X1 terminal is directly opposite the H1 terminal. Distribution transformers with a rating of 200kVA and greater and with a primary voltage of 8,600 volts or more are generally subtractive.

The transformer may be internally connected for additive polarity or subtractive polarity. Additive polarity can be checked as follows:

- Connect (jumper) adjacent high- and low-voltage terminals.
- Apply 240V to the transformer primary (H1 and H2).

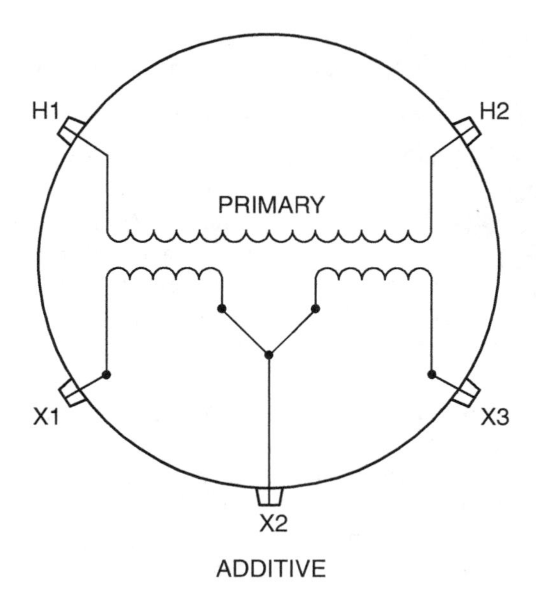

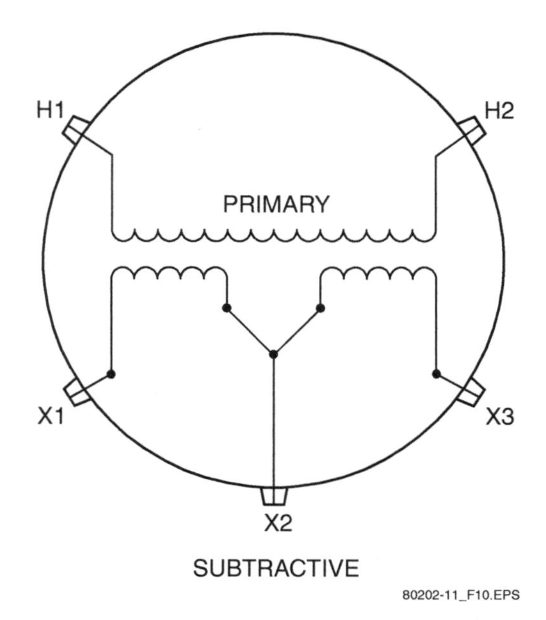

Figure 10 Single-phase transformer terminal designations.

- A voltmeter placed across the other adjacent high-and low-voltage terminals should read the sum of the high-and low-voltage windings (*Figure 11*).
- With additive polarity, the low-voltage terminal on the right when facing the high-voltage side of the transformer should be marked X1.

Subtractive polarity can be checked as follows:

- Connect (jumper) adjacent high- and low-voltage terminals.
- Apply 240V to the transformer primary (H1 and H2).
- A voltmeter placed across the other adjacent high-and low-voltage terminals should read the difference in the high- and low-voltage windings (*Figure 12*).
- With subtractive polarity, the low-voltage terminal on the left when facing the low-voltage side of the transformer should be marked X1.

Per industry standards, additive polarity is standard on all single-phase transformers sized 200kVA and smaller and having high-voltage windings of 8,600V and below. All other transformers are subtractive polarity. Polarity is important when connecting transformers in parallel. Always connect H1 to H1, X1 to X1, and so on.

GOING GREEN

Insulating Oil

Electrical grade mineral oil is widely used as insulating oil in transformers. However, it is not as kind to the environment as some other oils. Newer ester-based insulating oil is biodegradable and can be a net CO_2 absorber over the course of its life. Cooper Power Systems has a transformer that contains pure peanut oil.

WARNING!

Always wear the appropriate PPE and use the correct voltmeter when checking transformer polarity.

4.2.0 Common Transformer Connections

There are many ways that transformers can be connected in a distribution system. In this sec-

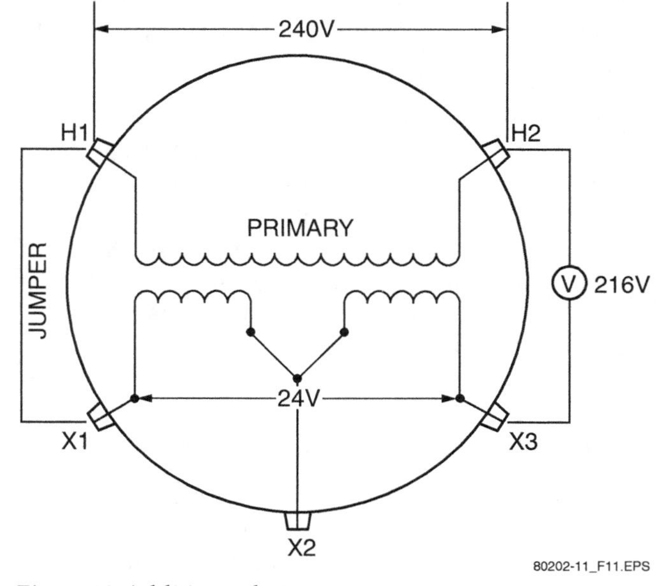

Figure 11 Additive polarity.

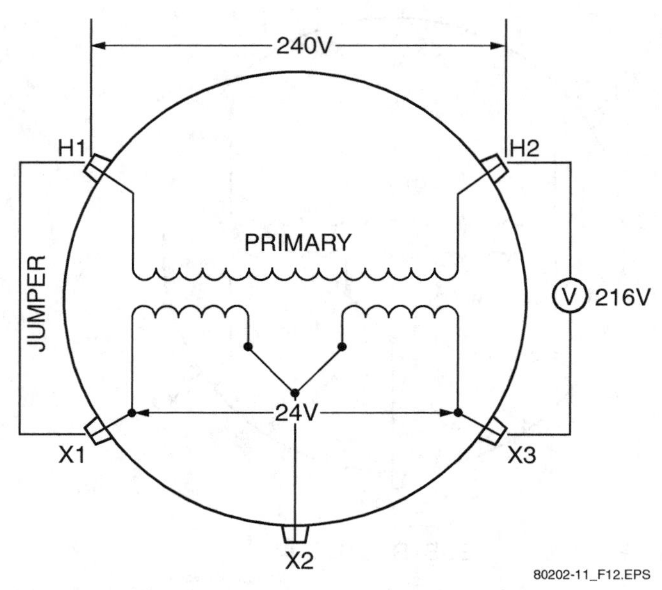

Figure 12 Subtractive polarity.

tion, only the most common connection schemes are shown. The ground connections must be made first when hooking-up a transformer. They are removed last when disconnecting it.

4.3.0 Single-Phase Light and Power Systems

One single-phase transformer is connected to serve homes and small businesses. It is the most common distribution scheme in use today. It is known as a 120/240V, single-phase system (*Figure 13*). Primary voltage is connected to the high-voltage bushings (H1 and H2). In this system, the transformer secondary center tap is grounded. Voltage between one of the secondary bushings (X1 or X3) and the grounded neutral (X2) is 120V. Voltage across the secondary bushings (X1 and X3) is 240V.

> **NOTE**
> If the transformer is equipped with a single high-voltage bushing, ground the transformer case before connecting the primary high-voltage conductor to the H1 high-voltage bushing. The H2 connector on the primary coil is internally connected to the case ground terminal in the transformer.

4.4.0 Three-Phase Power Systems

There are advantages and disadvantages of the different types of three-phase power distribution schemes. Today, the three-phase, four-wire wye-wye system (*Figure 14*) is widely used. It offers increased power-carrying capacity and better voltage regulation.

> **GOING GREEN**
> **Amorphous Metal**
> Amorphous metal has a non-crystalline structure that gives it unique magnetic properties. When this metal is used in the core of a transformer, losses are reduced up to 60 percent over laminated steel cores. The end result is a more energy-efficient transformer.

Three single-phase transformers are connected in the wye-wye scheme. Each H1 high-voltage bushing is connected to one of the high-voltage primary conductors. The H2 high-voltage bushings are all connected in parallel to the primary grounded neutral conductor. On the secondary side, conductors attached to each X1 bushing connect to different secondary conductors. The X2 secondary bushings are all connected in parallel with the secondary grounded neutral conductor. By industry standards, the X3 secondary bushing is not used in this scheme. Voltage between any individual secondary conductor and neutral should be 120V. Voltage between any two secondary conductors should be 208V. Transformers with a single high-voltage bushing can also be used in a three-phase wye-wye scheme by connecting each individual high-voltage primary conductor to each high-voltage bushing, and parallel-connecting the ground terminals on each transformer to the primary side neutral conductor.

> **NOTE**
> The three single-phase transformers used in a wye-wye scheme should be matched in power, voltage, and impedance.

> **On Site**
> **Low-Voltage Surge Protection**
> Some manufacturers equip their transformers with low-voltage surge protection. One type of protector is a spark gap that is installed inside the transformer between the low-voltage bushings (X1 and X2) and ground. Any surge in the low-voltage side causes the spark gap to conduct the surge spike to ground.

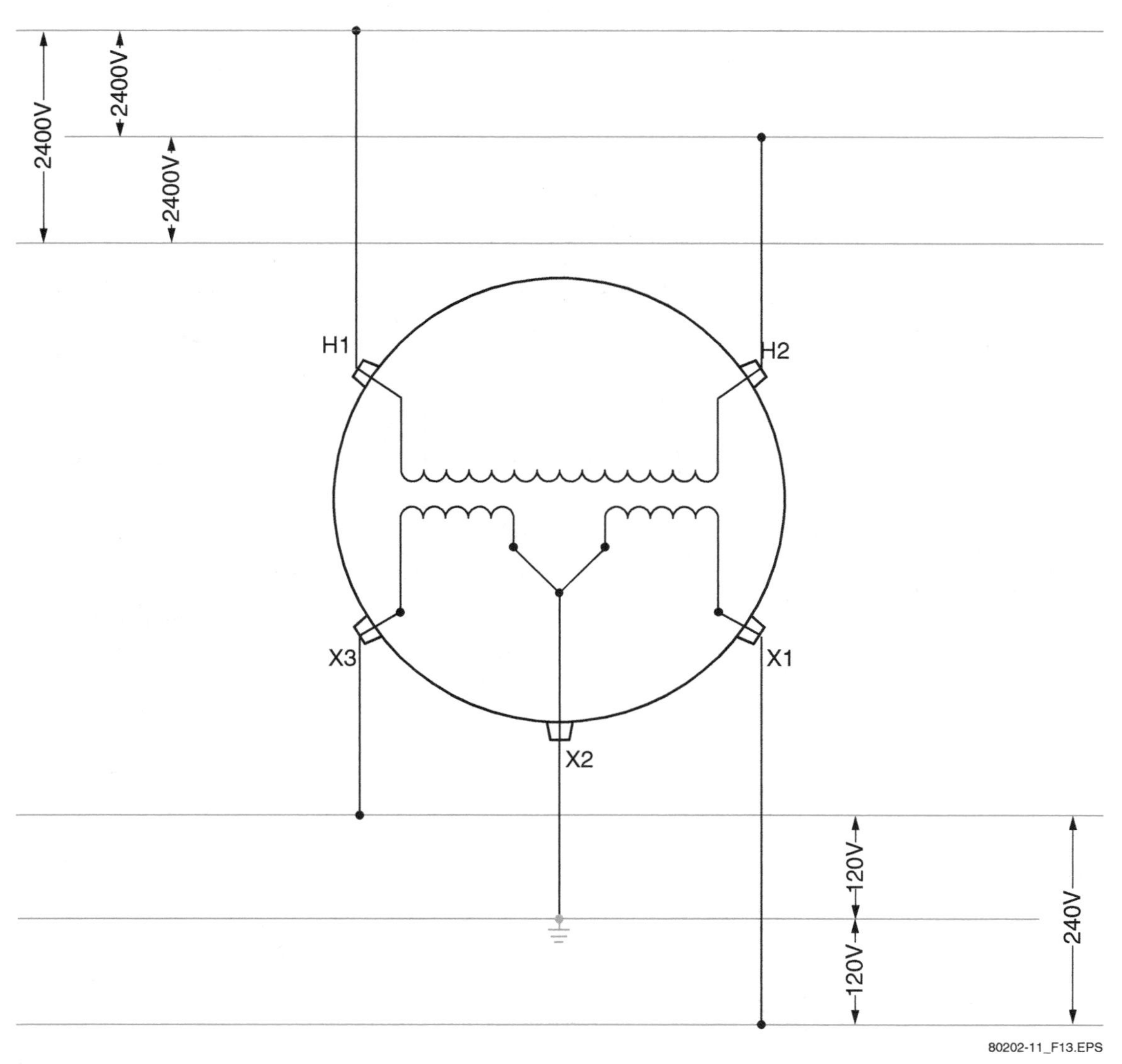

Figure 13 Single-phase transformer connection.

4.5.0 Connecting Transformers

Primary conductors are connected to the H1 and H2 terminals of the transformer. Secondary conductors are connected to the X1, X2, and X3 terminals. The utility will specify the type of terminals to be installed on transformers they order. Popular terminal styles include eyebolts and spades. The terminal material must be compatible with the conductors that will be connected to the terminal. Tin-plated terminals are often specified so that copper or aluminum conductors can be connected. Copper or aluminum terminals can also be specified. Terminal sizes are based on transformer power rating in kVA. As transformer kVA ratings increase, the terminals must be larger to handle larger diameter conductors.

4.5.1 Transformer Primary Terminals

Two-piece clamp-type terminals (eyebolts) are commonly used to connect primary conductors to H1 and H2 terminals. The clamp is shaped like the Greek letter omega (Ω) and is tightened with a nut. To install a conductor, the nut is backed off to loosen the clamp. Then the conductor is placed through a hole in the terminal. The nut is then tightened to clamp the bare conductor in place. *Figure 15* shows a clamp-type high-voltage terminal with two copper conductors clamped in place.

4.5.2 Transformer Secondary Terminals

The same clamp-type terminals used on the primary terminals are used on the secondary terminals (*Figure 16*). If multiple secondary conductors must be attached to the terminals, a spade terminal (*Figure 17*) is used. A nut, bolt, and lock washer of the same material as the spade terminal is used

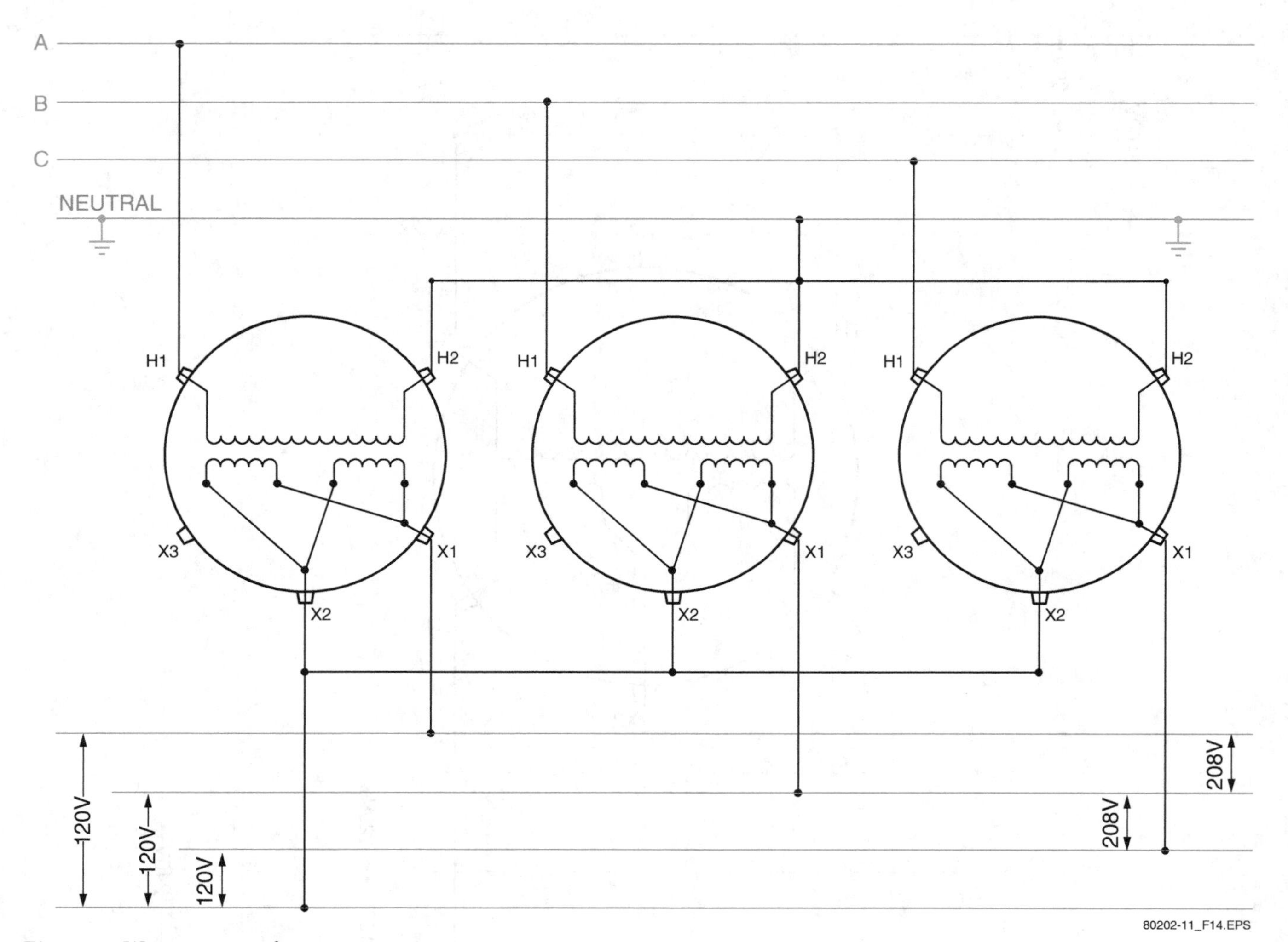

Figure 14 Wye-wye transformer connections.

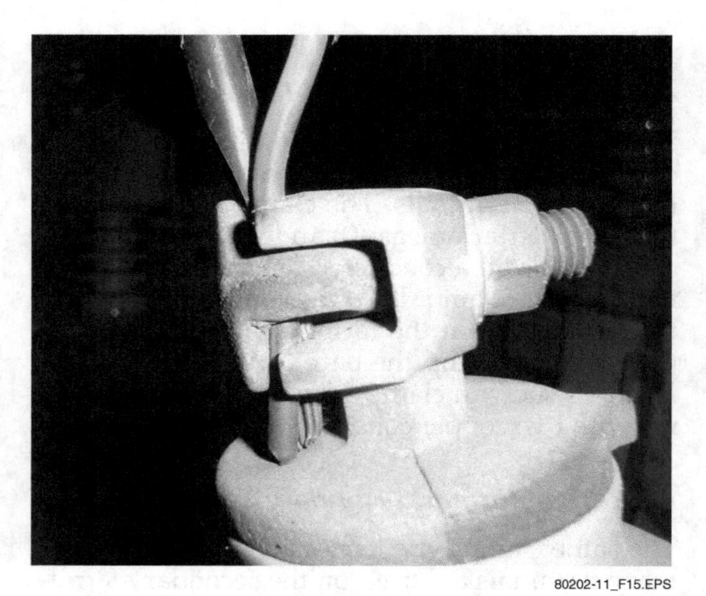

Figure 15 Primary clamp-type terminal.

Figure 16 Secondary clamp-type terminals.

80202-11_F17.EPS

Figure 17 Secondary spade terminals.

to connect a conductor with a ring terminal to a spade terminal. Insulated copper pin terminals are available that can be crimped on the ends of aluminum conductors that are connected to copper clamp-type secondary terminals.

NOTE

Always ensure that a conductor is clean and free of corrosion before connecting it to a transformer terminal. Apply oxide inhibitor to aluminum conductors.

5.0.0 Load Management and Protective Devices

An aerial distribution system contains devices designed to protect against surges, regulate voltage, and allow workers to isolate components or parts of the system.

Devices designed to protect against current surges include the following:

- Recloser
- Lightning arrestor
- Fuse

Devices designed to regulate voltage include the following:

- Transformer taps
- Field (pole-mounted) voltage regulators
- Capacitor

Devices designed to isolate components include the following:

- Reclosers
- Breakers
- Switches
- Sectionalizers

5.1.0 Current Surge Protection Devices

A current surge caused by a lightning strike or a short circuit in the primary or secondary side of the distribution system can damage system components. Several devices are available to prevent current surges or limit their damage.

5.1.1 Lightning Arrestors

A lightning arrestor (*Figure 18*) is a form of relief valve. When a lightning strike occurs, the arrestor provides a low-resistance path to ground for the excess current and voltage. After the surge passes, the arrestor acts as an open circuit so that normal line current and voltage is not grounded. The most common lightning arrestor in use today is the metal-oxide varistor (MOV).

This type of resistor offers a very high resistance to normal system current and voltage. When a surge voltage spike occurs, the resistance of the varistor breaks down at a predetermined voltage and conducts the surge to ground. Once the surge passes, the resistance of the varistor returns to a high value. Lightning arrestors are standard equipment on self-protected transformers.

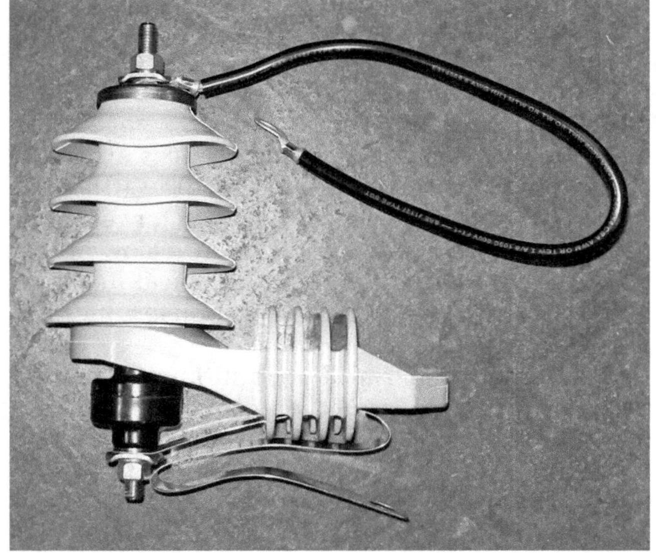

80202-11_F18.EPS

Figure 18 MOV lightning arrestor.

GOING GREEN

The Smart Grid

The present power distribution system in the United States (the grid) is often overloaded when demand is high, and has excess power when demand is low. The Smart Grid is in its early stages of being implemented. It will use digital technology and advanced communications to better manage electrical loads and the generation and distribution of power.

80202-11_SA02.EPS

5.1.2 Fused Cutouts

Fuses are designed to open a distribution primary circuit and prevent transformer damage if an overload or short circuit occurs. Fuses are rated based on current and response time (time required to open). A letter code is used for response time. Depending on the application, a slow or fast response time may be required. Typical fuse link letter designations and their response times are shown in *Table 2*.

Fuses used in aerial distribution systems are mounted in a cutout (*Figure 19*). When a fault occurs, the fusible link within the fuse blows out, vaporizing the link. The cutout is made so that the fuse is mounted about 20 degrees from vertical. This offset helps the fuse drop from the cutout when the fuse blows. The cutout is typically equipped with an eye or hook that enables the cutout to be opened with a hotstick or loadbreaking tool. Cutout types include the following:

- Enclosed
- Open
- Open-link

Table 2 Fuse Link Type and Response Time

Fuse Link	Response Time
Type K	Fast
Type N	Fast
Type T	Slow
Type H	Very Slow
Type D	Very Slow
Type S	Very Slow
Type C	Slow

80202-11_T02.EPS

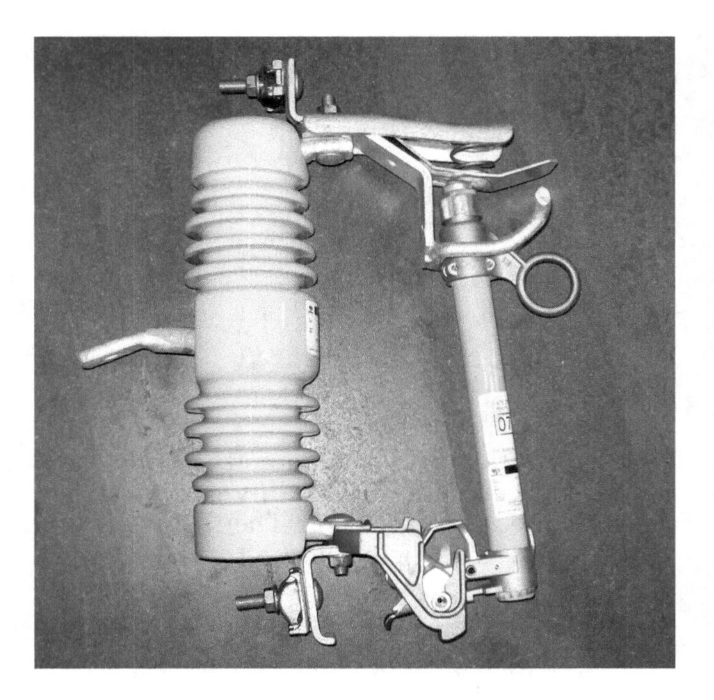

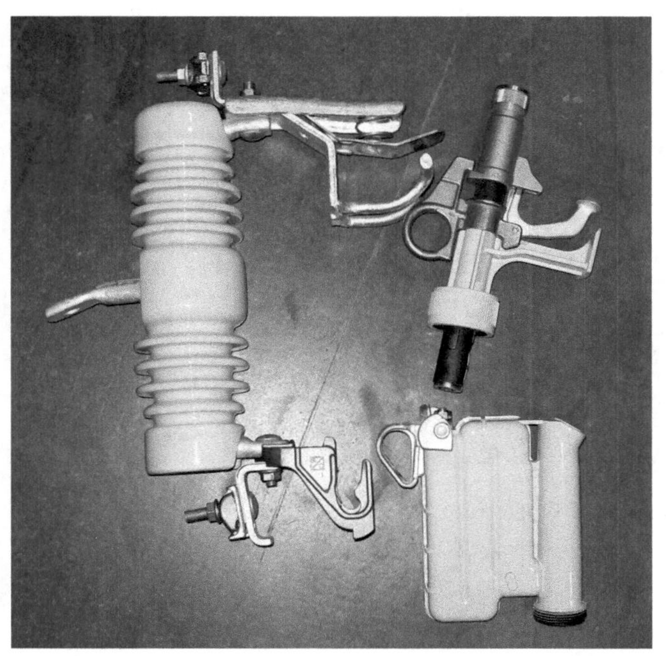

80202-11_F19.EPS

Figure 19 Cutout with fuse installed.

5.1.3 Enclosed Cutout

The enclosed cutout houses the fuse clips and fuse holder within an enclosure. A hinged door supports the fuse holder. When the fuse blows, the door drops open to indicate that a fault has occurred (*Figure 20*). Enclosed cutouts are used in distribution systems where the voltage does not exceed 7,200V. Current ratings are 50A, 100A, and 200A.

5.1.4 Open Cutout

The open cutout is widely used in distribution systems and can be used with all distribution system voltages. When the fuse blows, it drops from the cutout to indicate that a fault has occurred (*Figure 21*). Current ratings are 100A and 200A.

5.1.5 Open-Link Cutout

An open link cutout, sometimes called a flipper fuse, relies on spring tension to open the circuit when the fuse blows (*Figure 22*). This type of cutout cannot handle large fault currents, so it is not widely used.

5.1.6 Cutout Selection

The type of cutout used is based on distribution system voltage and the current rating of the fuse. The use of enclosed cutouts and open-link cutouts is limited. The enclosed cutout can only be used in systems where the voltage does not exceed 7,200V. Open-link cutouts cannot be used where large fault currents may occur. That leaves the open cutout as the cutout of choice in most applications. It can be used with all system voltages and has current ratings up to 200A.

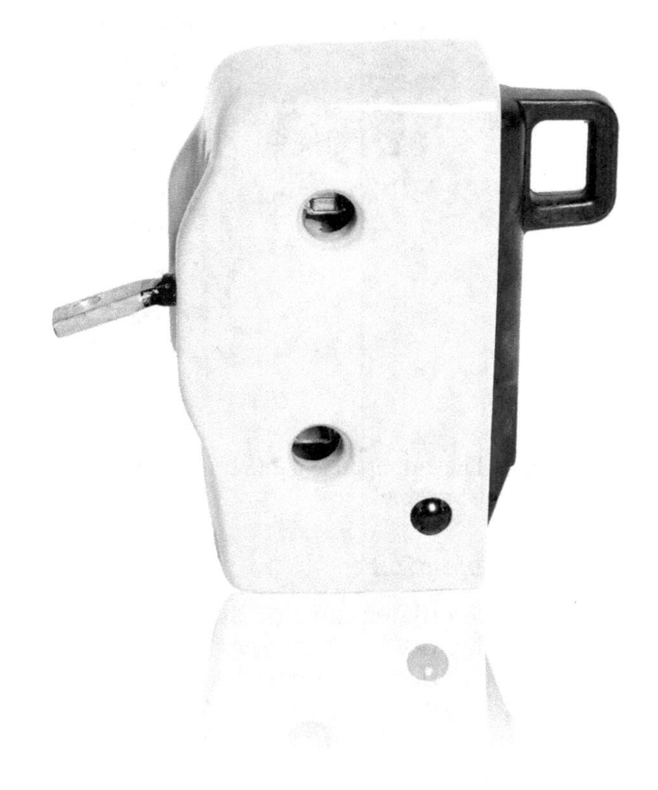

80202-11_F20.EPS

Figure 20 Enclosed cutout indicating blown fuse.

80202-11_F21.EPS

Figure 21 Open cutout with blown fuse.

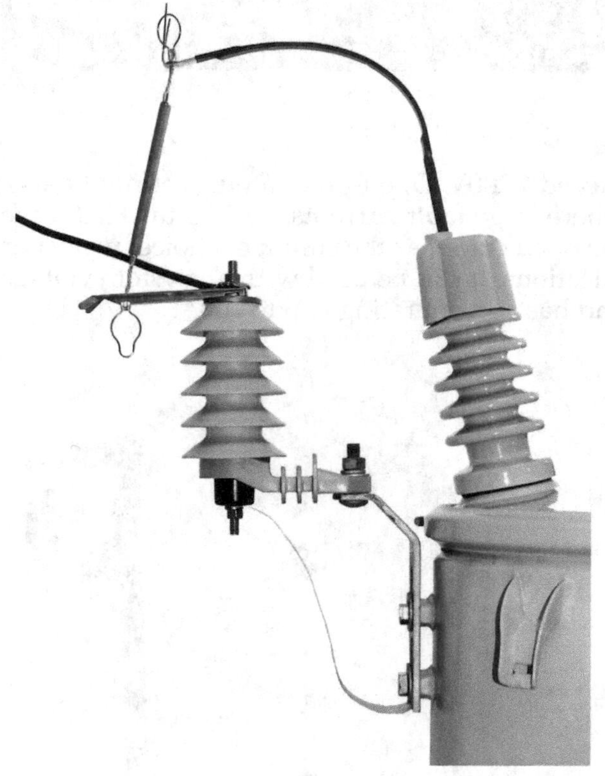

80202-11_F22.EPS

Figure 22 Open-link cutout attached to transformer high-voltage bushing.

5.1.7 Fuse Selection

An engineer calculates the fuse size based on the transformer size (kVA), its current rating, and the wiring configuration (wye or delta). *Table 3* shows recommended fuse sizes for 4,800V systems that are wye-connected. The fuses listed protects the transformer between 200 and 300 percent of the rated load.

Table 3 Recommended Transformer Fuse Sizes

Transformer Size (KVA)	Rated AMPs*	Fuse Link Rating*
5	1.04	1 H
10	2.08	3 H
25	5.2	6
50	10.4	12
100	20.8	25
500	104.2	140

*4800V wye-connected systems

80202-11_T03.EPS

Fuse selection is as much an art as it is a science. Local conditions and utility policy are factors that affect fuse selection. The **fusing ratio** is commonly used to help select a fuse. It is found by dividing the current-carrying capacity of the fuse by the transformer full load current.

FR = fuse current carrying capacity ÷ transformer full load current

A high fusing ratio increases the overload allowance for the transformer and increases service continuity. However, it may result in more transformer overloads and failures. A low fusing ratio produces opposite results. Transformers are less prone to failure, but service continuity is not as good.

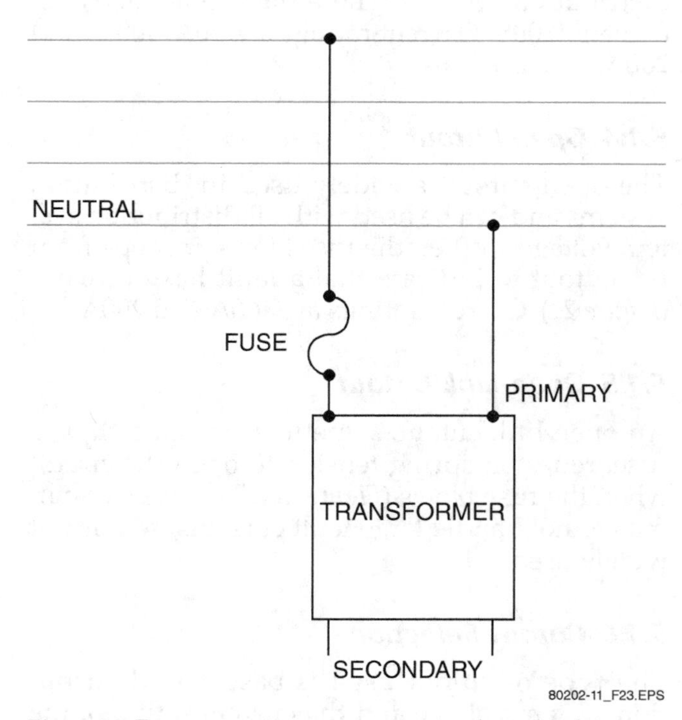

80202-11_F23.EPS

Figure 23 Fuse protecting a wye-connected single-phase transformer.

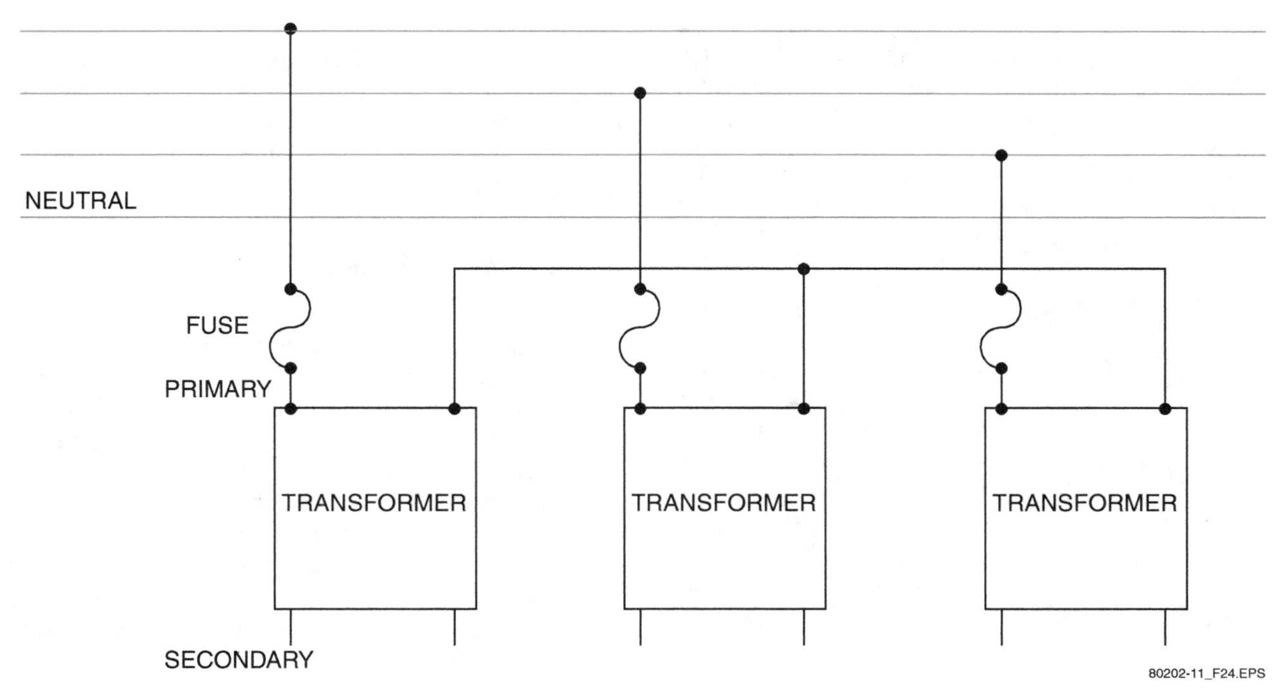

Figure 24 Fuses protecting three wye-connected transformers.

The fuse is installed based on the wiring configuration of the transformer. *Figure 23* shows a single fuse protecting a wye-connected single-phase transformer. Three fuses protecting three wye-connected transformers are shown in *Figure 24*.

5.2.0 Voltage Regulators

It is important to maintain the voltage supplied to customers within a range that allows lighting, appliances, and machinery to function properly. For example, voltage supplied to homes is standardized at 126V maximum and 114V minimum for a nominal 120V secondary service voltage. There are several devices and methods for regulating voltage.

5.2.1 Transformer Taps

A tap in the transformer primary (*Figure 25*) allows voltages above or below the rated primary voltage to be applied to the transformer. This lets the utility deliver a fixed voltage to the customer, even if the primary voltage is higher or lower than what is should be. Taps are typically set at increments of 2½ percent above and below the rated primary voltage. Normally, there are two increments above and two below the rated primary voltage. Taps can only be changed with the transformer de-energized. A rotary tap-changing switch (*Figure 26*) is used to select the correct tap. It can be located inside or outside the transformer shell.

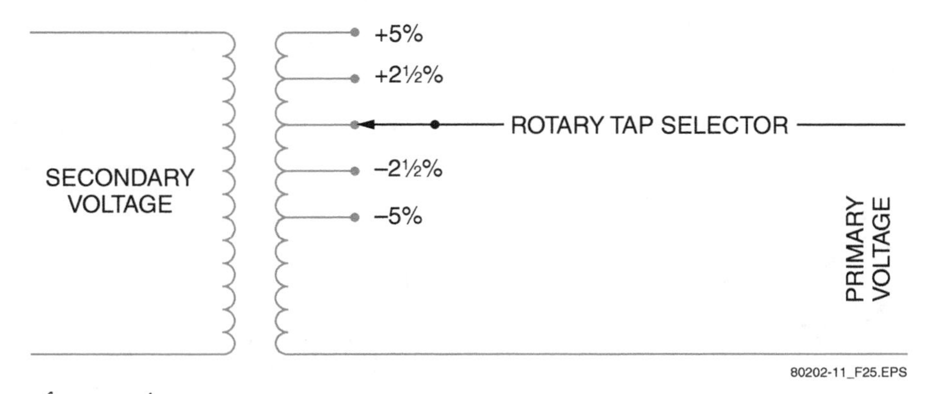

Figure 25 Tapped transformer primary.

80202-11_F26.EPS

Figure 26 Tap-changing switch.

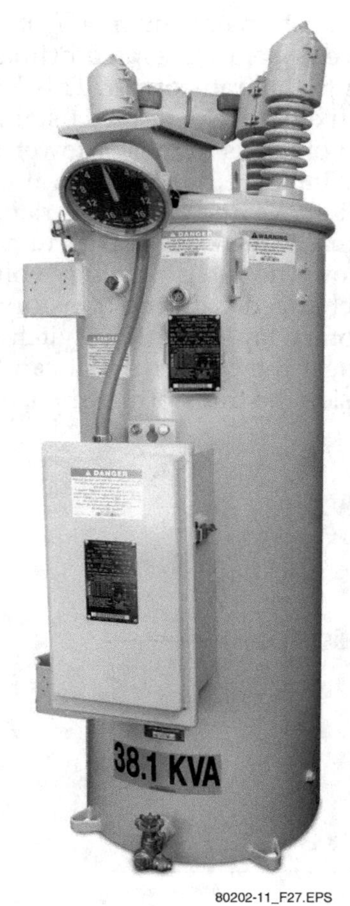

80202-11_F27.EPS

Figure 27 Feeder voltage regulator.

NOTE

Rotary tap-changing switches that can be operated from outside the transformer shell are safer and more convenient than those inside the shell. For safe operation, they often require a special key.

5.2.2 Feeder Voltage Regulators

A **feeder voltage regulator** (*Figure 27*), sometimes called a step voltage regulator, maintains a constant voltage in distribution feeder circuits even as loads change. Voltage regulators can be three-phase or single-phase. A motor-driven tap-changing switch can make changes in increments as small as ⅝ percent. Electronic switches are also used to select the correct number of secondary winding turns. Electronic switches eliminate arcing that can occur with mechanical switching. A voltage regulator bypass switch is used to bypass a voltage regulator.

WARNING!

Before operating the bypass switch, the voltage regulator must be in the neutral position.

5.2.3 Capacitors

In an alternating current circuit that contains a capacitor, current always leads voltage (*Figure 28*). Capacitors in a distribution system are used to offset the lagging current the line must supply to inductive loads such as motors and transformers. Capacitors are installed in banks (*Figure 29*) and can be switched in and out automatically or manually. Some banks are connected at all times.

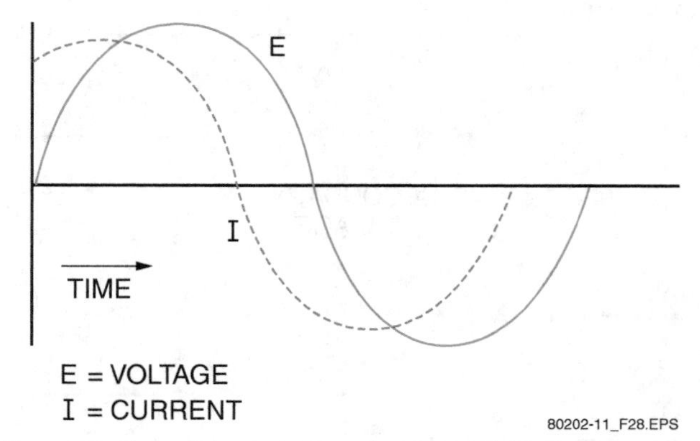

80202-11_F28.EPS

Figure 28 Capacitor current and voltage characteristics.

On Site

Tapped Aerial Transformer Use

Transformers with taps are not as widely used as they once were. In fact, ANSI no longer specifies taps in basic standard transformers. There are cases where tapped aerial transformers still make sense including:

- A utility may have different distribution voltages within their system. Stocking one type of tapped transformer would reduce their need for different types of replacement transformers.
- A manufacturer may provide one line of tapped transformers that it can sell to several utilities that have different distribution voltages.
- Transformers that serve heavy power loads may require taps to compensate for full-load voltage drops.

80202-11_F29.EPS

Figure 29 Capacitor bank.

WARNING!

Capacitors are electrical storage devices. Even after a capacitor is isolated and the circuit is de-energized, a residual charge may remain in the capacitor that can deliver a fatal shock. Before working on any capacitor, bleed off any charge according to the manufacturer's instructions before touching the capacitor.

5.3.0 Isolating Devices

There are times when line workers need to isolate components or sections of the line to make repairs or perform routine maintenance. A variety of devices are available to do this.

5.3.1 Air Break Switches

An **air break switch** (*Figure 30*) contains blade and fixed contacts equipped with arcing horns. When the switch opens, the horns spread to lengthen the arc until it breaks. Some air break switches use other methods to control the arc. Three-phase air break switches are mounted on a pole. They are operated together as a group by way of a crank handle that extends from the switch to the base of the pole. A hook stick is used to operate some air break switches. It is common for air break switches to be operated by remote control from a central location. Air break switches are used to open the circuit in an energized line. A **disconnect switch** (*Figure 31*) is an air break switch that does not have arcing horns. This type of switch must be opened with the line de-energized. A load break tool can be used to open a disconnect switch with the line still energized.

80202-11_F30.EPS

Figure 30 Three-pole air break switch.

On Site

Voltage Regulator Names

Tap-changing voltage regulators are known by several other names. Other names include the following:

- Line-drop compensator
- On-load tap changer
- Auto-boost regulator
- Mechanical tap changer

80202-11_F31.EPS

Figure 31 Disconnect switch.

5.3.2 Air Break Switch Installation

Before installing any switch, read the manufacturer's installation instructions. Three-phase air break switches are assembled on the ground, hoisted into place, and clamped directly to the pole. Crossarm braces are required. The switch is lifted into position using safe rigging practices. The connectors are wire-brushed, and oxide inhibitors are applied to both before connecting conductors to the switch (*Figure 32*).

Disconnect switches require little if any assembly before installation. They are typically installed on single or double crossarms and can be mounted in a vertical or underhung position (*Figure 33*). Both positions allow the switch to be opened with a hook stick. The switch body is bolted to a backstrap located on the opposite side of the crossarm. The connectors are wire-brushed, and oxide inhibitors are applied to both before connecting conductors to the switch.

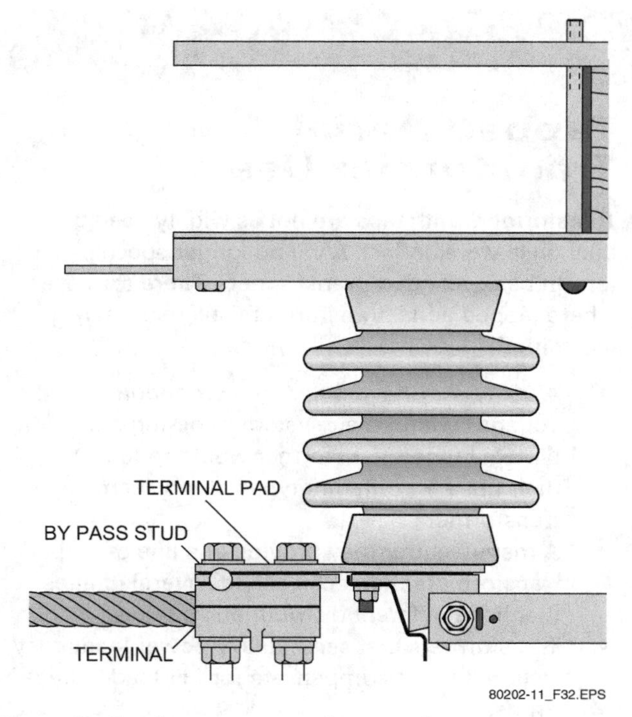

Figure 32 Conductor connected to switch.

5.3.3 Oil Switches

An **oil switch** (*Figure 34*) has its contacts immersed in oil to help quench any arc that forms as the contacts open. Oil switches can be opened and closed manually, or can open automatically like a circuit breaker.

5.3.4 Reclosers

A recloser can be compared to a circuit breaker. When a fault occurs, it resets itself after a preset time. A magnetic solenoid in the recloser opens contacts to stop current flow if fault current occurs. Once current flow stops, the solenoid de-energizes to restore current flow. If the fault is still present, the recloser opens again. The advantage of a recloser is that it can deal with temporary faults without causing a longer outage that might occur if a fuse were used in the same circuit. If the fault is not temporary, the recloser locks the circuit open after a certain number of reset attempts. Reclosers are used in single- and three-phase circuits. The contacts can be immersed in oil (*Figure 35*) or the contacts can be in a vacuum (*Figure 36*). The oil or vacuum helps suppress any arc that forms when the recloser opens. Single- and three-phase versions are available.

5.3.5 Sectionalizers

A sectionalizer (*Figure 37*) is a type of switch that is used with reclosers to isolate faults in distribu-

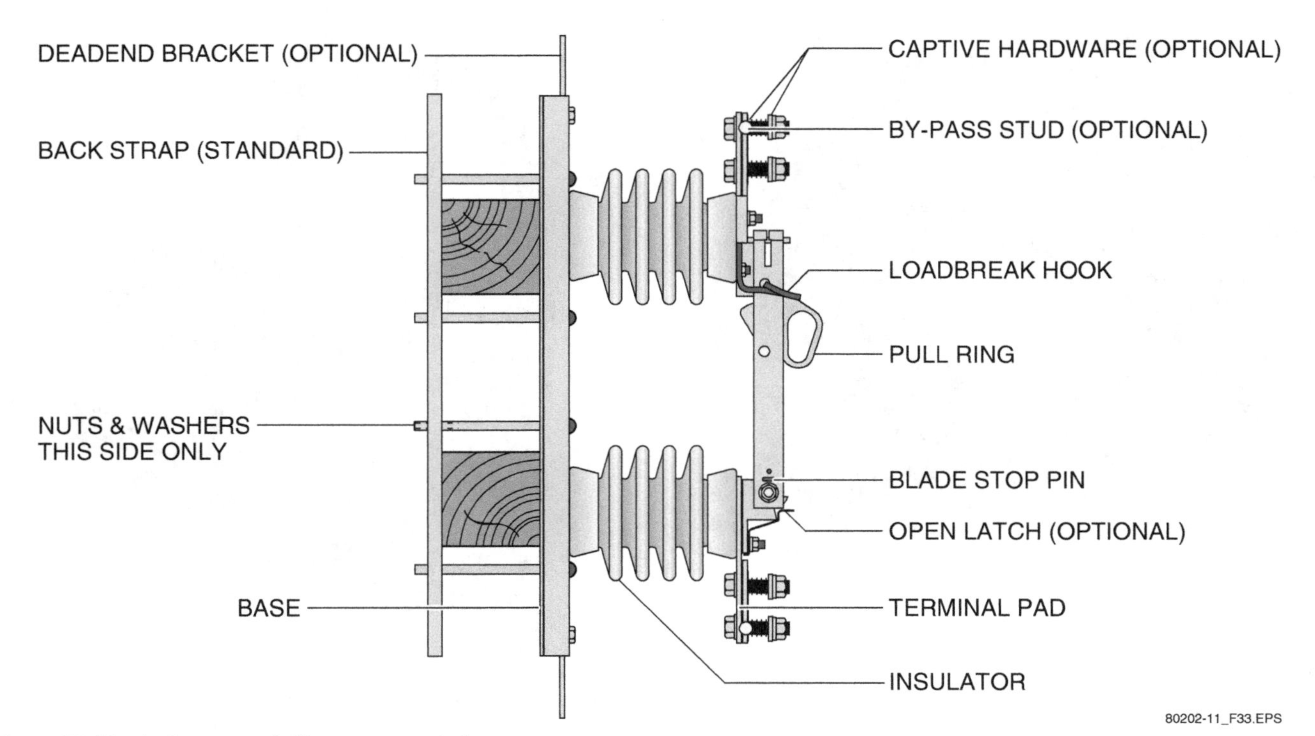

Figure 33 Vertical-mounted disconnect switch.

Figure 34 Oil switch.

Figure 35 Oil-type recloser.

tion circuits. A recloser opens and then closes a circuit when a transient fault occurs. A one-time transient fault would not affect the sectionalizer. If a fault continues to occur that results in many openings and resets of the circuit, the sectionalizer activates to isolate (sectionalize) the faulty section of line. The sectionalizer counts the number of resets and opens the circuit and locks it out after a certain number of resets are reached. The sectionalizer must be reset manually. Modern sectionalizers often use electronic circuits to monitor reclosers and open the sectionalizer.

80202-11_F36.EPS

Figure 36 Vacuum recloser.

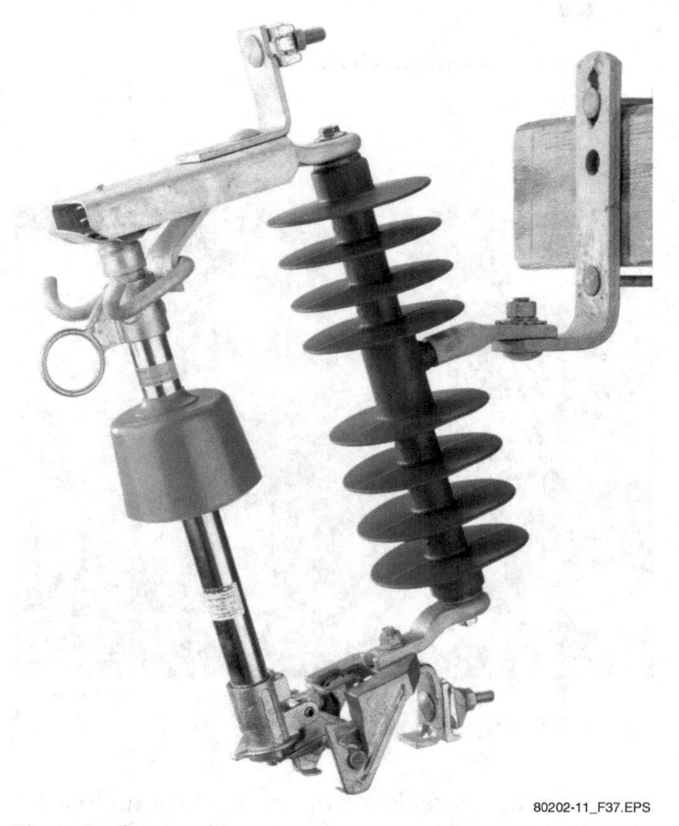

80202-11_F37.EPS

Figure 37 Sectionalizer.

5.3.6 Fault Indicators

Whenever an outage occurs, it is critical to quickly isolate the problem so that power can be restored. A **faulted circuit indicator (FCI)** provides line workers with a visual indication that a fault has occurred in a line. This helps reduce troubleshooting time. Fault indicators that clip onto conductors (*Figure 38*) are widely used. They are spring-

80202-11_F38.EPS

Figure 38 Clip-on fault indicator.

80202-11_F39.EPS

Figure 39 Line worker installing fault indicators.

loaded and are attached by pressing them against the primary conductor using a hook stick (*Figure 39*). Spring tension holds them in place (*Figure 40*). The clip can handle a wide range of conductor sizes. A single device can handle a wide range of distribution system voltages and currents. The indicators are electronic devices that use induction to detect the current flowing through a conductor. If the device does not detect current (an outage), a bright light in the device comes on to alert line workers of a problem. Some indicators can tell the difference between a temporary fault and a permanent fault. All have the ability to automatically reset after power is restored.

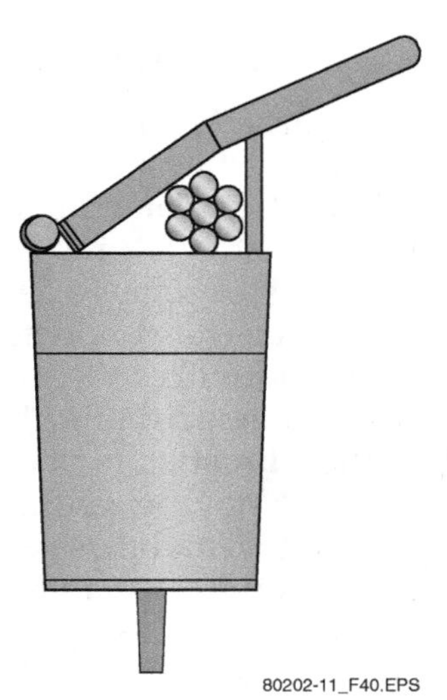

80202-11_F40.EPS

Figure 40 Fault indicator attached to conductor.

6.0.0 Insulated Tools

OSHA regulations require that power line workers use insulated tools when working on or near energized circuits. Insulated hand tools are individually tested and certified by the manufacturer to be suitable for specific working conditions. They must also be periodically inspected to ensure that they remain safe to use. Generally, the maximum rated voltage for insulated hand tools, such as pliers and screwdrivers, is 1,000 volts AC and 1,500 volts DC. However, power line workers must obey the minimum approach distance guidelines, which means there will be many times they cannot get close enough to their specific work area to use traditional hand tools. To enable power line workers to maintain the components on energized lines, a variety of long-reach insulated tools are available. Requirements for live-line tools are found in *OSHA Standard 29 CFR 1910.269(j)*. Live-line tools made of fiberglass-reinforced plastic (FRP) must be designed and built to withstand test voltage of 100,000 volts per foot of length for five minutes per *ASTM F711, Standard Specification for Fiberglass-Reinforced Plastic (FRP) Rod and Tube Used in Live Line Tools.* OSHA requires that live-line tools be cleaned and inspected daily for defects daily.

In addition to using approved insulated tools, it is necessary to make sure that these tools are properly maintained. This includes the following:

- Keeping tools clean and dry.
- Handling insulated tools with care.
- Inspecting insulation before each use. If you doubt the integrity of the insulation, destroy the tool or have it retested.
- Following the manufacturer's temperature recommendations for use.
- Having a qualified person inspect and recertify tools annually for safe use.
- Storing the tools in a protective sleeve or tube to prevent damage and to protect against condensation and dust.
- Using required personal protective equipment in addition to with insulated tools.

6.1.0 Hot Sticks

The hot stick (*Figure 41*) is probably the most important tool available to the power line worker. It is an insulated tool that protects the line worker from electric shock and arc burns while working on live lines. It is designed to accept a large number of specialized attachments that are used to perform specific work tasks such as testing voltage, applying tie wires, pruning tree limbs, replacing fuses, and opening and closing switches. A telescoping version of the hot stick, called an extendo stick, allows workers to perform tasks

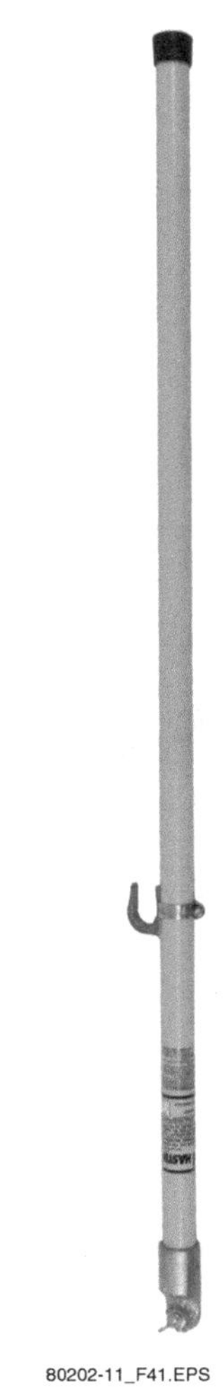

80202-11_F41.EPS

Figure 41 Hot stick.

from ground level. This eliminates the need to climb a pole or use a bucket in some cases.

Hot sticks are made in different lengths, from a few feet long up to telescoping types (extendo poles) that can be extended out to 40' long. The hot stick not only insulates the worker from the energized conductor, it also provides physical separation from the device being operated. This helps reduce the chance of burns that might result from electrical arcing if there is a malfunction of the device being operated.

Many hot sticks are equipped with a universal tool head (*Figure 42*) that allow various tools to be attached the end of the stick. These universal head sticks have a swivel head design that allows tools to be easily connected at the proper angle to efficiently perform a particular job.

80202-11_F42.EPS

Figure 42 Hot stick with universal head.

80202-11_F43.EPS

Figure 43 Shotgun stick.

6.2.0 Clamp Stick (Shotgun Stick)

A clamp stick (*Figure 43*), commonly referred to as a shotgun stick, contains an easy-to-control hook on the end of an insulated hot stick that can be extended and retracted to grab objects. These tools are used to apply or remove devices such as clamps on electrical lines or conductors. The stick includes a support rod with an insulated head at one end and a movable actuating handle at the other end. The shotgun stick is used most often for installing and removing grounds and mechanical jumpers, but it can be used with various end fittings to open and close a variety of overhead and underground circuits.

Shotgun sticks can also be equipped with a universal tool head (*Figure 44*) at the bottom end of the stick. This head allow various tools to be connected.

6.3.0 Extendo Stick

Extendo sticks (*Figure 45*) are telescoping sticks that offer maximum convenience for line workers because they can be used to perform many overhead tasks from ground level. These poles have universal end fittings that accept a variety of attachments so that the power line worker can disconnect switches, replace cutout tubes, remove pole covers, prune trees, and perform many other overhead tasks. These sticks retract to approximately 4' or 5' for easy storage and portability.

Heavy-duty, spring-loaded plastic buttons are used to keep the extended portions of the stick securely locked into position. As each section is extended and slightly rotated, the buttons pop into place, providing a secure lock of the each extended section of the pole.

Disconnect tools (*Figure 46*) are used to open switches and enclosed cutouts. They are also used for installing and removing open-link fuse links.

6.4.0 Load Break Tool

A load break tool (*Figure 47*) is used to quickly and safely open energized disconnects, cutouts, power fuses, capacitor banks, and fuse limiters. The load break tool accomplishes this by containing the external electrical charge that occurs when a fuse under a load is opened. A load break

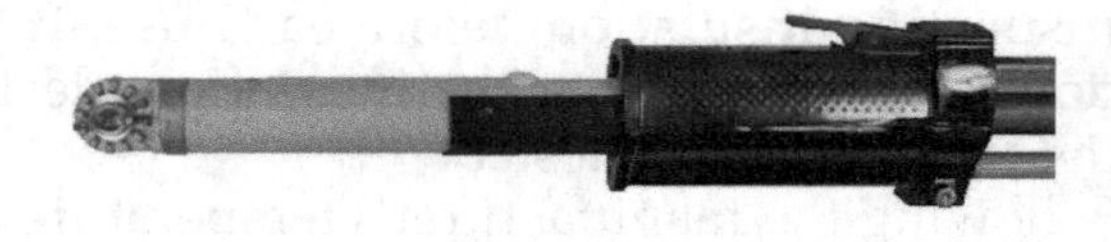

80202-11_F44.EPS

Figure 44 Shotgun stick with universal head.

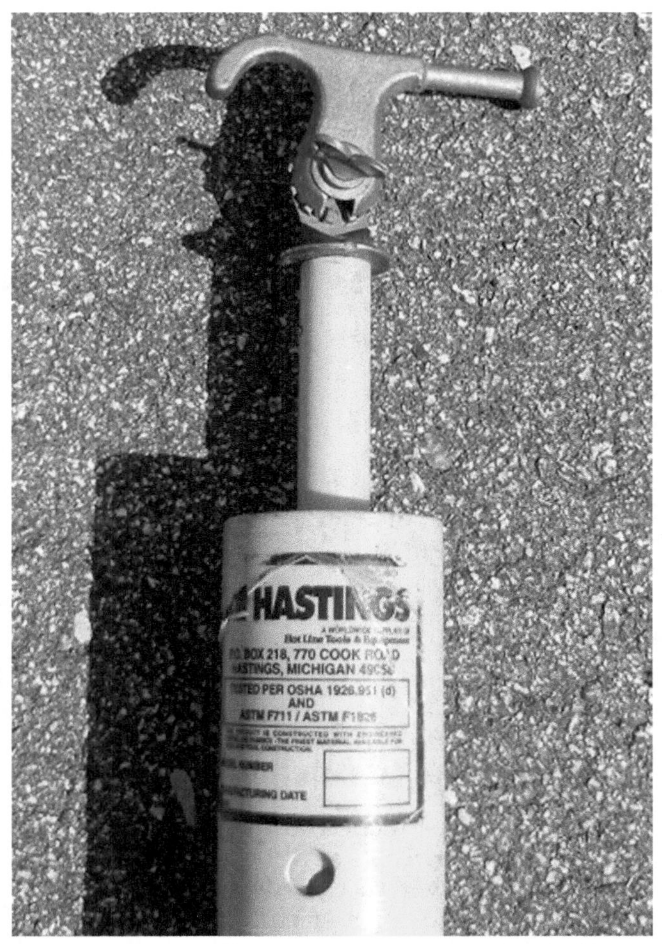

80202-11_F45.EPS

Figure 45 Telescoping extendo stick.

80202-11_F46.EPS

Figure 46 Disconnect attachments.

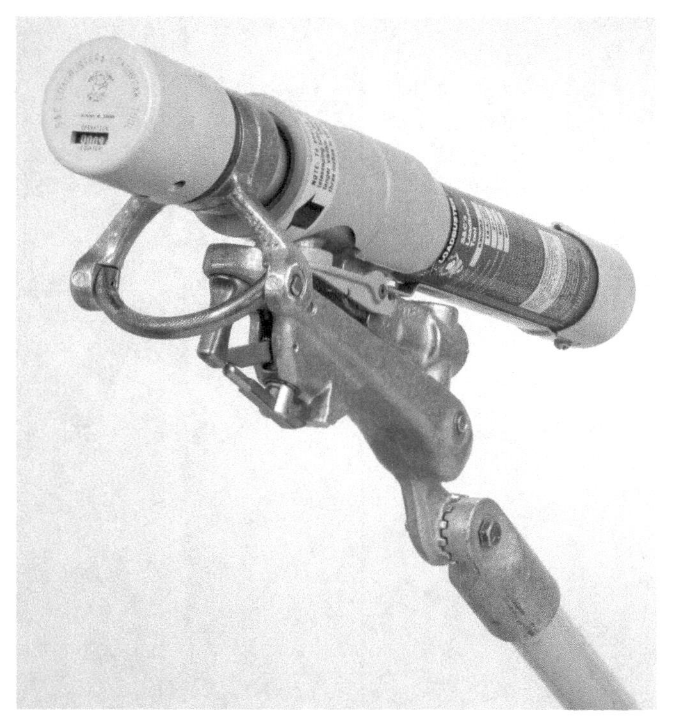

80202-11_F47.EPS

Figure 47 Load break tool.

tool has a built-in current-limiter that enables it to be used to disconnect load currents up to 900A.

The tool is positioned across the front of the device to be opened, with the tool's anchor placed on the attachment hook on the far side of the device. The device's pull-ring is engaged, with the pull-ring hook on the tool. A firm, steady, downward pull is used to extend the tool to its maximum length. This opens the device and diverts the current through the load arc-extinguishing chamber, or silencer. At the same time, the tool's internal operating spring is charged. Then, at a predetermined time during the operating stroke, its internal trigger trips, the charged operating spring is released, the internal circuits are separated, and the circuit is positively interrupted. Before using it again, the tool must be reset, and then it is ready to break the next circuit. *Figure 48* shows how the circuit is energized using the load break tool.

Use of the load break tool may vary based on company policy.

7.0.0 Overhead Street Lights

Overhead street lights can be mounted on dedicated poles or they can be mounted on the same wood poles used in overhead power distribution systems. Line workers often have to install overhead street lights on wood poles. This section focuses on street lights installed on wood poles.

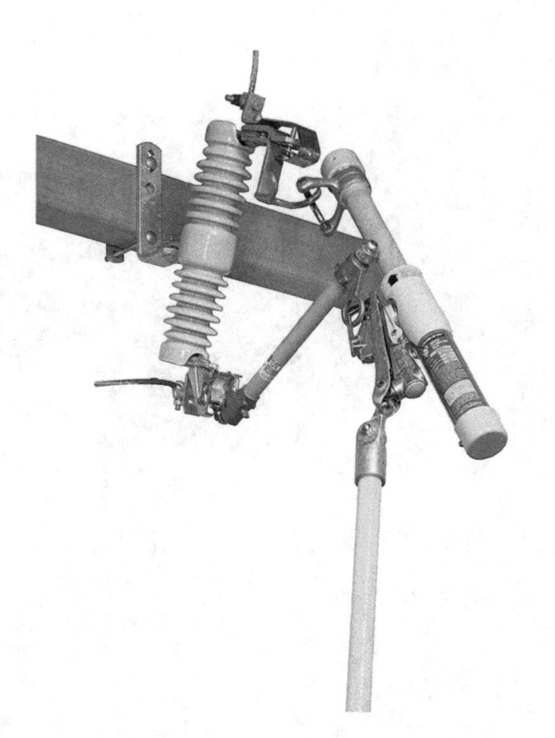
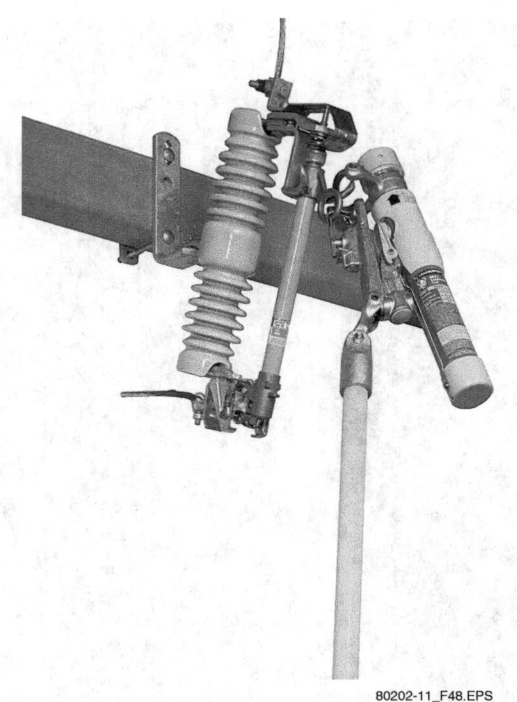

80202-11_F48.EPS

Figure 48 Load break tool operation.

7.1.0 Street Light Power Supply

Street lights are powered by AC voltage that ranges from 120V to 480V. In urban areas, voltage is often supplied through underground cables from a pad-mounted transformer. In rural areas and places that have overhead service, power is supplied through overhead cables. In many areas, the utility owns the street lights. It charges the municipality a flat rate that includes the cost of the power, plus maintenance and upkeep of the lights. Metered power for street lights is not widely used.

7.2.0 Pole Requirements

A street light may be installed on a dedicated pole (*Figure 49*) or on a pole that supports primary and secondary conductors. If the pole supports primary conductors, a minimum distance of 3 feet must be maintained from the base of the mast arm to the lowest primary conductor. The height of the luminaire above the road and the distance that it hangs over the road are a function of road width and the size and type of the luminaire used. Mast arms of different lengths allow the luminaire to be correctly placed over the road.

> **NOTE**
> The cobra-head luminaire is widely used for street lighting. It gets it name because it looks like a cobra in a ready-to-strike pose.

7.3.0 Luminaire Installation

The luminare is attached to the end of the mast arm. The mast arm acts as a conduit for the wires that supply power to the luminaire. The wiring is checked for proper connections and correct length. The correct lamp, which is the actual light source, is installed before the mast with attached luminaire is installed. A base on which the mast arm will be attached must be secured to the pole at the proper height above ground. The mast arm base is clamped around the pole (*Figure 50*). Other types of bases are bolted through the pole.

80202-11_F49.EPS

Figure 49 Street light on dedicated pole.

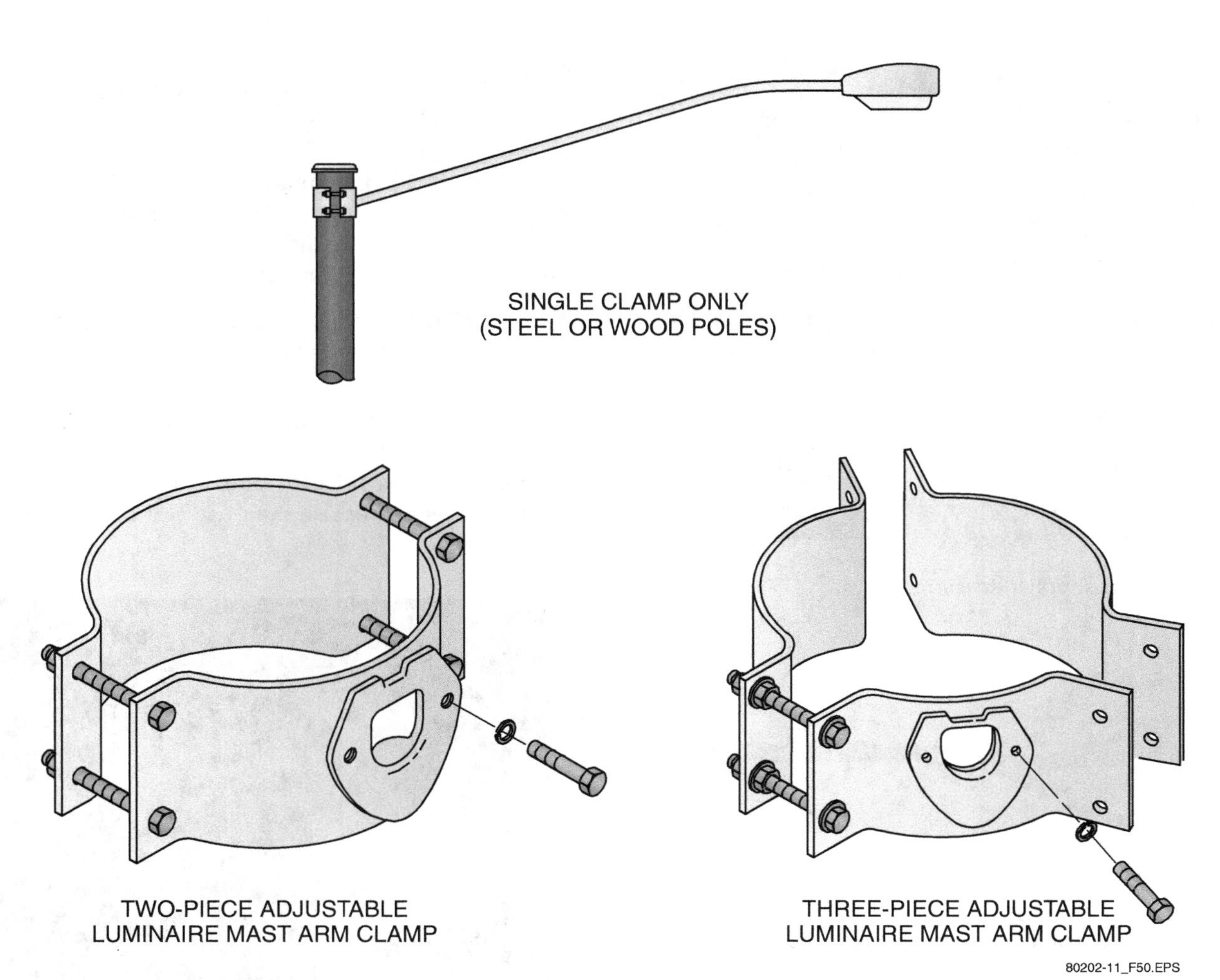

Figure 50 Clamp-on mast arm base.

NOTE

Very long mast arms may require a brace for added support.

The base and the mast arm must be grounded using a No. 6 AWG copper wire connected to a driven ground rod at the base of the pole. The luminaire assembly is hoisted up the pole using a gin pole and block and tackle or other safe rigging practice, and then attached to its base. Conductors are routed out the bottom of the base, allowing enough slack to form a drip loop. The conductors are attached to the overhead secondary cables. *Figure 51* shows a pole-mounted street light with power supplied through an overhead cable. The neutral conductor supports the cable and is attached to a spool-type insulator (*Figure 52*).

CAUTION

Never stand below any object that is being hoisted into place.

7.4.0 Luminaire Control

To conserve energy, street lights are only turned on after dark. A photocell on the top of the luminaire (*Figure 53*) controls operation. It is simple and reliable and is the most widely used device to turn street lights on and off. At dusk, a photocell senses a lack of light and applies power. At dawn, it senses increased light and removes power.

80202-11_F51.EPS

Figure 51 Power supplied to a street light.

80202-11_F52.EPS

Figure 52 Neutral conductor supporting cable.

80202-11_F53.EPS

Figure 53 Photocell control.

GOING GREEN

LED Street Lights

Luminaires that use LEDs for illumination are now available. They offer significant benefits over conventional lamps including:

- 60 to 80 percent reduction in energy costs
- Longer life than conventional lamps
- Equal light output

LEDs are more expensive than conventional lamps and that is hindering their wider use. However, as the price of power increases and the cost of LEDs decreases, their use will become more widespread.

80202-11_SA03.EPS

SUMMARY

Transformers play a key role in a power distribution system. At the power station, they are used to raise the voltage to a very high level so that power can be moved a great distance with low losses. Transformers lower the power at substations, and lower it again to a level that can be used by the customer. Pole-mounted (aerial) transformers are available in a variety of power levels. They also are available with different levels of protection from current surges and lightning strikes. Transformers can be adapted to a variety of voltage distribution systems based on how they are connected. Polarity must be observed when connecting transformers.

Single-phase transformers are widely used in both single- and three-phase distribution systems. Three single-phase transformers can be connected for a three-phase system. The most common three-phase system in use is the wye-wye scheme.

Fuses and lightning arrestors are used protect transformers from current surges. Voltage in a distribution system is maintained with tapped transformers, feeder voltage regulators, and capacitors. A variety of switches are used to isolate components or sections of the line to make repairs or perform routine maintenance. Fault indicators can help workers find a fault so that power can be restored sooner.

The street lights in cities are often owned by the local utility. Power line workers may be required to install and maintain street lights.

Review Questions

1. At a power plant substation, an output transformer is used to _____.
   a. increase voltage
   b. reduce voltage
   c. increase current
   d. modulate frequency

2. Which distribution system voltage is most likely to be found in a lightly populated rural area?
   a. 2,400V
   b. 4,800V
   c. 7,200V
   d. 500kV

3. The temperature-rise rating of an aerial transformer is related to the temperature of the transformer's _____.
   a. high-voltage bushing
   b. windings
   c. outer shell
   d. insulating oil

4. Single-phase transformers can only be used to supply single-phase power to customers.
   a. True
   b. False

5. Which statement about single-phase transformers is *correct*?
   a. Single-phase transformers must have two high-voltage bushings.
   b. The insulating oil used in single-phase transformers is highly flammable.
   c. A secondary circuit breaker with external handle is standard on all single-phase transformers.
   d. Single-phase transformers must have at least one high-voltage bushing.

6. When facing the low-voltage side of a subtractive transformer, the secondary terminal on the right should always be _____.
   a. X3
   b. X1
   c. H1
   d. H2

7. The standard polarity of single-phase transformers sized 200kVA and smaller is _____.
   a. positive
   b. negative
   c. subtractive
   d. additive

8. On a single-phase transformer serving a home, the voltage measured across transformer terminals X1 and X2 should be _____.
   a. 120V
   b. 240V
   c. 264V
   d. 2,400V

9. The most common lightning arrestor used with transformers is a(n) _____.
   a. spark gap
   b. metal-oxide varistor
   c. full-wave rectifier
   d. oil-filled capacitor

10. Which cutout can be used with all distribution system voltages?
    a. Open-link
    b. Enclosed
    c. Open
    d. Locked

11. A high fusing ratio can increase service continuity.
    a. True
    b. False

12. Which statement about tapped transformer primaries is *correct*?
    a. The taps are always changed inside the transformer shell.
    b. All single-phase transformers have primary winding taps.
    c. Tap positions are changed with the transformer de-energized.
    d. Tap switch contacts are immersed in oil to prevent contact arcing.

13. Capacitors are used in a power distribution system to _____.
    a. protect transformers from lightning strikes
    b. bring additional inductance into the system
    c. change the polarity of transformer windings
    d. offset lagging current supplied to motors

14. Which of these isolating device has arcing horns?
    a. Air break switch
    b. Capacitor
    c. Recloser
    d. Oil switch

15. A recloser is similar to a(n) _____.
    a. disconnect switch
    b. fuse
    c. circuit breaker
    d. thermostat

16. Sectionalizers automatically reset after they open.
    a. True
    b. False

17. What electrical principle is used to operate clip-on fault indicators?
    a. Rectification
    b. Modulation
    c. Capacitance
    d. Inductance

18. The live-line tool that has a built-in current limiter is the _____.
    a. extendo stick
    b. hot stick
    c. load break tool
    d. shotgun stick

19. A street light is attached to a pole using a_____.
    a. cable
    b. mast arm
    c. luminaire
    d. crossarm

20. Which method or device is widely used to turn street lights on and off?
    a. Manual operation
    b. Photocell
    c. Time-delay relay
    d. Contactor

Trade Terms Introduced in This Module

Additive polarity: Internal construction of a transformer that allows current in two adjacent primary and secondary terminals to flow in opposite directions. It can be detected if the voltage measured across unconnected high- and low-voltage transformer terminals is greater than the voltage applied across the high-voltage terminals.

Air break switch: A switch used in power distribution systems that can be opened while the line is energized. Arc horns on the switch help break the arc that forms when the energized switch is opened.

Amorphous metal: A metal with a noncrystalline structure that gives it unique magnetic properties. When this metal is used in the core of a transformer, losses are reduced up to 60 percent over laminated steel cores.

Basic impulse level (BIL): A measure of a transformer's ability to withstand a current surge such as what might occur when lightning strikes.

Bushing: A terminal on a transformer where high- and low-voltage conductors are connected. Bushings are often made of porcelain.

Completely self-protected (CSP) transformer: A transformer with built-in secondary current overload protection. The type of protective equipment that is standard on a CSP transformer has to be added on to other transformers. This type of transformer is being replaced with conventional transformers.

Disconnect switch: A type of air break switch that is used to isolate a component. A disconnect must be opened with the line de-energized. For that reason, it does not contain arc horns.

Faulted circuit indicator: A device that provides line workers with a visual indication that a fault has occurred in a line.

Feeder voltage regulator: A regulator used to maintain a constant voltage in distribution feeder circuits even as loads change. A motor-driven tap-changing switch in the secondary windings is used to make changes as small as ⅝ percent increments. Sometimes called a step-voltage regulator.

Fused cutout: A type of switch that contains a fuse cartridge.

Fusing ratio: A ratio that reflects the relationship between the current-carrying capacity of a fuse and the transformer full-load current.

Lightning arrestor: A device used to protect transformers and other distribution components from damage caused by lightning-induced current surges.

Luminaire: A lighting unit consisting of a lamp, together with parts designed to distribute the light, protect the lamp, and connect it to the power source.

Mast arm: The structure that extends the luminaire out from the pole and serves as a conduit for the wiring that powers the lamp.

Metal-oxide varistor: A resistive device commonly used in lightning arrestors. The varistor offers a high resistance to normal system current and voltage. When a surge occurs, the resistance of the varistor breaks down at a preset voltage and conducts the surge to ground. Once the surge passes, the varistor returns to a high value.

Oil switch: A type of disconnect switch used to open energized lines. Switch contacts are immersed in oil to suppress the arc that forms when the switch opens. Oil switches can be opened manually or automatically.

Recloser: An overload device that operates like a circuit breaker. It resets itself after a preset time when a fault occurs. A magnetic solenoid in the recloser opens contacts to stop current flow if fault current occurs. Once current flow stops, the solenoid de-energizes to restore current flow. If the fault is still present, the recloser opens again. If the fault is not temporary, the recloser locks the circuit open after a predetermined number of reset attempts.

Sectionalizer: A type of switch that is used with reclosers to isolate faults in distribution circuits. If a fault continues to occur that results in many openings and resets of a recloser, the sectionalizer activates to isolate (sectionalize) the faulty section of line. Sectionalizers must be reset manually.

Self-protected (SP) transformer: A transformer that has limited protective features such as a lightning arrestor.

Shell: The metal housing containing the transformer. Sometimes called the tank.

Shell-type construction: A type of transformer construction in which the core surrounds the transformer windings instead of the coils surrounding the core.

Subtractive polarity: Internal construction of a transformer that allows current in two adjacent primary and secondary terminals to flow in the same direction. It can be detected if the voltage measured across unconnected high- and low-voltage transformer terminals is less than the voltage applied across the high-voltage terminals.

Transformer taps: Points on a transformer's windings that can be connected to vary voltage. The taps increase or decrease the number of windings available in the transformer, thus varying voltage. Taps can be in the primary or secondary windings.

Additional Resources

This module presents thorough resources for task training. The following resource material is suggested for further study.

The Lineman's and Cableman's Handbook, 11th Edition. New York, NY: McGraw-Hill.
National Electrical Safety Code C2-2007. New York, NY: Institute of Electrical and Electronics Engineers.
Electrical Power Distribution and Transmission. Upper Saddle River, NJ: Prentice Hall.

Figure Credits

International Renewable Energy Facilitation Company-www.inrefco.com & JDW & Associates, Module opener and SA02

Topaz Publications, Inc., Figures 1, 2, 6, 15–17, 19 (right photo), 21, 29, 45, 49, 51, and 53

Cooper Power Systems, Table 1 and Figure 22

Howard Industries, Inc., Figures 27 and SA01

ERMCO, Figure 3

Mike Wilson, Figures 4, 5, 18, 19 (left photo), 26, 31, 34, and 35

GE Consumer & Industrial, Figure 7

John Traister, Figure 8 (top illustration)

ABB North America, Figure 20

Hubbell Power Systems, Figures 30, 32, 33, 36, and 37

Photo courtesy of Schweitzer Engineering Laboratories, Inc., Figures 38–40

Hastings, Figures 41, 43, and 44

Salisbury Electrical Safety, Figures 42 and 46

S&C Electric Company, Figures 47 and 48

Pelco Products, Inc., Figure 50

Jim Mitchem, Figure 52

LEDtronics, Inc., www.LEDtronics.com, SA03

NCCER CURRICULA — USER UPDATE

NCCER makes every effort to keep its textbooks up-to-date and free of technical errors. We appreciate your help in this process. If you find an error, a typographical mistake, or an inaccuracy in NCCER's curricula, please fill out this form (or a photocopy), or complete the online form at **www.nccer.org/olf**. Be sure to include the exact module ID number, page number, a detailed description, and your recommended correction. Your input will be brought to the attention of the Authoring Team. Thank you for your assistance.

Instructors – If you have an idea for improving this textbook, or have found that additional materials were necessary to teach this module effectively, please let us know so that we may present your suggestions to the Authoring Team.

NCCER Product Development and Revision
13614 Progress Blvd., Alachua, FL 32615

Email: curriculum@nccer.org
Online: www.nccer.org/olf

❑ Trainee Guide ❑ AIG ❑ Exam ❑ PowerPoints Other ________________

Craft / Level: Copyright Date:

Module ID Number / Title:

Section Number(s):

Description:

Recommended Correction:

Your Name:

Address:

Email: Phone:

80203-11

Cable and Conductor Installation and Removal

Module Three

Trainees with successful module completions may be eligible for credentialing through NCCER's National Registry. To learn more, go to **www.nccer.org** or contact us at **1.888.622.3720**. Our website has information on the latest product releases and training, as well as online versions of our *Cornerstone* newsletter and Pearson's product catalog.

Your feedback is welcome. You may email your comments to **curriculum@nccer.org,** send general comments and inquiries to **info@nccer.org**, or fill in the User Update form at the back of this module.

V.1 12/11

80203-11
Cable and Conductor Installation and Removal

Objectives

When you have completed this module, you will be able to do the following:

1. Install cables and conductors.
2. Describe how to remove cables and conductors.
3. Splice and terminate cables and conductors.
4. Explain how to select and size a conductor for a given application.
5. Operate cable-pulling equipment.

Performance Tasks

Under the supervision of your instructor, you should be able to do the following:

1. Install cables and conductors.
2. Splice and terminate cables and conductors.
3. Operate cable-pulling equipment.

Trade Terms

Alumoweld
Aluminum conductor steel reinforced (ACSR)
Armor rods
Conductor splice
Copperweld
Crimp-on splice
Full-tension splice
Loading district
Messenger
Puller/tensioner
Repair sleeve
Service drop
Splice shunt
Tensile strength
Tension stringing method
Western Union tie

Industry Recognized Credentials

If you're training through an NCCER-accredited sponsor you may be eligible for credentials from NCCER's Registry. The ID number for this module is 80203-11. Note that this module may have been used in other NCCER curricula and may apply to other level completions. Contact NCCER's Registry at 888.622.3720 or go to nccer.org for more information.

Contents

Topics to be presented in this module include:

1.0.0 Introduction ... 1
2.0.0 Safety ... 1
3.0.0 Cables and Conductors ... 2
3.1.0 Characteristics of Conductors ... 2
3.1.1 Conductor Conductivity ... 2
3.1.2 Conductor Strength ... 2
4.0.0 Selecting Conductors ... 4
4.1.0 Line Current and Voltage ... 4
4.2.0 Snow and Ice Loads ... 4
4.3.0 Wind Loads ... 4
4.4.0 Ambient Temperature ... 8
4.5.0 Material Cost ... 9
5.0.0 Installing Cables and Conductors ... 9
5.1.0 Stringing Conductors ... 9
5.2.0 Dead-End Conductors ... 10
5.3.0 Tensioning and Sagging Conductors ... 10
5.4.0 Tie Conductors to Insulators ... 11
5.4.1 Insulator Types ... 12
5.4.2 Conductor Ties ... 12
5.5.0 Installing Overhead Cables ... 14
6.0.0 Repair and Replace Cables and Conductors ... 15
6.1.0 Full-Tension Splice ... 17
6.2.0 Full-Tension Crimp-On Splice ... 17
6.3.0 Conductor Crimp-On Splice ... 18
6.4.0 Splice Shunt ... 18
6.5.0 Repair Sleeve ... 19
6.6.0 Automatic Splice ... 19
7.0.0 Cable-Pulling Equipment ... 19

Figures and Tables

Figure 1 Aerial cable use 3
Figure 2 Comparison of wire sizes 4
Figure 3 Loading districts 7
Figure 4 Conductor spacer 7
Figure 5 Air flow spoiler 7
Figure 6 Temperature effect on conductors 8
Figure 7 Bullwheel tensioner 9
Figure 8 Pulling machine 9
Figure 9 Stringing block 10
Figure 10 Basket grips installed on conductors 10
Figure 11 Installing bands on the Kellems grip 11
Figure 12 Spider® system 12
Figure 13 Dead-end clamp 12
Figure 14 Dead-end insulator 12
Figure 15 Tensioning and sagging conductors 13
Figure 16 Dynamometer. 13
Figure 17 Sag chart 13
Figure 18 Sag scope 14
Figure 19 Insulator 14
Figure 20 Plastic clamp-style insulator 14
Figure 21 Conductor ties 14
Figure 22 Prefabricated tie kit 15
Figure 23 Prefabricated tie kit on top-tie insulator 15
Figure 24 Prefabricated tie kit on side-tie insulator 15
Figure 25 Western Union tie 16
Figure 26 Correct placement of armor rods 16
Figure 27 Energizing new conductors 16
Figure 28 Service drop 16
Figure 29 Conductor splice 17
Figure 30 Full tension splice (non-crimp) 18
Figure 31 Crimping tool 18
Figure 32 Full-tension crimp-on splice 18
Figure 33 Conductor crimp-on splice 19
Figure 34 Splice shunt 19
Figure 35 Repair sleeve 19
Figure 36 Automatic splice 20
Figure 37 Tension stringing conductors 20
Figure 38 Puller/tensioner 20

Table 1 Aluminum Conductor Types 3
Table 2 ACSR Physical Properties 5
Table 3 ACSR Electrical Properties 6
Table 4 Loading District Values 7
Table 5 Tie Wire Sizes and Applications 15

1.0.0 INTRODUCTION

Power is supplied to customers through an aerial distribution system that contains transformers, conductors, and different devices to protect and/or isolate parts of the system. The conductors and cables that tie the system together can be damaged by accidents or weather. Upgrades to the system may require that new conductors and cables be installed and/or existing ones replaced. Power line workers are tasked to install and repair cables and conductors. In this module, you will learn how to install, repair, and remove conductors and cables.

2.0.0 SAFETY

All power line workers are required to wear personal protective equipment (PPE) to prevent on-the-job injuries.

- *Clothing* – Clothing should be made of natural fibers, such as cotton or wool, and treated to be arc-resistant and fire-retardant. Avoid clothing made of certain synthetic fibers, such as rayon, nylon, and polyester, which could burn or melt and cling to skin. Pant legs should extend over the top of shoes or boots. Never wear shorts on the job. Shirts should be long-sleeved to protect the skin from the sun, to minimize injuries caused by wood splinters, and to provide protection from chemicals used to treat the wood against rot. On most job sites, workers are required to wear a bright-colored shirt or vest to increase their visibility.
- *Footwear* – Only safety-toe leather shoes or boots should be worn on the job. High-top boots have the added advantage of greater support for the ankles. Many companies require dielectrically rated boots or overshoes. Never wear sandals or canvas-type shoes on the job.
- *Eye protection* – Safety glasses with side shields in accordance with *ANSI Standard Z87.1* must be worn at all times while on the job. If a power tool such as a chain saw is used on the job, a full face-shield may be necessary in addition to the safety glasses.
- *Hearing protection* – If power tools or noisy machinery such as a chain saw, trencher, or air compressor are used on the job, approved earmuffs or earplugs are generally required.
- *Dust mask* – If a wood pole is cut or drilled, a dust mask must be worn. The preservatives in the resulting sawdust may be harmful if inhaled.
- *Gloves* – Working on wood poles used in aerial distribution systems can be rough on hands. Leather work gloves provide protection from rope burns, wood splinters, and other hazards.
- *Rubber gloves and sleeves* – Power line workers often work near energized lines or equipment. Rubber gloves and sleeves protect against electrical shock. Use an air and water test on the gloves and roll the sleeves in each direction to look for damage. Inspect all other PPE, and tools according to the manufacturer's recommendations to ensure they are safe to use.
- *Safety helmet/hard hat/hard cap* – An approved *ANSI Standard Z89.1 Class E* safety helmet made of a non-conducting plastic material, such as polypropylene or polycarbonate, must be worn at all times while on the job site. The helmet protects the head from injuries caused by bumps and falling objects, and it provides shade from the sun.

In addition to wearing the proper PPE, follow these safety precautions when working with cables, conductors, and their related devices:

- Avoid installing or removing aerial equipment in wet or bad weather except in emergencies.
- Use protective covers, line hose, rubber blankets, and shields around conductors and other components when working near energized lines.
- Before starting any job, inspect and test rubber gloves and sleeves using the water and/or air method. Inspect all PPE and tools to ensure they are safe to use.
- Only use tools and equipment that are approved for power line use.
- Use fall protection equipment when working at heights above six feet (four feet for unqualified climbers).
- If an insulated lift bucket is used for bare hand work, test the insulating arm for excessive leakage current each day before it is used. A non-barehand bucket requires an annual boom leakage test.

On Site

Metal Purity

The copper or aluminum used to make conductors must have a very high purity. Contaminants in the metal can reduce its conductivity. Copper is a metal that is heavily recycled. However, recycled copper is not used to make conductors because it is too costly to refine it to the required purity level.

- Have a qualified observer at all job sites to ensure that minimum approach distances to energized equipment are observed, safe work practices are being followed, and correct PPE and tools are being used.

> **NOTE**
>
> A qualified observer is a member of the work crew with the primary duty of ensuring that clearances to energized equipment are maintained, safe work practices are followed, and appropriate PPE and protective shields are being used. That person is familiar with all aspects of installation and construction of distribution systems and is trained in first aid. While performing the duties of a qualified observer, that person cannot perform any other tasks.

3.0.0 CABLES AND CONDUCTORS

Conductors and/or cables are used to carry electric power. It might seem that the two terms are interchangeable, but there is a difference. In power distribution, the term *cable* usually means a conductor that is insulated. A conductor, on the other hand, does not have any insulation. Conductors are used in aerial power distribution systems. Cables are used in underground power distribution systems, but have limited use in aerial systems. A typical use of aerial cable is for **service drop** conductors from a pole to a customer (*Figure 1*). The focus of this module is on conductors and cables used in aerial distribution systems. Cables used in underground distribution systems are covered in a later module.

3.1.0 Characteristics of Conductors

Conductors used in power distribution systems must have characteristics that enable them to deliver power in a safe and efficient manner. Desirable characteristics include good conductivity and high strength.

3.1.1 Conductor Conductivity

A conductor is a material that allows the easy flow (conductivity) of electric current. Most metals are good conductors. For example, gold and silver have excellent conductivity but their high cost precludes their use in distribution systems. Instead, less costly metals such as copper, aluminum, and steel, are used. Copper is very close to silver in conductivity. Copper is cheap and plentiful, making it an excellent choice.

Aluminum is another widely used conductor material. Its conductivity is not as good as copper, but that disadvantage is outweighed by its lighter weight and lower cost. Steel has strength but much lower conductivity than copper or aluminum and can rust if not treated. A steel conductor that is coated with copper is called a **copperweld** conductor. Steel that is coated with aluminum is an **alumoweld** conductor. The copper or aluminum coats the steel and increases its conductivity. The coating also prevents rust. A popular conductor material is **aluminum conductor steel reinforced (ACSR)**. This conductor has a steel core that is coated with aluminum. The core is then wrapped with aluminum conductors. This combination provides the strength of steel, corrosion resistance, and low cost. Modern distribution system conductors tend to be aluminum because it offers good conductivity, good strength, and a much lower cost than copper.

Table 1 lists common types of bare aluminum conductors used for overhead service. To obtain the same conductivity, an aluminum conductor must have 1.66 times the cross-sectional area of a copper conductor. However, even at this larger size, the less-costly aluminum still has 75 percent of the **tensile strength** but only 55 percent of the weight of copper. That is why aluminum is the material of choice for conductors.

Table 1 Aluminum Conductor Types

Type	Construction
AAAC	All aluminum alloy conductor
AAC	All aluminum conductor
ACAR	All aluminum conductor alloy reinforced
ACCC®	Aluminum conductor composite core
ACSS	Aluminum conductor steel supported
ACS/TW	Aluminum conductor steel supported (Trapezoidal)
ACSR	Aluminum conductor steel reinforced
ACSR/SD	Aluminum conductor steel reinforced (Self-daming)
ACSR/TW	Compact aluminum conductor trapezoidal wire steel

80203-11_T01.EPS

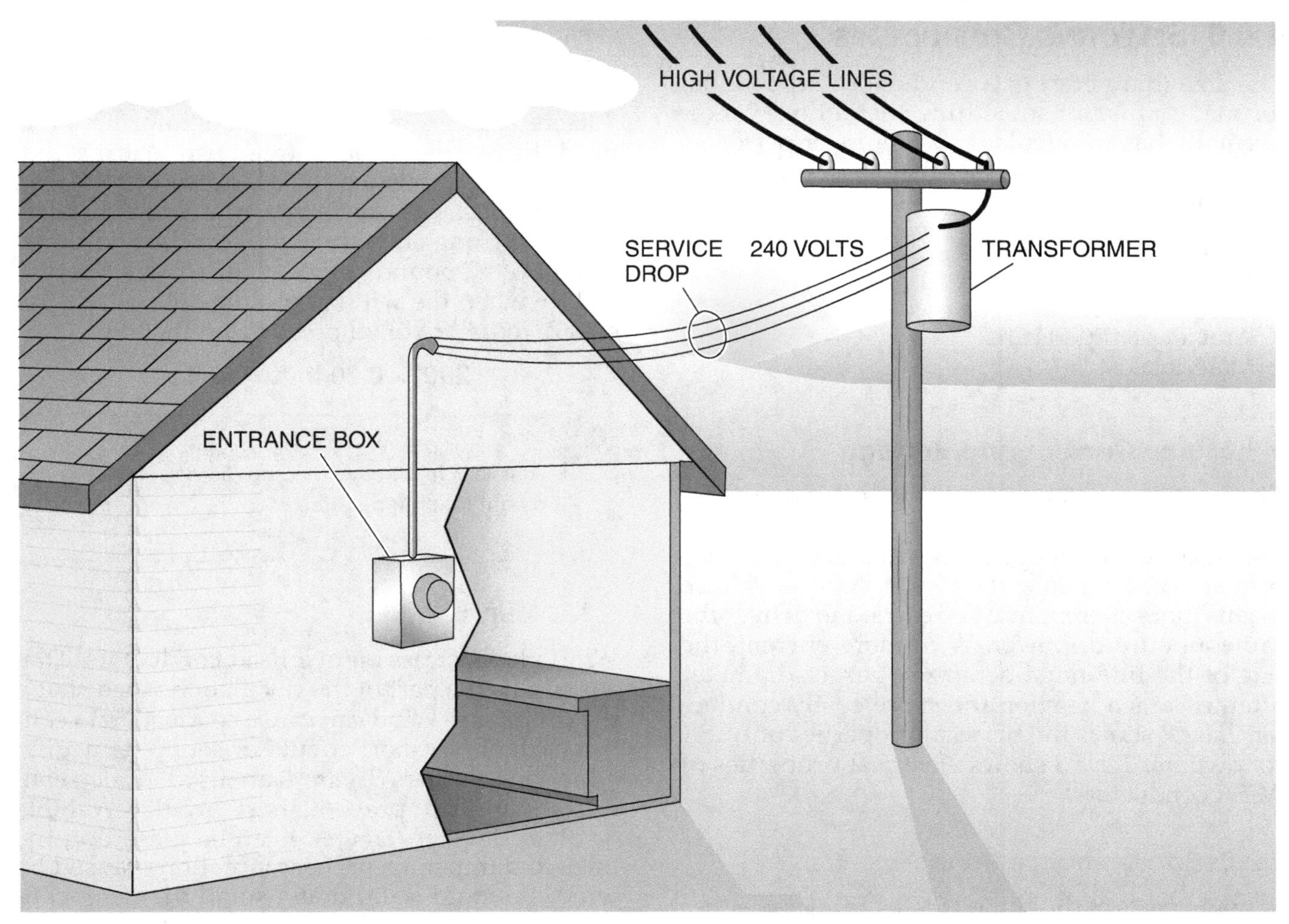

Figure 1 Aerial cable use.

3.1.2 Conductor Strength

A conductor must have the strength to be able to withstand the stresses placed upon it. It must have the strength to support its own weight and the added weight of ice or snow. Other forces that can stress a conductor include wind, and expansion and contraction caused by temperature changes. There are several factors that contribute to the strength of a conductor including tensile strength and the diameter of the conductor.

Tensile strength can be defined as the ability of a material to resist being pulled apart. Different metals have different tensile strengths. For example, copper has a higher tensile strength than aluminum. When selecting a material for a conductor, select one with enough tensile strength to handle the loads placed upon it without failing. The diameter of a conductor affects its strength. For a given material, a larger-diameter conductor is stronger than one with a smaller diameter. However, increased diameter means more weight.

NOTE

The strength of ACSR compared to aluminum without a steel core offers utilities several benefits. Since ACSR is stronger, the distance between poles can be increased. This translates to fewer poles and less time spent installing the poles and their related hardware.

Conductor (wire) sizes are expressed in numbers. The most common number system is the American Wire Gauge (AWG). Generally, the larger the gauge number, the smaller the diameter of the wire will be (*Figure* 2). Smaller diameter conductors tend to be solid, while sizes No. 6 AWG and larger are stranded. Stranded conductors are more flexible and easier to handle than solid conductors. Conductors are sometimes sized using circular mils. A circular mil is a circle with a diameter of 1 mil (0.001 in). When a conductor size is 250 kcmil, the cross-sectional area of the conductor(s) that carries the current is 250,000 circles having a diameter of 0.001 inch.

4.0.0 Selecting Conductors

The size (diameter) of a conductor and conductor material is not something that a line worker normally has to calculate. An engineer performs that task and bases the calculation on the following factors:

- Line current and voltage
- Weight of ice or snow
- Wind loads
- Ambient temperature
- Material cost

4.1.0 Line Current and Voltage

The voltage and current that a line carries has a direct impact on the size of the conductor. A higher primary voltage means that current in the primary conductors will be lower. A lower voltage means more current must be carried to deliver the same amount of power. With more current, the size of the line must be larger. Larger diameter often means a heavier and more costly conductor. *Table 2* shows the physical properties of ACSR conductors. *Table 3* shows electrical properties of ACSR conductors.

4.2.0 Snow and Ice Loads

In many parts of the United States, winter weather can wreak havoc on overhead power lines. Freezing rain can coat lines, adding enough weight to cause them to fail. The US Bureau of Standards has divided the continental United States into three loading districts. They are labeled light, medium, and heavy to represent different winter weather conditions (*Figure 3*). In each loading district, an ice thickness, wind pressure, and temperature are specified which are used to develop constants in pounds per foot of length of conductor. *Table 4* lists the values and constants for the three loading districts.

Here is an example of how this information is used. In North Carolina (medium loading district), poles supporting the conductors are placed 200 feet apart. Conductors spanning that distance must be strong enough to support the additional weight of 40 pounds of ice at an air temperature of 15°F when the wind is placing a 4 pounds per square foot horizontal pressure on the line.

$$200' \times 0.20 \text{ lb/ft} = 40 \text{ lbs}$$

NOTE

Alaska is in a heavy loading district. Hawaii is in a light loading district.

4.3.0 Wind Loads

Wind places stress on overhead conductors. That stress must be part of the calculation when sizing the conductor. Wind can cause conductors to contact each other (short circuit) or set up damaging vibrations. Spacers (*Figure 4*) are used to keep conductors apart to prevent short circuits. Weights or other devices (*Figure 5*) can be placed on the lines to dampen or prevent vibrations caused by wind. It would seem that a small diameter conductor would not offer that much resistance to wind. That is incorrect. Let's see how much area a ½-inch (0.500) conductor spanning 200 feet presents to the wind. First find the conductor diameter in feet by dividing the diameter by 12. Then multiply the conductor diameter in feet by the length of the conductor:

$$0.500'' \div 12 = 0.042'$$
$$0.042' \times 200' = 8.4 \text{ sqft}$$

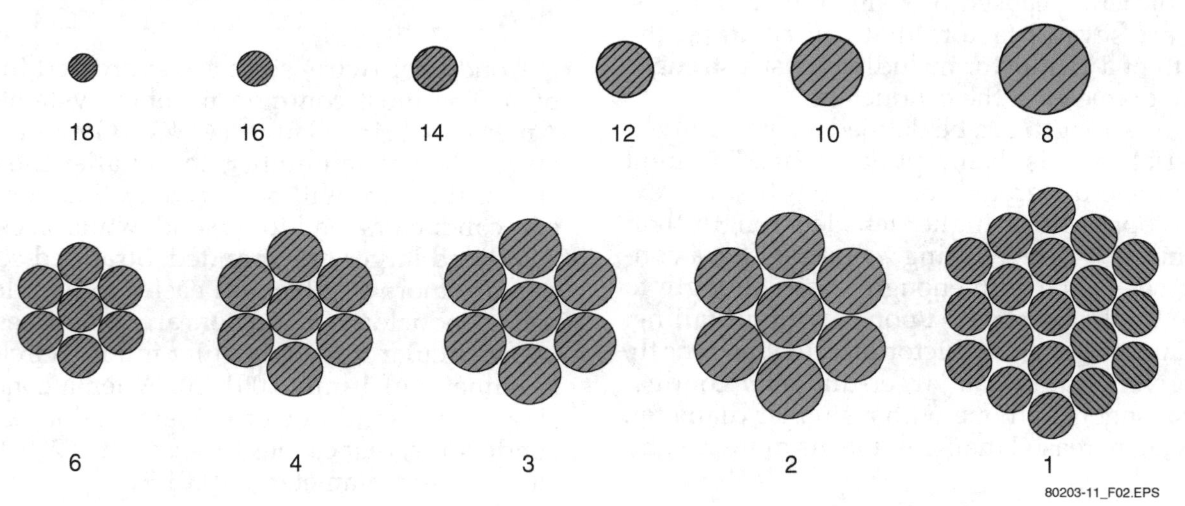

Figure 2 Comparison of wire sizes.

ACSR

Aluminum Conductor Steel Reinforced

ASTM: B230, Specification for Aluminum 1350-H19 Wire for Electrical Purposes
B232, Specification for Concentric-Lay-Stranded Aluminum Conductors, Coated Steel Reinforced (ACSR)
B498, Specification for Zinc-Coated (Galvanized) Steel Core Wire for Aluminum Conductors Steel Reinforced (ACSR)

Physical Properties

CODE WORD	SIZE AWG or kcmil	STRANDING Number & Diameter (In.) Aluminum	Steel	Nominal Diameter (In.) Complete Conductor	Steel Core	Rated Strength (Lbs.)	WEIGHT (Lbs./1000 Ft.) Total	Aluminum	Steel	PERCENT OF TOTAL WEIGHT Aluminum	Steel
TURKEY	6	6 x 0.0661	1 x 0.0661	0.198	0.0661	1190	36.0	24.4	11.6	67.8	32.2
SWAN	4	6 x 0.0834	1 x 0.0834	0.250	0.0834	1860	57.4	39.0	18.4	67.9	32.1
SWANATE	4	7 x 0.0772	1 x 0.1029	0.257	0.1029	2360	67.0	39.0	28.0	58.2	41.8
SPARROW	2	6 x 0.1052	1 x 0.1052	0.316	0.1052	2850	91.2	61.9	29.3	67.9	32.1
SPARATE	2	7 x 0.0974	1 x 0.1299	0.325	0.1299	3640	106.6	61.9	44.7	58.1	41.9
ROBIN	1	6 x 0.1181	1 x 0.1181	0.354	0.1181	3550	115.0	78.1	36.9	67.9	32.1
RAVEN	1/0	6 x 0.1327	1 x 0.1327	0.398	0.1327	4380	145.2	98.6	46.6	67.9	32.1
QUAIL	2/0	6 x 0.1489	1 x 0.1489	0.447	0.1489	5300	182.8	124.1	58.7	67.9	32.1
PIGEON	3/0	6 x 0.1672	1 x 0.1672	0.502	0.1672	6620	230.5	156.4	74.1	67.9	32.1
PENGUIN	4/0	6 x 0.1878	1 x 0.1878	0.563	0.1878	8350	290.8	197.4	93.4	67.9	32.1
WAXWING	266.8	18 x 0.1217	1 x 0.1217	0.609	0.1217	6880	289.1	249.9	39.2	86.4	13.6
PARTRIDGE	266.8	26 x 0.1013	7 x 0.0788	0.642	0.236	11300	366.9	251.3	115.6	68.5	31.5
MERLIN	336.4	18 x 0.1367	1 x 0.1367	0.684	0.1367	8700	364.8	315.3	49.5	86.4	13.6
LINNET	336.4	26 x 0.1137	7 x 0.0884	0.720	0.265	14100	462.0	316.5	145.5	68.5	31.5
ORIOLE	336.4	30 x 0.1059	7 x 0.1059	0.741	0.318	17300	526.4	317.7	208.7	60.4	39.6
CHICKADEE	397.5	18 x 0.1486	1 x 0.1486	0.743	0.1486	9900	431.0	372.5	58.5	86.4	13.6
IBIS	397.5	26 x 0.1236	7 x 0.0961	0.783	0.288	16300	546.0	374.1	171.9	68.5	31.5
LARK	397.5	30 x 0.1151	7 x 0.1151	0.806	0.345	20300	621.8	375.2	246.6	60.4	39.6
PELICAN	477	18 x 0.1628	1 x 0.1628	0.814	0.1628	11800	517.3	447.1	70.2	86.4	13.6
FLICKER	477	24 x 0.1410	7 x 0.0940	0.846	0.282	17200	613.9	449.4	164.5	73.2	26.8
HAWK	477	26 x 0.1354	7 x 0.1053	0.858	0.316	19500	655.3	448.9	206.4	68.5	31.5
HEN	477	30 x 0.1261	7 x 0.1261	0.883	0.378	23800	746.4	450.4	296.0	60.4	39.6
OSPREY	556.5	18 x 0.1758	1 x 0.1758	0.879	0.1758	13700	603.3	521.4	81.9	86.4	13.6
PARAKEET	556.5	24 x 0.1523	7 x 0.1015	0.914	0.304	19800	716.1	524.3	191.8	73.2	26.8
DOVE	556.5	26 x 0.1463	7 x 0.1138	0.927	0.341	22600	765.2	524.2	241.0	68.5	31.5
EAGLE	556.5	30 x 0.1362	7 x 0.1362	0.953	0.409	27800	870.7	525.4	345.3	60.4	39.6
PEACOCK	605	24 x 0.1588	7 x 0.1059	0.953	0.318	21600	778.8	570.1	208.7	73.2	26.8
SWIFT	636	36 x 0.1329	1 x 0.1329	0.930	0.1329	13800	642.8	596.0	46.8	92.7	7.3
KINGBIRD	636	18 x 0.1880	1 x 0.1880	0.940	0.1880	15700	689.9	596.3	93.6	86.4	13.6
ROOK	636	24 x 0.1628	7 x 0.1085	0.977	0.326	22600	818.2	599.1	219.1	73.2	26.8
GROSBEAK	636	26 x 0.1564	7 x 0.1216	0.990	0.365	25200	874.2	599.0	275.2	68.5	31.5
EGRET	636	30 x 0.1456	19 x 0.0874	1.019	0.437	31500	987.2	600.5	386.7	60.8	39.2
FLAMINGO	666.6	24 x 0.1667	7 x 0.1111	1.000	0.333	23700	857.9	628.2	229.7	73.2	26.8
STARLING	715.5	26 x 0.1659	7 x 0.1290	1.051	0.387	28400	983.7	674.0	309.7	68.5	31.5
REDWING	715.5	30 x 0.1544	19 x 0.0926	1.081	0.463	34600	1109.3	675.3	434.0	60.8	39.2
COOT	795	36 x 0.1486	1 x 0.1486	1.040	0.1486	16800	803.6	745.1	58.5	92.7	7.3
TERN	795	45 x 0.1329	7 x 0.0886	1.063	0.266	22100	895	749	146	83.7	16.3
CUCKOO	795	24 x 0.1820	7 x 0.1213	1.092	0.364	27900	1023	749	274.0	73.2	26.8
CONDOR	795	54 x 0.1213	7 x 0.1213	1.092	0.364	28200	1022	748	274.0	73.2	26.8
DRAKE	795	26 x 0.1749	7 x 0.1360	1.108	0.408	31500	1093	749	344	68.5	31.5
MALLARD	795	30 x 0.1628	19 x 0.0977	1.140	0.489	38400	1233.9	750.7	483.2	60.8	39.2
RUDDY	900	45 x 0.1414	7 x 0.0943	1.131	0.283	24400	1013	848	165	83.7	16.3
CANARY	900	54 x 0.1291	7 x 0.1291	1.162	0.387	31900	1158	848	310	73.2	26.8
CORNCRAKE	954	20 x 0.2184	7 x 0.0971	1.165	0.291	25600	1074	899	175	83.7	16.3
REDBIRD	954	24 x 0.1994	7 x 0.1329	1.196	0.399	33500	1228	899	329	73.2	26.8
TOWHEE	954	48 x 0.1410	7 x 0.1097	1.175	0.329	28500	1123	899	224	80.1	19.9
RAIL	954	45 x 0.1456	7 x 0.0971	1.165	0.291	25900	1075	899	176	83.7	16.3
CARDINAL	954	54 x 0.1329	7 x 0.1329	1.196	0.399	33800	1227.1	898.4	328.7	73.2	26.8
ORTOLAN	1033.5	45 x 0.1515	7 x 0.1010	1.212	0.303	27700	1163	973	190	83.7	16.3
CURLEW	1033.5	54 x 0.1383	7 x 0.1383	1.245	0.415	36600	1329	973	356	73.2	26.8
BLUEJAY	1113.0	45 x 0.1573	7 x 0.1049	1.259	0.315	29800	1254	1049	205	83.7	16.3
FINCH	1113.0	54 x 0.1436	19 x 0.0862	1.293	0.431	39100	1430	1054	376	73.7	26.3
BUNTING	1192.5	45 x 0.1628	7 x 0.1085	1.302	0.326	32000	1342	1123	219	83.7	16.3
GRACKLE	1192.5	54 x 0.1486	19 x 0.0892	1.338	0.446	41900	1531	1128	403	73.7	26.3
SKYLARK	1272.0	36 x 0.1880	1 x 0.1880	1.316	0.1880	26400	1286	1192	94	92.7	7.3
BITTERN	1272.0	45 x 0.1681	7 x 0.1121	1.345	0.336	34100	1432	1198	234	83.7	16.3
PHEASANT	1272.0	54 x 0.1535	19 x 0.0921	1.382	0.461	43600	1634	1205	429	73.7	26.3
DIPPER	1351.5	45 x 0.1733	7 x 0.1155	1.386	0.347	36200	1521	1273	248	83.7	16.3
MARTIN	1351.5	54 x 0.1582	19 x 0.0949	1.424	0.475	46300	1735	1279	456	73.7	26.3
BOBOLINK	1431.0	45 x 0.1783	7 x 0.1189	1.427	0.357	38300	1611	1348	263	83.7	16.3
PLOVER	1431.0	54 x 0.1628	19 x 0.0977	1.465	0.489	49100	1838	1355	483	73.7	26.3
LAPWING	1590.0	45 x 0.1880	7 x 0.1253	1.504	0.376	42200	1790	1498	292	83.8	16.3
FALCON	1590.0	54 x 0.1716	19 x 0.1030	1.545	0.515	54500	2042	1505	537	73.7	26.3
CHUKAR	1780.0	84 x 0.1456	19 x 0.0874	1.602	0.437	51000	2072	1685	387	81.3	18.7
MOCKINGBIRD	2034.5	72 x 0.1681	7 x 0.1122	1.681	0.337	46800	2163	1929	234	89.2	18.7
BLUEBIRD	2156.0	84 x 0.1602	19 x 0.0961	1.762	0.481	60300	2508	2040	468	81.3	18.7
KIWI	2167.0	72 x 0.1735	7 x 0.1157	1.735	0.347	49800	2301	2052	249	89.2	10.8
THRASHER	2312.0	76 x 0.1744	19 x 0.0814	1.802	0.407	56700	2523	2188	335	86.7	13.3
JOREA	2515.0	76 x 0.1819	19 x 0.0850	1.880	0.425	61700	2749	2383	366	86.7	13.3
High Strength ACSR											
GROUSE	80.0	8 x 0.1000	1 x 0.1670	0.367	0.1670	5200	148.8	74.9	73.9	50.3	49.7
PETREL	101.8	12 x 0.0921	7 x 0.0921	0.461	0.276	10400	253.8	95.9	157.9	37.8	62.2
MINORCA	110.8	12 x 0.0961	7 x 0.0961	0.481	0.288	11300	276.3	104.4	171.9	37.8	62.2
LEGHORN	134.6	12 x 0.1059	7 x 0.1059	0.530	0.318	13600	335.5	126.8	208.7	37.8	62.2
GUINEA	159.0	12 x 0.1151	7 x 0.1151	0.576	0.345	16000	396.3	149.7	246.6	37.8	62.2
DOTTEREL	176.9	12 x 0.1214	7 x 0.1214	0.607	0.364	17300	440.9	166.6	274.3	37.8	62.2
DORKING	190.8	12 x 0.1261	7 x 0.1261	0.631	0.378	18700	475.7	179.7	296.0	37.8	62.2
COCKHIN	211.3	12 x 0.1327	7 x 0.1327	0.664	0.398	20700	526.8	199.0	327.8	37.8	62.2
BRAHMA	203.2	16 x 0.1127	19 x 0.0977	0.714	0.489	28400	674.6	191.4	483.2	28.4	71.6

80203-11_T02.EPS

Table 3 ACSR Electrical Properties

ACSR

ALCAN

Aluminum Conductor Steel Reinforced

Electrical Properties

CODE WORD	SIZE & STRANDING: AWG or kcmil	SIZE & STRANDING: Aluminum/ Steel	RESISTANCE: DC (Ohms/1000 Ft.) @20°	RESISTANCE: AC-60-HZ(Ohms/1000 Ft.) @25° C	AC-60-HZ @50° C	AC-60-HZ @75° C	60 HZ REACTANCE 1 FOOT EQUIVALENT SPACING: Capacitive (Megohms-1000 Ft.)	Inductive (Ohms/1000 Ft.) @25° C	Inductive @50° C	Inductive @75° C
TURKEY	6	6/1	0.6419	0.6553	0.750	0.8159	0.7513	0.1201	0.1390	0.1439
SWAN	4	6/1	0.4032	0.4119	0.4794	0.5218	0.7149	0.1152	0.1314	0.1369
SWANATE	4	7/1	0.3989	0.4072	0.4633	0.5165	0.7102	0.11533	0.1239	0.1303
SPARROW	2	6/1	0.2534	0.2591	0.3080	0.3360	0.6785	0.1100	0.1235	0.1277
SPARATE	2	7/1	0.2506	0.2563	0.2966	0.3297	0.6737	0.1081	0.1176	0.1206
ROBIN	1	6/1	0.2011	0.2059	0.2474	0.2703	0.6600	0.1068	0.1191	0.1224
RAVEN	1/0	6/1	0.1593	0.1633	0.1972	0.2161	0.6421	0.1040	0.1138	0.1163
QUAIL	2/0	6/1	0.1265	0.1301	0.1616	0.1760	0.6241	0.1017	0.1117	0.1135
PIGEON	3/0	6/1	0.1003	0.1034	0.1208	0.1445	0.6056	0.0992	0.1083	0.1095
PENGUIN	4/0	6/1	0.0795	0.0822	0.1066	0.1157	0.5966	0.0964	.01047	0.1053
								Inductive (Ohms/1000 Ft.)	GMR (Ft.)	
WAXWING	266.8	18/1	0.0644	0.0657	0.0723	0.0788	0.576	0.0934	0.0197	
PARTRIDGE	266.8	26/7	0.0637	0.0652	0.0714	0.0778	0.565	0.0881	0.0217	
MERLIN	336.4	18/1	0.0510	0.0523	0.0574	0.0625	0.560	0.0877	0.0221	
LINNET	336.4	26/7	0.0506	0.0517	0.0568	0.0619	0.549	0.0854	0.0244	
ORIOLE	336.4	30/7	0.0502	0.0513	0.0563	0.0614	0.544	0.0843	0.0255	
CHICKADEE	397.5	18/1	0.0432	0.0443	0.0487	0.0528	0.544	0.0856	0.0240	
IBIS	397.5	26/7	0.0428	0.0438	0.0481	0.0525	0.539	0.0835	0.0265	
LARK	397.5	30/7	0.0425	0.0434	0.0477	0.0519	0.533	0.0824	0.0277	
PELICAN	477.0	18/1.	0.0360	0.0369	0.0405	0.0441	0.528	0.0835	0.0263	
FLICKER	477.0	24/7	0.0358	0.0367	0.0403	0.0439	0.524	0.0818	0.0283	
HAWK	477.0	26/7	0.0357	0.0366	0.0402	0.0438	0.522	0.0814	0.0290	
HEN	477.0	30/7	0.0354	0.0362	0.0389	0.0434	0.517	0.0803	0.0304	
OSPREY	556.5	18/1	0.0309	0.0318	0.0348	0.0379	0.518	0.0818	0.0284	
PARAKEET	556.5	24/7	0.0307	0.0314	0.0347	0.0377	0.512	0.0801	0.0306	
DOVE	556.5	26/7	0.0305	0.0314	0.0345	0.0375	0.510	0.0795	0.0313	
EAGLE	556.5	30/7	0.0300	0.0311	0.0341	0.0371	0.505	0.0786	0.0328	
PEACOCK	605.0	24/7	0.0282	0.0290	0.0378	0.0347	0.505	0.0792	0.0319	
SWIFT	636.0	36/1	0.0267	0.0281	0.0307	0.0334	0.509	0.0806	0.0300	
KINGBIRD	636.0	18/1	0.0269	0.0278	0.0306	0.0332	0.507	0.0805	0.0301	
ROOK	636.0	24/7	0.0268	0.0277	0.0300	0.0330	0.502	0.0786	0.0327	
GROSBEAK	636.0	26/7	0.0267	0.0275	0.0301	0.0328	0.500	0.0780	0.0335	
EGRET	636.0	30/19	0.0266	0.0273	0.0299	0.0326	0.495	0.0769	0.0351	
FLAMINGO	666.6	24/7	0.0256	0.0263	0.0290	0.0314	0.498	0.0780	0.0335	
STARLING	715.5	26/7	0.0238	0.0244	0.0269	0.0292	0.490	0.0767	0.0355	
REDWING	715.5	30/19	0.0236	0.0242	0.0267	0.0290	0.486	0.0756	0.0372	
COOT	795.0	36/1	0.0217	0.0225	0.0247	0.0268	0.492	0.0780	0.0335	
TERN	795.0	45/7	0.0216	0.0225	0.0246	0.0267	0.488	0.0764	0.0352	
CUCKOO	795.0	24/7	0.0215	0.0223	0.0243	0.0266	0.484	0.0763	0.0361	
CONDOR	795.0	54/7	0.0215	0.0222	0.0244	0.0265	0.484	0.0759	0.0368	
DRAKE	795.0	26/7	0.0214	0.0222	0.0242	0.0263	0.482	0.0756	0.0375	
MALLARD	795.0	30/19	0.0213	0.0220	0.0241	0.0261	0.477	0.0744	0.0392	
RUDDY	900.0	45/7	0.0191	0.0200	0.0218	0.0237	0.479	0.0755	0.0374	
CANARY	900.0	54/7	0.0190	0.0197	0.0216	0.0235	0.474	0.0744	0.0392	
CORNCRAKE	954.0	20/7	0.0180	0.0188	0.0206	0.0224	0.474	0.0751	0.0378	
REDBIRD	954.0	24/7	0.0179	0.0186	0.0204	0.0221	0.470	0.0742	0.0396	
TOWHEE	954.0	48/7	0.0180	0.0188	0.0205	0.0223	0.473	0.0745	0.0391	
RAIL	954.0	45/7	0.0180	0.0188	0.0206	0.0223	0.474	0.0748	0.0385	
CARDINAL	954.0	54/7	0.0179	0.0186	0.0205	0.0222	0.470	0.0757	0.0404	
ORTOLAN	1033.5	45/7	0.0167	0.0175	0.0191	0.0208	0.468	0.0739	0.0401	
CURLEW	1033.5	54/7	0.0165	0.0172	0.0189	0.0201	0.464	0.0729	0.0420	
BLUEJAY	1113.0	45/7	0.0155	0.0163	0.0178	0.0193	0.462	0.0731	0.0416	
FINCH	1113.0	54/19	0.0154	0.0161	0.0176	0.0191	0.458	0.0702	0.0436	
BUNTING	1192.5	45/7	0.0144	0.0152	0.0167	0.0181	0.456	0.0723	0.0431	
GRACKLE	1192.5	54/19	0.0144	0.0151	0.0165	0.0179	0.452	0.0710	0.0451	
SKYLARK	1272.0	36/1	0.0135	0.0145	0.0159	0.0173	0.455	0.072	0.0427	
BITTERN	1272.0	45/7	0.0135	0.0144	0.0157	0.0170	0.451	0.072	0.0445	
PHEASANT	1272.0	54/19	0.0135	0.0142	0.0155	0.0169	0.447	0.070	0.0466	
DIPPER	1351.5	45/7	0.0127	0.0136	0.0148	0.0161	0.447	0.071	0.0459	
MARTIN	1351.5	54/19	0.0127	0.0134	0.0147	0.0159	0.442	0.070	0.0480	
BOBOLINK	1431.0	45/7	0.0120	0.0129	0.0141	0.0152	0.442	0.070	0.0472	
PLOVER	1431.0	54/19	0.0120	0.0127	0.0134	0.0151	0.438	0.069	0.0495	
LAPWING	1590.0	45/7	0.0108	0.0117	0.0127	0.0138	0.434	0.069	0.0498	
FALCON	1590.0	54/19	0.0108	0.0116	0.0126	0.0137	0.430	0.068	0.0521	
CHUKAR	1780.0	84/19	0.0097	0.0106	0.0115	0.0125	0.424	0.067	0.0534	
MOCKINGBIRD	2034.5	72/7	0.0085	0.0096	0.0104	0.0112	0.416	0.066	0.0553	
BLUEBIRD	2156.0	84/19	0.0080	0.0090	0.0098	0.0105	0.409	0.065	0.0588	
KIWI	2167.0	72/7	0.0080	0.0092	0.0099	0.0106	0.411	0.068	0.0570	
THRASHER	2312.0	76/19	0.0075	0.0086	0.0092	0.0100	0.405	0.065	0.0600	
JOREA	2515.0	76/19	0.0069	0.0081	0.0087	0.0093	0.399	0.064	0.0621	
							60 HZ REACTANCE 1 FOOT EQUIVALENT SPACING			
							Capacitive (Megohms-1000 Ft.)	Inductive (Ohms/1000 Ft.) @25° C	@50° C	@75° C
HIGH STRENGTH ACSR										
GROUSE	80.0	8/1	0.2065	0.2110	0.2362	0.2612	0.6547	0.1047	0.1129	0.1150
PETREL	101.8	12/7	0.1583	0.1625	0.2072	0.2394	0.6193	0.1019	0.1161	0.1282
MINORCA	110.8	12/7	0.1454	0.1491	0.1932	0.2233	0.6125	0.1017	0.1176	0.1269
LEGHORN	134.6	12/7	0.1198	0.1233	0.1638	0.1894	0.5972	0.0998	0.1148	0.1227
GUINEA	159.0	12/7	0.1014	0.1045	0.1426	0.1653	0.5845	0.0979	0.1117	0.1189
DOTTEREL	176.9	12/7	0.0911	0.0945	0.1301	0.1513	0.5760	0.0970	0.1102	0.1169
DORKING	190.8	12/7	0.0845	0.0875	0.1229	0.1424	0.5697	0.0956	0.1093	0.1150
COCHIN	211.3	12/7	0.0763	0.0792	0.1125	0.1311	0.5618	0.0945	0.1074	0.1129
BRAHMA	203.2	16/19	0.0764	0.0790	0.1089	0.1348	0.5507	0.0934	0.1047	0.1121

Notes:

1. DC reesistance is based on 16.946 ohn-cmil/ft. 61.2% IACS for 1350 wires and 129.64 ohm-cmil/ft. 8% IACS for the steel core at 20° C with stranding increment as per ASTM B232.

80203-11_T03.EPS

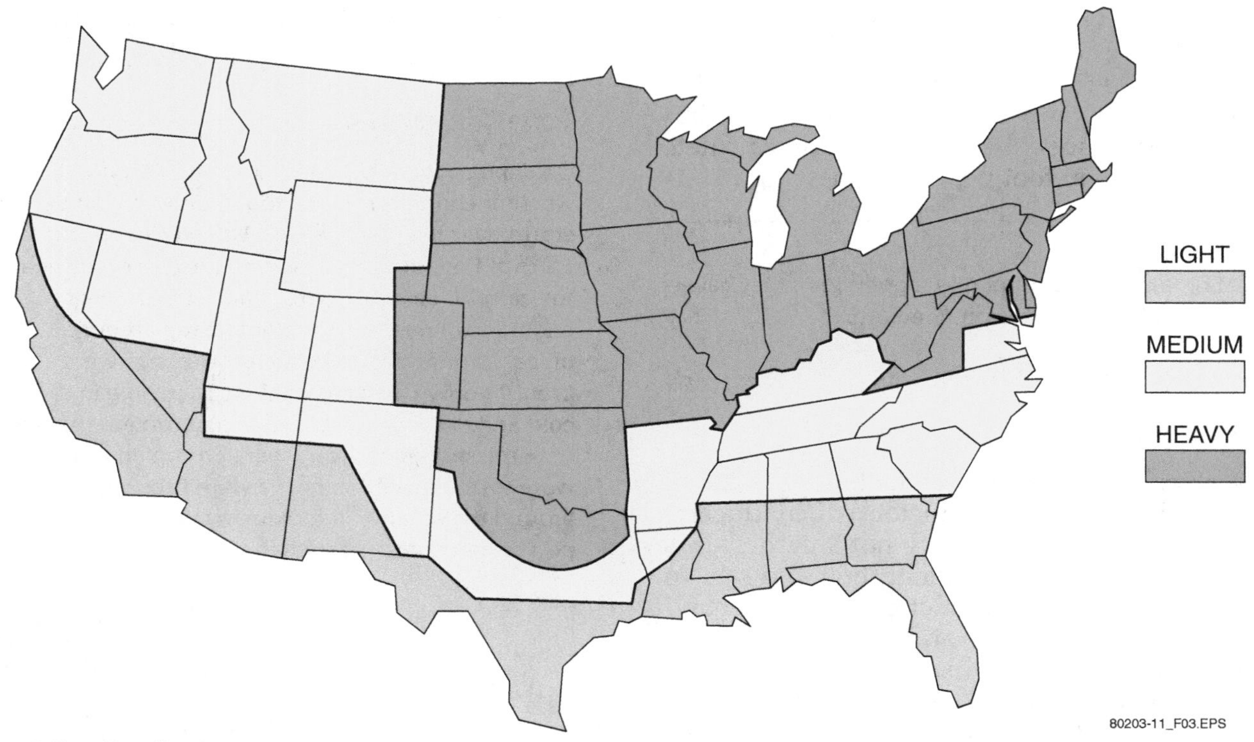

Figure 3 Loading districts.

Table 4 Loading District Values

Loading District	Ice Thickness	Horizontal Wind Pressures	Temperature	Constant
Heavy	0.5"	4 lb./sq. ft	0°F	0.30 lb./ft.
Medium	0.25"	4 lb./sq. ft.	15°F	0.20 lb./ft.
Light	0.0"	9 lb./sq. ft	30°F	0.05 lb./ft.

80203-11_T04.EPS

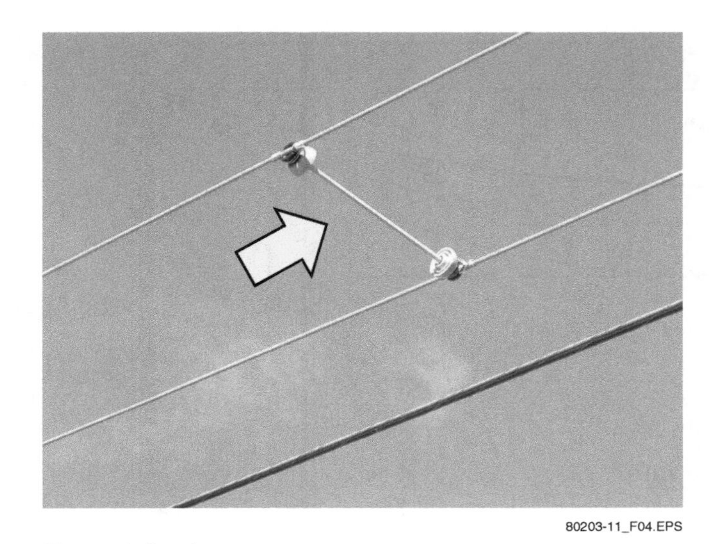

Figure 4 Conductor spacer.

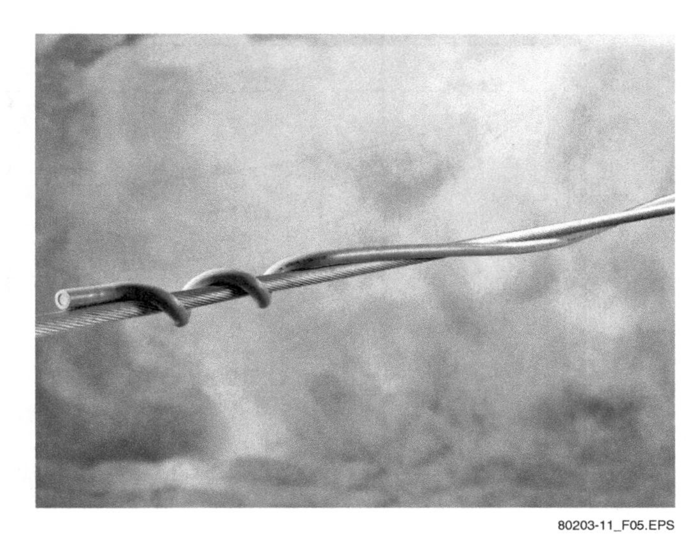

Figure 5 Air flow spoiler.

An area of 8.4 square feet is like having a 2' × 4' sheet of plywood attached to the conductor between each 200-foot span. Imagine the stress that the line would be under if a 40 mile per hour wind was blowing against it. An engineer can calculate the actual stress in foot-pounds using a formula.

> **NOTE**
> Some manufacturers offer conductors that have a built-in line-dampening feature.

4.4.0 Ambient Temperature

The ambient temperature in which the conductor is installed affects the conductor. All conductors expand and contract. In very hot weather, this can cause the conductor to lengthen and sag. In very cold weather, the conductor can contract and tighten the line. Expansion and contraction factors and local temperature is used to help determine what type of material to use for conductors. *Figure 6* shows the effect that temperature can have on a conductor that spans 200 feet. Temperature also affects conductivity. As temperature rises, the resistance of any conductor increases. Resistance decreases as temperature drops. A conductor that is used in a desert climate may have to be sized larger to compensate for reduced conductivity at higher temperatures.

4.5.0 Material Cost

When sizing and selecting a conductor, material cost is always a major factor. Utilities are always looking to improve their bottom line. That means

On Site

Ice Storm

On January 10, 1998, the right combination of weather conditions came together to form a perfect storm over the northeastern United States and parts of Canada. Rain fell at air temperatures near freezing for several hours. That caused tree limbs and power lines to accumulate large amounts of ice. Hundreds of miles of power lines were brought down, leaving millions of people in the cold and dark. So severe was the problem that steel transmission-line towers were crumpled as if they were toys. Power line crews were brought in from around the country. It took many weeks before power was completely restored.

80203-11_SA01.EPS

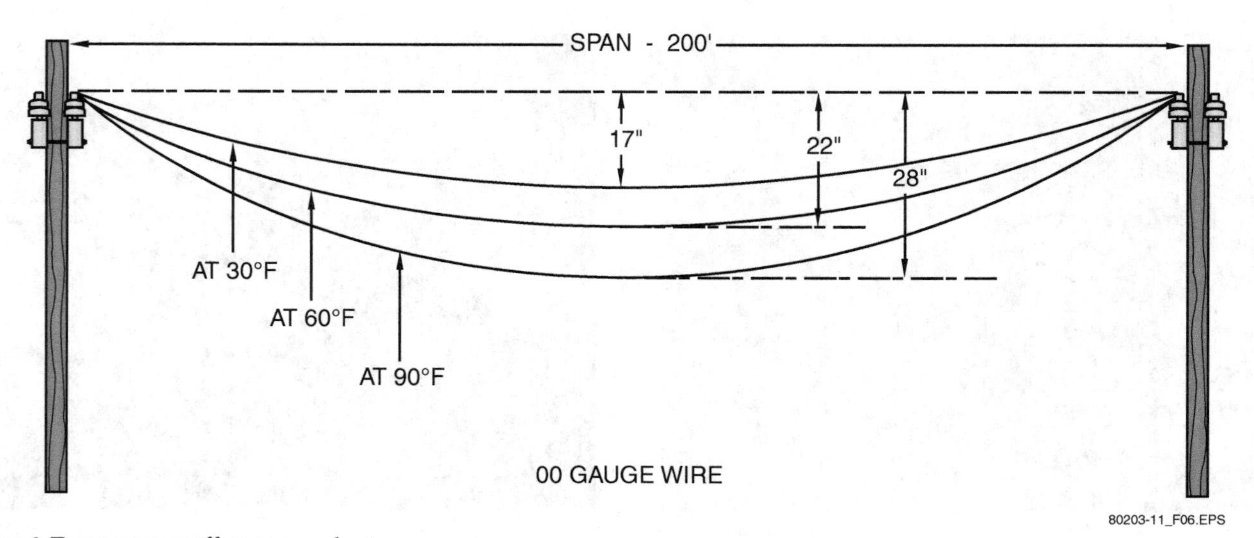

Figure 6 Temperature effect on conductors.

using less costly materials that deliver the desired performance. In recent years, worldwide demand has driven the price of copper to unheard-of levels. That, in turn, has caused all products that contain copper, including cables and conductors, to rise. This is driving utilities to use aluminum-based conductors.

5.0.0 Installing Cables and Conductors

Cables and conductors must be installed with care to prevent damage to the conductors. Steps for installing a primary conductor include the following:

- Pull conductor from the supply reel
- Raise conductor to installation height
- Dead-end conductor to the first pole
- Tension, sag, and dead-end conductor to the last pole
- Tie-in new conductors to insulators

5.1.0 Stringing Conductors

The tension stringing method is the preferred method for paying off a conductor from the supply reel. This method prevents conductors from touching the ground or sagging into nearby energized lines. A bullwheel tensioner (*Figure 7*) is located at the conductor supply reel. A drum-style pulling machine (*Figure 8*) is located at the end of the line (note the grounding mat in the foreground). Both machines provide tension.

The tension method requires that stringing blocks (*Figure 9*) be mounted on the top of the cross-arms. These pulleys must be sized to the conductor. They allow the conductor to be guided smoothly between poles. Top-mounted blocks provide a more stable pull than suspended stringing blocks.

NOTE

The slack/layout method requires that the conductor be pulled off the reel and laid on the ground before being raised onto the pole. This method can damage the conductors. It should only be used if the tension method cannot be used.

A basket grip (*Figure 10*), also called a Kellems grip, is used to attach the conductor to a pulling rope. Bands are usually applied to both ends of the grip keep it in place during the pull. *Figure 11* shows the sequence. Do not use a bulldog grip or Chicago-style grip to attach the conductor to

80203-11_F07.EPS

Figure 7 Bullwheel tensioner.

80203-11_F08.EPS

Figure 8 Pulling machine.

a pulling rope. Both types will damage the conductor. A swivel is used to prevent twisting. The pulling rope is passed through stringer blocks to the pulling machine. The rope, in turn, pulls the conductor off the reel and over the stringer blocks without touching the ground. A smaller diameter rope called a finger line or pilot line is tied to the heavier pulling rope and used to draw it up over

On Site

Deadly Static Electricity

Air moving over an object can cause a static charge to build up. On low-humidity days, wind blowing over a de-energized transmission or distribution line can build up a substantial static charge if that line is not grounded. Line workers have been killed by the static charge on a de-energized line.

80203-11_F09.EPS

Figure 9 Stringing block.

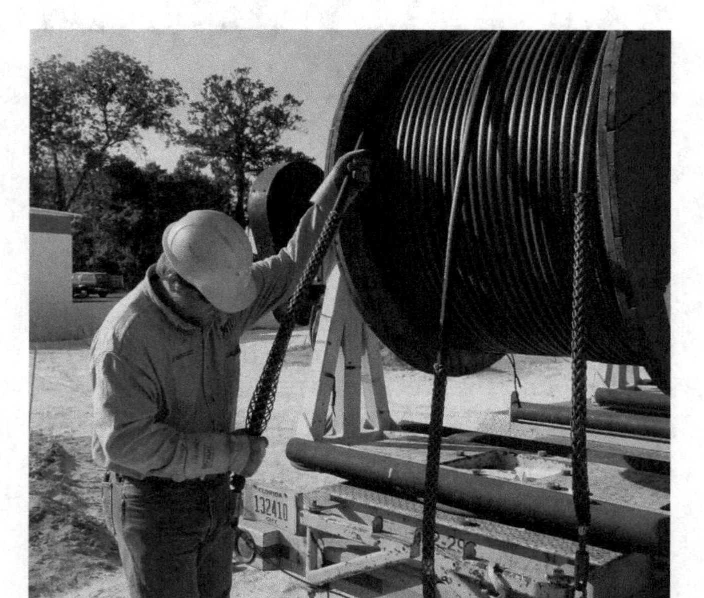

80203-11_F10.EPS

Figure 10 Basket grips installed on conductors.

the stringer block. Many utilities use the Spider® System (*Figure 12*) to better manage the pilot lines. Once the conductor is raised to the correct height on the pole, it must be dead-ended.

WARNING!

Conductors being installed may pick up a static charge or an induced voltage from a nearby conductor. They can also become accidentally energized. To protect against these voltages, always install a running ground on any new conductor being installed.

5.2.0 Dead-End Conductors

Dead-ending is the process of securing the terminating ends of conductors at the poles on each end of the section of new line. Once the line is strung between the poles, it is dead-ended at the first pole using a dead-end clamp (*Figure 13*) or a helical wire wrap connected to a dead-end insulator (*Figure 14*). After the line is dead-ended at the first pole, the length of conductor can be tensioned and sagged. Once that is done, the other end of the line can be dead-ended to the last pole.

5.3.0 Tensioning and Sagging Conductors

The line to be tensioned and sagged is dead-ended at the first pole with the conductor supported at each of the other poles by stringing blocks (*Figure 15*). A pulling machine at the other end of the conductor pulls it to apply tension and bring the line up to the proper sag. Once the proper sag is achieved, the line is snubbed off to an anchor and left to set for at least two hours to creep. During that time, tension equalizes and sag becomes evenly distributed between the poles. The conductor stretches from its own weight and expands or contracts based on the ambient temperature.

A dynamometer (*Figure 16*) measures pulling force when installed in the pull line. It is the preferred method used to measure tension and adjust sag. A sag chart (*Figure 17*) is then used to find the correct sag when pull tension, conductor size, span length, and temperature are known. To check the extent of sag, use a sag scope (*Figure 18*). This device is similar to a telescopic sight on a rifle. It is attached to an end pole and used to sight along the conductors. It can confirm that sag is correct or it can be used to make adjustments to sag. Once sag is set, confirm that the minimum ground clearance for conductors is maintained between all poles.

NOTE

An analog or digital dynamometer can be used to measure tension. The dynamometer is often built into the operator's console on the bullwheel tensioner.

A. BANDING TOOL WITH BAND

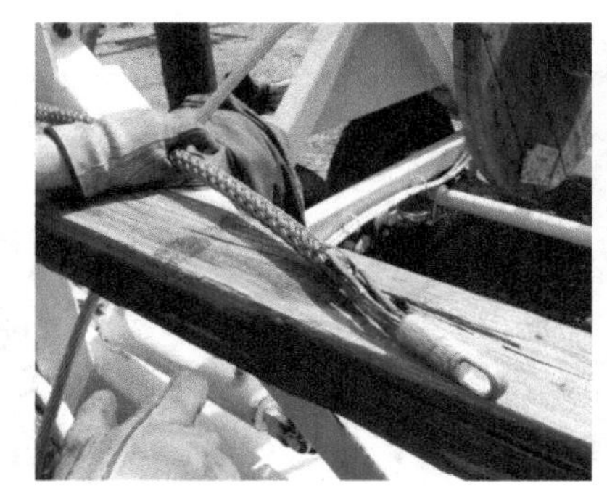

B. SWIVEL END

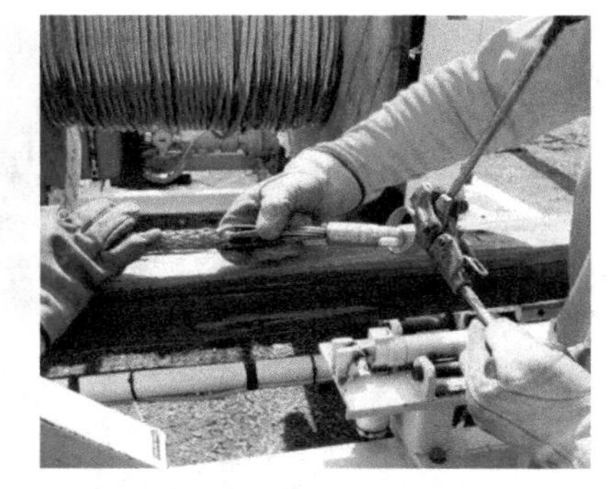

C. POSITIONING THE FIRST BAND

D. TIGHTENING THE BAND

E. REMOVING THE BANDING TOOL

F. FIRST BAND IN PLACE

G. SWIVEL END BANDED AND TAPED

H. CABLE END BANDED AND TAPED

80203-11_F11.EPS

Figure 11 Installing bands on the Kellems grip.

Creep in new conductors can be compensated by either over-tensioning or pre-stressing. With over-tensioning, the conductor is tensioned 5 to 10 percent over the required tension using a dynamometer. Over time, the conductor creeps back down to the correct tension and sag height. Pulling the conductor slightly higher than the desired sag height can also over-tension the line.

Pre-stressing greatly over-tensions the conductor for up to 30 minutes. Over-tensioning should never exceed 50 percent. A dynamometer is used to measure tension. After tension is released, the relaxed conductor is pulled to the correct sag height.

NOTE

Always compensate for temperature when adjusting sag.

5.4.0 Tie Conductors to Insulators

At each pole, the conductor is secured to one or more insulators with clamps or special wire ties.

5.4.1 Insulator Types

Insulators have grooves in the top or side of the body in which the conductor is cradled before it

Figure 12 Spider® system.

is tied in place (*Figure 19*). Clamp-style insulators (*Figure 20*) do not require ties. They are tightened to grip the conductor and are quicker and easier to use than other types. Insulators can be placed on top of a cross-arm or mounted on the side of the pole with a standoff. Insulators have always been made of porcelain; however, plastic insulators are now available.

5.4.2 Conductor Ties

Ties of various lengths are used to attach the conductor to an insulator (*Figure 21*). The ties are twisted and made of bare metal so that they can be easily wrapped around the conductor and the insulator. Plastic ties are used for cables. The tie wire AWG size is based on the AWG size of the conductor (*Table 5*). The length of the tie wire is also based on conductor size. For example, a 4/0 bare conductor requires a 54-inch tie. Prefabricated tie kits for insulators are available and

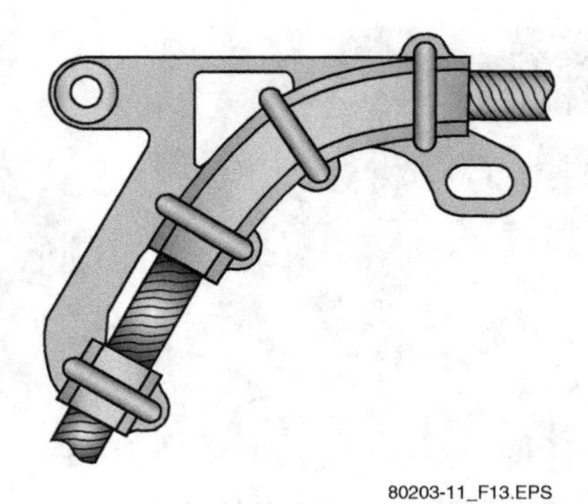

Figure 13 Dead-end clamp.

DEAD-END CLAMP

DEAD-END INSULATOR

80203-11_F14.EPS

Figure 14 Dead-end insulator.

widely used (*Figure 22*). A tie that is properly made should provide a secure binding between conductor, insulator, and tie. It should also reinforce the conductor near the insulator, and prevent chafing. When tying a conductor, always do the following:

- Use annealed (soft) metal ties.
- Use a tie of the correct length and strength.
- Apply the tie by hand. Tools may cut or nick the tie.
- Cut off any excess after the tie is made.
- Discard used ties. Always apply a new tie.

NOTE

Use aluminum ties with aluminum conductors. Use copper ties with copper conductors.

CAUTION

To prevent cuts, always wear gloves when applying ties.

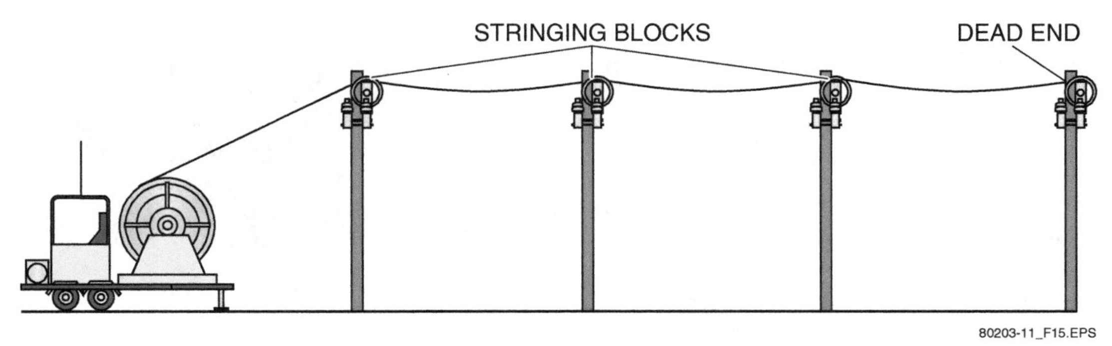

Figure 15 Tensioning and sagging conductors.

Figure 16 Dynamometer.

Apply a prefabricated tie kit to a top-tie insulator as follows (*Figure 23*):

- Center the neoprene pad in the insulator groove between the conductor and the insulator.
- Ensure that both legs of the tie are parallel to the conductor.
- Rotate the tie ends around the insulator so that the neoprene pad forms an S across the top of the insulator.
- Wrap both legs of the tie around the conductor and cut off any excess.

Apply a prefabricated tie kit to a side-tie insulator as follows (*Figure 24*):

- Center the neoprene pad around the conductor so that both rest against the side of the insulator.
- Form a loop with the tie and place the loop on the side of the insulator that is opposite the conductor.
- Cross the loop on the conductor side of the insulator.
- Wrap both legs of the tie around the conductor and cut off any excess.

CONDUCTOR SAG & TENSION DATA

BARE ALUMINUM – 1/0 AAAC (AZUSA)				
RATED STRENGTH: 4,460 LBS				
AMPACITY: 160A (NOMINAL) 235A (EMERGENCY OVERLOAD)				
RULING SPAN: 150 FT				
	INITIAL		FINAL	
AMBIENT TEMPERATURE (°F)	TENSION (LBS)	SAG (FT)	TENSION (LBS)	SAG (FT)
50	1,277	0.25	894	0.36
60	1,177	0.27	785	0.41
70	1,077	0.30	678	0.48
80	977	0.33	574	0.56
90	877	0.37	476	0.68
100	778	0.42	387	0.83
MAXIMUM OPERATING CONDITIONS	–	–	120	2.70

80203-11_F17.EPS

Figure 17 Sag chart.

If prefabricated tie kits are not available, the conductor can be tied to the insulator with a standard tie using a **Western Union tie** (*Figure 25*). This popular and simple way to apply a tie is done as follows:

- Place the middle of the tie on the side of the insulator opposite the conductor.
- Loop the tie around the insulator with each side of the loop underneath the conductor.
- Wrap the two ends of the tie around the conductor for about seven turns.
- Cut off any excess.

Armor rods, sometimes called line guards, are spiral rods that are placed around the conductor on either side of the insulator. Their purpose is to stiffen the conductor so that it does not bend or flex at that critical point. *Figure 26* shows the correct placement or armor rods around a conductor.

> **NOTE**
> Ties, armor rods, and other conductor tie accessories are often color-coded so that they can be easily matched to a conductor type.

Conductors must be energized after they are installed. Do this by clamping a jumper between each new conductor and an energized conductor (*Figure 27*).

5.5.0 Installing Overhead Cables

Cables are used for the service drop between the utility pole and the customer's electric service (*Figure 28*). For a home, the service drop consists of two insulated conductors and one bare conductor that is the neutral. The bare conductor also

80203-11_F19.EPS

Figure 19 Insulator.

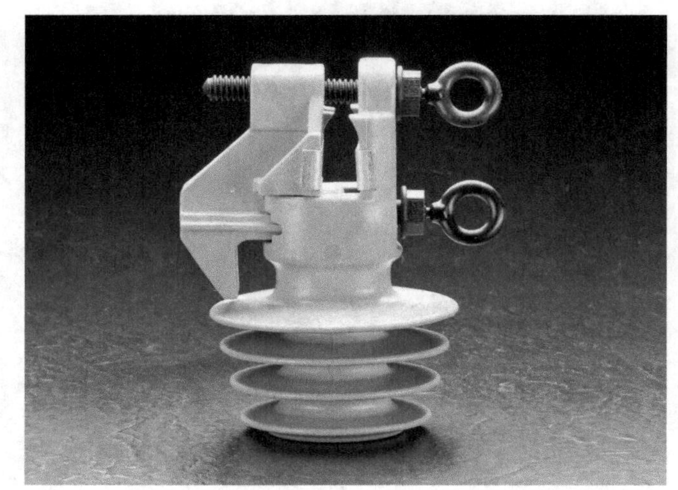

80203-11_F20.EPS

Figure 20 Plastic clamp-style insulator.

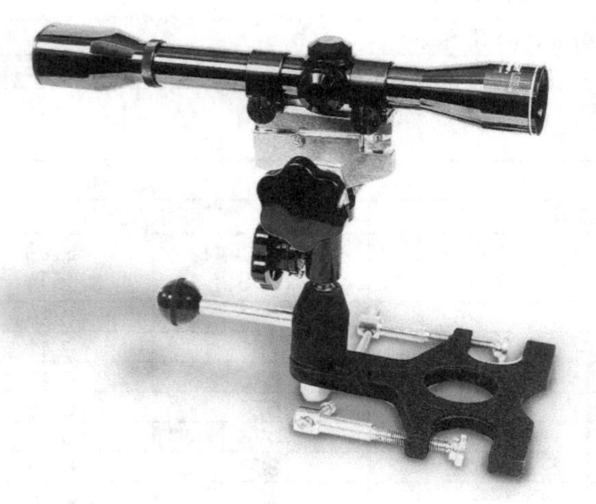

80203-11_F18.EPS

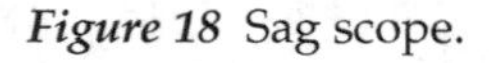

Figure 18 Sag scope.

80203-11_F21.EPS

Figure 21 Conductor ties.

Table 5 Tie Wire Sizes and Applications

Tie Wire Size AWG	Conductor Size AWG
#6	No. 4 AWG and smaller
#4	No. 1 AWG to No. 4 AWG
#2	No. 1/0 AWG and larger

80203-11_T05.EPS

serves as the messenger. After the correct length of conductor is laid on the ground, a line worker dead-ends the cable at the service mast.

Another line worker raises the other end of the service drop up the utility pole and dead-ends the conductors. This procedure is done by hand to prevent overstressing the service mast. Service drops that have to be removed are cut and dropped to the ground under controlled and safe conditions.

NOTE

Triplex is the name commonly used for residential service drop cable.

6.0.0 Repair and Replace Cables and Conductors

Conductors and cables that become damaged have to be repaired or replaced. Conductors are normally repaired. Replacement is only done if the damage is severe or if the distribution system is being upgraded due to an increased electrical load. Sections of conductor between spans can be cut and dropped to the ground under controlled and safe conditions. A worker in a bucket truck attaches a rope to the conductor before it is cut. Once cut, it can be lowered to the ground where it is rolled up by hand. The bucket truck and the conductor being removed must be properly grounded. Conductors that are removed from service cannot be re-used.

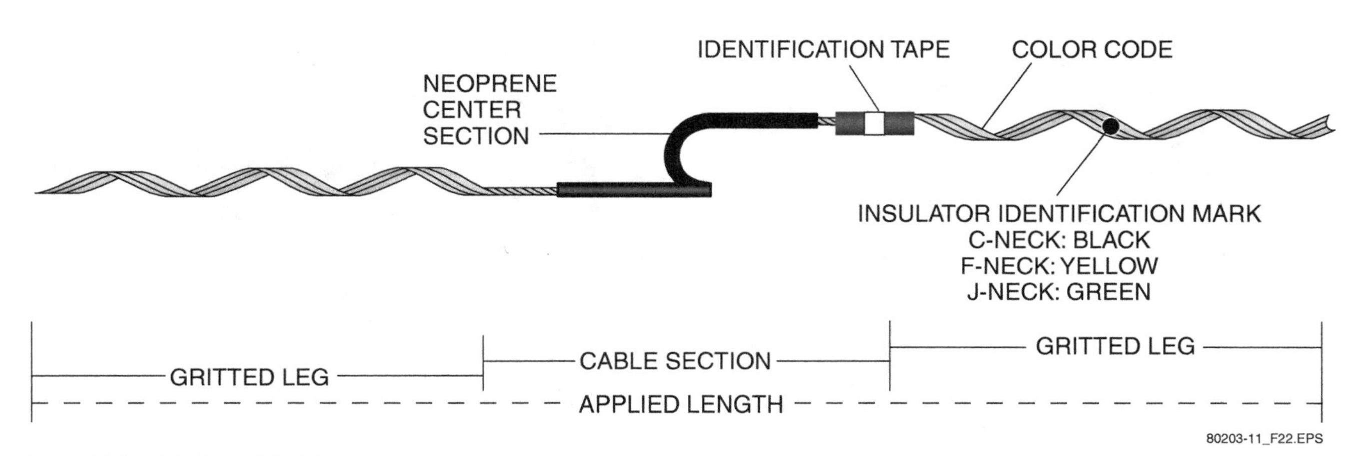

80203-11_F22.EPS

Figure 22 Prefabricated tie kit.

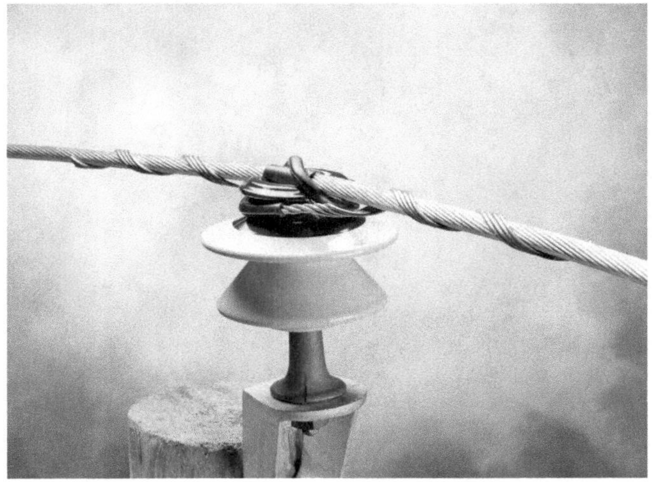

80203-11_F23.EPS

Figure 23 Prefabricated tie kit on top-tie insulator.

80203-11_F24.EPS

Figure 24 Prefabricated tie kit on side-tie insulator.

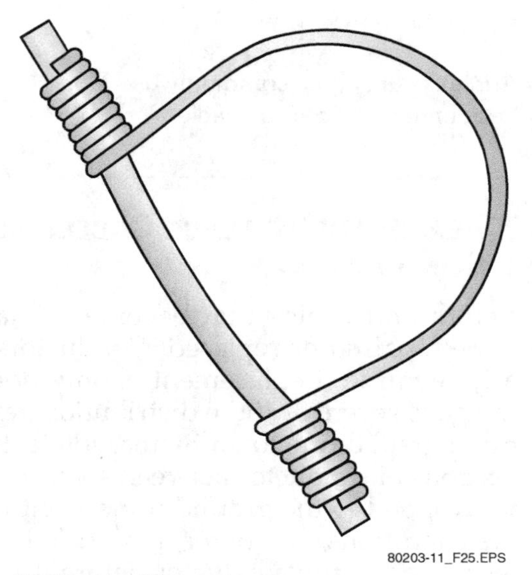

Figure 25 Western Union tie.

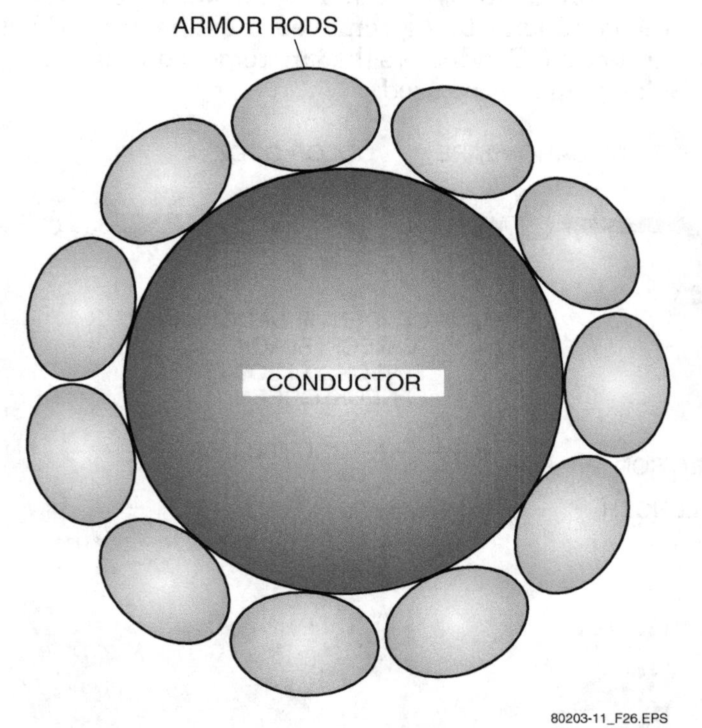

Figure 26 Correct placement of armor rods.

GOING GREEN

Conductor Recycling

Conductors that are removed from service can be recycled. The steel, copper, and aluminum that are recovered can be sold to provide the utility with a cash stream that adds to its bottom line.

Figure 27 Energizing new conductors.

Figure 28 Service drop.

Remove long sections spanning several poles by placing the conductors in stringing blocks and winding them on an empty reel. The process is the reverse of stringing new conductors. Connect a running ground to any conductors being removed and follow the same safety precautions for stringing new conductors. Replacement conductors are installed the same as stringing new conductors.

Many conditions can damage a conductor, including the following:

- Arcing and lightning strikes
- Stress and/or chafing caused by wind
- Tree limb damage
- Ice or snow damage
- Metal fatigue
- Corrosion
- Vandalism (gunfire)

If the damage is limited to a small area, which is often the case, repairs can be made using armor rods, splices, or **repair sleeves**. This section focuses on making localized repairs to ACSR conductors using various methods. The types of repair methods and devices include the following:

- **Conductor splice**
- **Full-tension splice**
- **Crimp-on splice**
- **Splice shunt**
- Repair sleeve

A conductor splice (*Figure 29*) is used to restore conductivity when the steel core of the conductor is intact. The outer layers of aluminum may be damaged to the point that conductivity is reduced. Before the conductor splice is installed, the aluminum conductors in the area where the splice will be applied must be brushed to remove any oxide coating. After that, a corrosion inhibitor is applied to the cleaned aluminum before the splice is wrapped around the conductor. This type of repair will not add much strength to the conductor. It is meant to increase conductivity.

6.1.0 Full-Tension Splice

A full-tension (non-crimp) splice (*Figure 30*) is used to restore conductivity and strength to a damaged ACSR conductor. An example might be chafing that wore through the conductor to the steel core, but left the core intact. This repair splice contains a core splice, filler rods, and conductor splice. Before starting the repair, the conductors on both sides of the repair must be evenly cut down to expose the core. The cut ends are taped to prevent unraveling of the conductors. The galvanized steel core splice is then wrapped around the steel core wire. Filler rods are applied over the core to restore the original outside diameter of the conductor. The conductors on both sides of the repair must be brushed to remove any oxide coating. After that, a corrosion inhibitor is applied to the cleaned aluminum before the outer splice is wrapped around the conductor. This type of repair restores the original strength and conductivity of the conductor.

6.2.0 Full-Tension Crimp-On Splice

Crimp-on splices are squeezed around the conductor with a crimping tool (*Figure 31*). When properly applied, they can restore full strength and conductivity to the conductor. Two types of crimp-on splices are available: a full-tension type, and a type meant to restore conductivity.

The full-tension type is used to splice a conductor in which the steel core wire is broken or badly damaged. The splice has two parts. An inner steel sleeve joins and strengthens the steel core. An outer aluminum sleeve joins the aluminum conductors and restores conductivity. To make the repair, the aluminum is cut back evenly on each side of the break to expose the core. Tape the cut ends of the aluminum to prevent unraveling of the conductors. Before joining the core wires, slide the aluminum sleeve over one of the conductors. Crimp the steel sleeve to join the core wire. The conductors on both sides must be brushed to re-

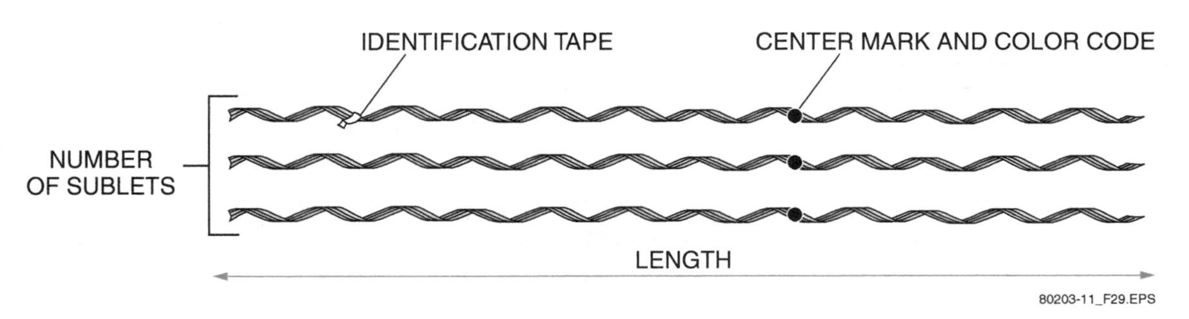

Figure 29 Conductor splice.

Figure 30 Full tension splice (non-crimp).

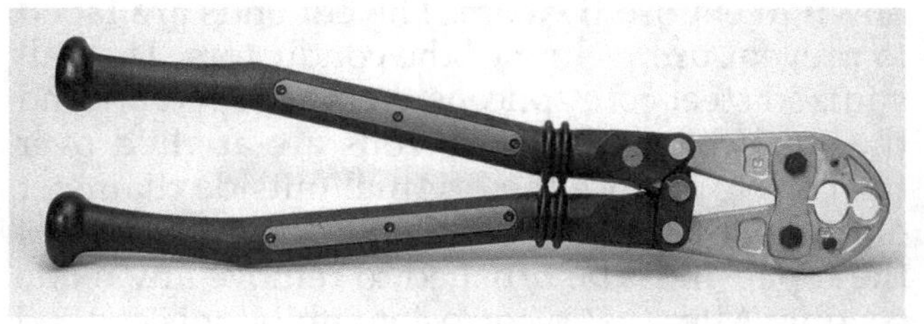

Figure 31 Crimping tool.

move any oxide coating. After that, a corrosion inhibitor is applied to the cleaned aluminum before the outer sleeve is slid into a center position and crimped in place around the conductor. *Figure 32* shows a cut away view of the completed repair. This type of repair restores the original strength and conductivity of the conductor.

6.3.0 Conductor Crimp-On Splice

A crimp-on splice (*Figure 33*) can be applied to restore conductivity. The sleeve must be placed on an open end of the conductor and slid into place. The conductor in the area where the splice will be applied must be brushed to remove any oxide coating. After that, a corrosion inhibitor is applied to the cleaned aluminum before the sleeve is slid into a center position and crimped in place around the conductor. This type of repair restores the original conductivity of the conductor. It is not meant to increase strength.

> **NOTE**
>
> Crimp-on splices often come coated with corrosion inhibitor.

6.4.0 Splice Shunt

Crimp-on splices should restore full strength and conductivity to a conductor. However, there may be cases where the original conductivity has degraded due to corrosion or other factors. A splice that is corroded or has reduced conductivity often shows up as a hot spot when the line is scanned with a thermal camera. A splice shunt (*Figure 34*) is applied over the crimp-on splice to restore conductivity and strength to the conductor. It is wrapped around the conductor and splice in the same manner as applying a conductor splice.

6.5.0 Repair Sleeve

It can be inconvenient to disconnect a conductor to slide on a crimp-on splice. For those cases, a repair sleeve (*Figure 35*) can be used. The sleeve can come in two halves that are placed over the damaged area and crimped in place. Another

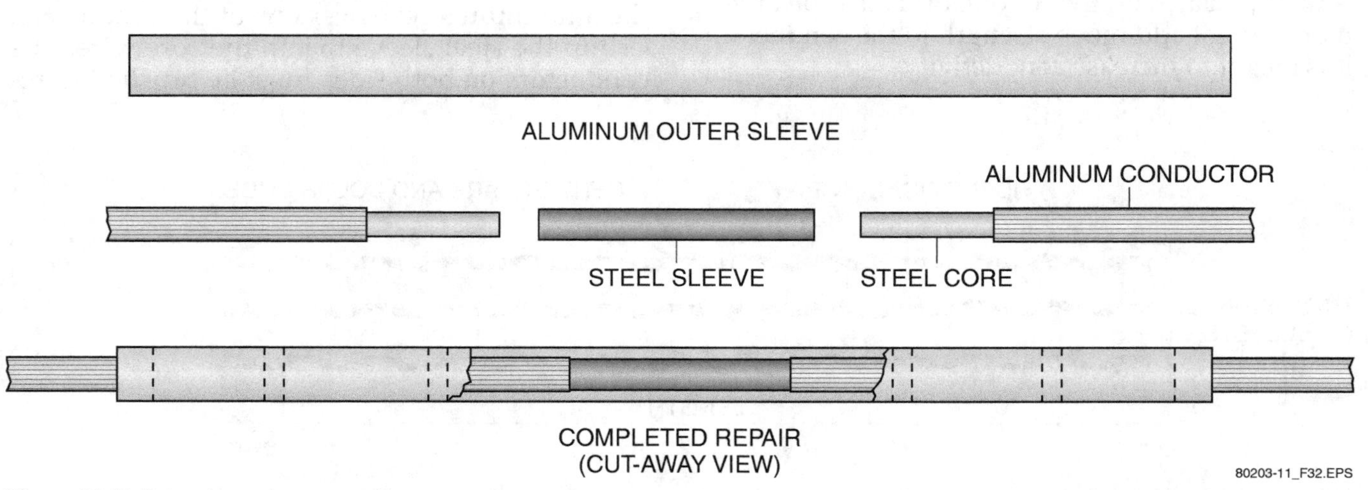

Figure 32 Full-tension crimp-on splice.

80203-11_F33.EPS

Figure 33 Conductor crimp-on splice.

80203-11_F34.EPS

Figure 34 Splice shunt.

version has a slit along its length that allows it to be placed over the conductor. Once in place, it is crimped over the conductor.

6.6.0 Automatic Splice

The full-tension automatic splice shown in *Figure 36* is used to make quick repairs on broken conductors. It functions like a Kellems grip, in that conductors can be slid into each end, but cannot be pulled out. Each end of the conductor is measured and marked so that it slides into the ends of the splice. The conductors are then inserted into the splice and the splice is pulled down to set the jaws of the splice.

7.0.0 Cable-Pulling Equipment

There are four major types of equipment used to pull distribution cables and conductors using the tension stringing method:

- Conductor reel and carriage
- Bullwheel tensioner
- Puller
- Take-up reel and carriage

The four pieces of equipment can be separate items, but they are often combined. The bullwheel tensioner and the conductor supply reel are placed on one trailer. The puller along with a

On Site

Pulling Rope

Rope used to pull conductors must be strong and resist stretching and abrasion. A product called Uniline® is widely used for pulling conductors. A ⅝"-diameter rope has a breaking strength of 16,800 pounds and less than 2 percent stretching. Uniline® ropes that have seen 20 years of service have been shown to retain 75 percent of their strength.

80203-11_SA02.EPS

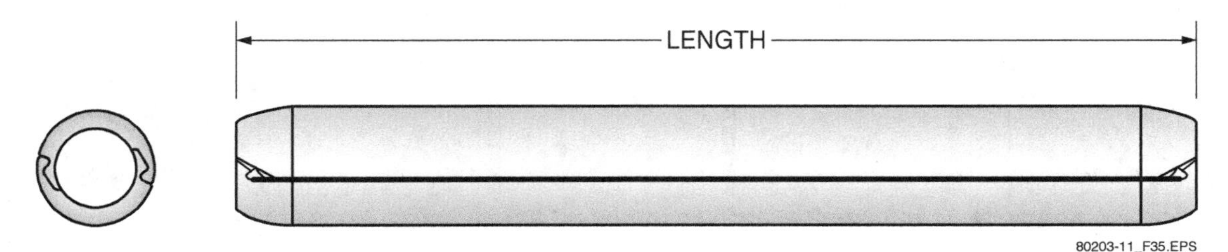

80203-11_F35.EPS

Figure 35 Repair sleeve.

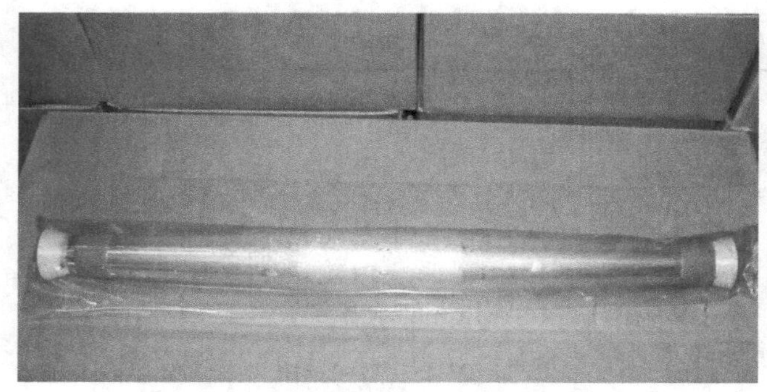

Figure 36 Automatic splice.

take-up reel is placed on another trailer. *Figure 37* shows the position of the two pieces of equipment for tension stringing. A **puller/tensioner** (*Figure 38*) is a dual-purpose machine that can be used as a puller or to apply tension. Two puller/tensioners can be used on a job to string conductors.

Safety has the highest priority when stringing conductors. Follow these safety guidelines when operating cable-pulling equipment:

- Install a grounding grid around and beneath all pulling equipment. Ground all pulling/tensioning equipment to the grounding grid.
- Install a running ground on all conductors being pulled. Connect the running ground to the grounding grid.
- Never allow workers to stand beneath conductors or ropes as they are being pulled. A snapped rope or cable can cause serious injuries.
- Only qualified persons are allowed to operate pulling/tensioning machines.

Figure 38 Puller/tensioner.

- Operators of pulling machines must always remain behind the protective cage when operating the machine.
- Establish and maintain constant two-way contact with all crew members when pulling conductors.

> **NOTE**
> Read the instructions before operating unfamiliar pulling equipment. Certification training is often required before a person is allowed to operate an unfamiliar machine. Most manufacturers provide training for their equipment.

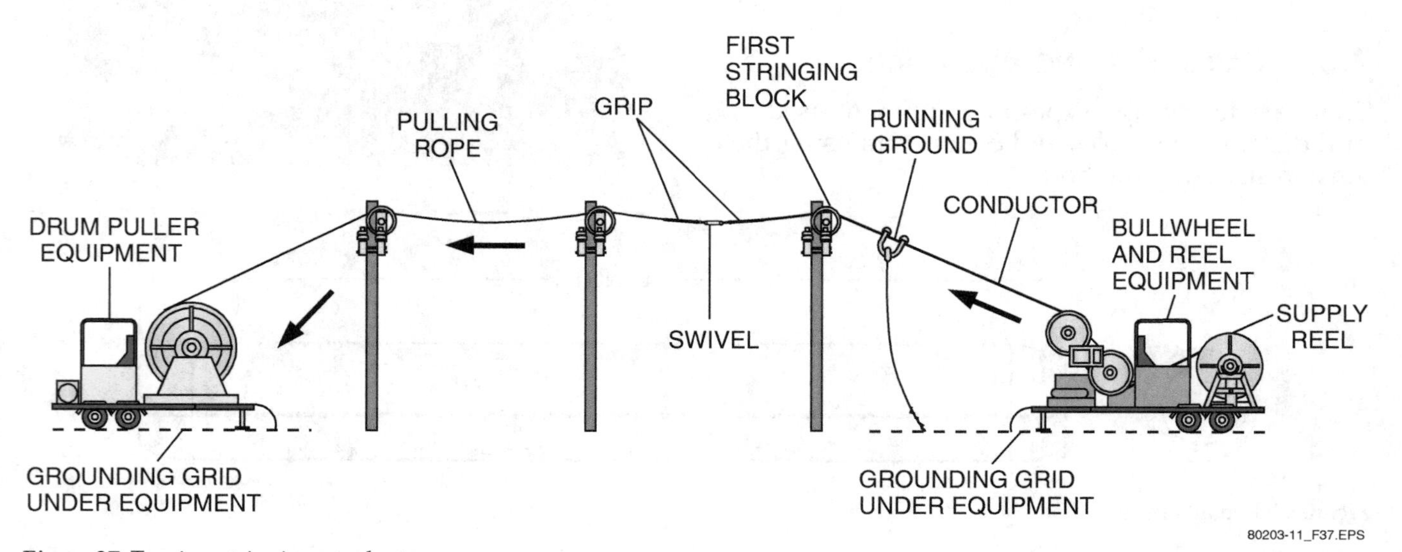

Figure 37 Tension stringing conductors.

The amount of tension to be applied during stringing is based on a formula that takes into account a number of factors, such as weight of the conductor, length of the span, and number of stringing blocks. An engineer supplies the tension requirement. The operator's console on modern pulling equipment is able to monitor and display pulling tension during the pull. If a problem occurs that increases tension, the operator can stop pulling until the problem is corrected. Some machines automatically stop if tension limits are exceeded.

SUMMARY

A common task of power line workers is installing and repairing cables and conductors. In power distribution, the term *cable* usually means a conductor that is insulated. A conductor, on the other hand, does not have any insulation. Conductors are used in aerial distribution systems. Cables are used in underground systems, but have limited use in aerial systems.

Conductors used in power distribution systems must be strong and have good conductivity. Aluminum conductors are widely used because they offer good strength and conductivity, and moderate cost. A popular conductor material is aluminum conductor steel reinforced (ACSR). This conductor has a strong steel core that is wrapped with aluminum conductors. When determining the size and type of conductor for a job, an engineer must consider line current and voltage, loads imposed by wind and icing, ambient temperature, and material cost.

The tension stringing method is the preferred method for stringing conductors between poles. It uses a bullwheel tensioner located at the conductor supply reel and a pulling machine located at the opposite end of the line. After the conductor is pulled it is secured at each pole to one or more insulators with clamps or special wire ties.

Conductors and cables that are damaged have to be repaired. Damage to small areas can be repaired using armor rods, splices, and repair sleeves. Conductors are replaced if they are badly damaged or if the system is being upgraded due to an increased load.

The types of equipment used to pull distribution conductors using the tension stringing method include the conductor reel, a bullwheel tensioner, a puller, and a take-up reel. Workers operating cable-pulling equipment must be trained and certified in their safe use.

Review Questions

1. In aerial power distribution systems, cable is _____.
   a. never used
   b. insulated
   c. widely used
   d. always made of copper

2. The conductor material most often used in modern distribution systems is _____ .
   a. copper
   b. copper-plated steel
   c. zinc-coated steel
   d. aluminum

3. A distribution system conductor that is larger than No. 6 AWG is likely to be _____.
   a. stranded
   b. copper
   c. insulated
   d. grounded

4. A smaller-diameter distribution system conductor can be used if the voltage carried by that conductor is increased.
   a. True
   b. False

5. Which statement about ice and snow loads on aerial conductors is *correct*?
   a. Ice and snow loads are only calculated for states above the Mason-Dixon line.
   b. The formula for calculating ice and snow loads is uniform across the United States.
   c. The formula for calculating ice and snow loads varies across the United States.
   d. Conductors coated with Teflon® are available that shed ice or snow.

6. What effect does temperature have on a stranded aluminum distribution conductor installed in North Dakota?
   a. No effect below –20°F or above 75°F.
   b. The conductor will loosen up (sag) in winter.
   c. Conductor resistance will increase in summer.
   d. Conductor resistance will increase in winter.

7. A bullwheel tensioner is located _____.
   a. at each stringing block
   b. on the pulling machine
   c. on all line trucks
   d. at the conductor supply reel

8. The conductor is attached to the pulling rope with a _____.
   a. bulldog grip
   b. Kellems grip
   c. Chicago-style grip
   d. Japanese finger

9. The tension on a conductor being pulled is measured with a _____.
   a. dynamometer
   b. bull wheel
   c. strain gauge
   d. pressure gauge

10. To compensate for creep when installing new conductors, the over-tensioning method over-tensions the conductor by _____.
    a. 40 to 50 percent
    b. 35 percent
    c. 20 percent
    d. 5 to 10 percent

11. Which statement about metal conductor ties is *correct*?
    a. Re-use ties no more than two times.
    b. Use pliers to apply metal ties.
    c. Use annealed metal ties.
    d. Use ties with clamp-type insulators.

12. Which statement about a residential service drop is *correct*?
    a. All conductors are bare.
    b. The bare conductor is the messenger.
    c. The neutral conductor is insulated.
    d. A pulling machine installs the cable.

13. The main purpose of a conductor splice is to strengthen the conductor.
    a. True
    b. False

14. What type of splice is used to repair a conductor with a broken core wire?
    a. Conductor splice
    b. Full-tension splice
    c. Full-tension splice (crimp-on)
    d. Conductor splice (crimp-on)

15. The main purpose of a splice shunt is to _____.
    a. restore conductivity
    b. repair the core wire
    c. replace the core wire
    d. prevent abrasion

Trade Terms Introduced in This Module

Alumoweld: Steel wire that is coated with aluminum to increase conductivity and prevent rust.

Aluminum conductor steel reinforced (ACSR): A widely used aluminum conductor that has an alumoweld steel core for added strength.

Armor rods: Spiraled metal rods that are placed around a conductor on both sides of the insulator. They are used to stiffen the conductor to prevent bending and flexing. They are often called line guards.

Conductor splice: A type of conductor repair used to restore conductivity. It does not add much strength to the conductor.

Copperweld: Steel wire that is coated with copper to increase conductivity and prevent rust.

Crimp-on splice: A type of conductor repair that is squeezed (crimped) around the conductor with a tool. Crimp-on splices can restore conductivity and/or strengthen the conductor.

Full-tension splice: A type of splice meant to restore strength and conductivity to a conductor. Crimp-on and non-crimp styles are available.

Loading districts: Areas of the United States established by the Bureau of Standards for the purpose of calculating snow and ice loads on power line conductors. Loading areas are designated light, medium, and heavy.

Messenger: A line that serves as a support for conductors. In a service drop, the neutral conductor serves as the messenger.

Puller/tensioner: A dual-purpose pulling machine that can be used as either a puller or a tensioner.

Repair sleeve: A type of crimp-on conductor repair that can be applied over a conductor without having to slip it on over the end of a conductor. They are available as two halves, or with a slit and are placed over the damaged area.

Service drop: The cables going between a utility pole and the customer's electrical service. For a residential service, a service drop consists of two insulated conductors and one bare conductor that serves as the neutral and the messenger.

Splice shunt: A repair meant to restore conductivity to an existing splice in a conductor. It is applied over the existing splice to form a parallel path (shunt) for current flow.

Tensile strength: The resistance of material to being pulled apart. Conductors must have enough tensile strength to withstand the various forces placed on them.

Tension stringing method: A method used to string conductors between poles. The conductor is pulled under tension over stringing blocks on each pole. The method keeps conductors off the ground to avoid damage. It is the preferred method for installing new conductors.

Western Union tie: A popular method used to tie conductors to insulators using a standard tie.

Additional Resources

This module presents thorough resources for task training. The following resource material is suggested for further study.

Power Distribution Engineering. New York, NY: Marcel Dekker.

Guide to Electrical Power Distribution Systems, 5th Edition. Tulsa, OK: PennWell Books.

Figure Credits

Sherman & Reilly, Inc., Module opener, Figures 7, 9, 12, 38, and SA02

ACSR Cable technical data reprinted with the permission of Alcan Cable, Tables 2 and 3

Topaz Publications, Inc., Figures 4, 21, 28, and 36

Photo, Courtesy of Preformed Lines Products, Figures 5, 19, 20, 22–24, 29, 30, and 34

Jim Mitchem, Figures 6, 13–15, 27, and 37

Dave Roszowski, SA01

Bruce Chesley, Figure 8

Tony Vazquez, Figure 10

Joe Holley, Figure 11

Dillon Force Measurement, Figure 16

Condux Tesmec, Figure 18

Burndy LLC, Figure 31

A.B. Chance Co./Hubbell Power Systems, Figure 35

NCCER CURRICULA — USER UPDATE

NCCER makes every effort to keep its textbooks up-to-date and free of technical errors. We appreciate your help in this process. If you find an error, a typographical mistake, or an inaccuracy in NCCER's curricula, please fill out this form (or a photocopy), or complete the online form at **www.nccer.org/olf**. Be sure to include the exact module ID number, page number, a detailed description, and your recommended correction. Your input will be brought to the attention of the Authoring Team. Thank you for your assistance.

Instructors – If you have an idea for improving this textbook, or have found that additional materials were necessary to teach this module effectively, please let us know so that we may present your suggestions to the Authoring Team.

NCCER Product Development and Revision
13614 Progress Blvd., Alachua, FL 32615
Email: curriculum@nccer.org
Online: www.nccer.org/olf

❑ Trainee Guide ❑ AIG ❑ Exam ❑ PowerPoints Other ______________________________

Craft / Level: Copyright Date:

Module ID Number / Title:

Section Number(s):

Description:

Recommended Correction:

Your Name:

Address:

Email: Phone:

80204-11

Underground Residential Distribution (URD) Systems

Module Four

Trainees with successful module completions may be eligible for credentialing through NCCER's National Registry. To learn more, go to **www.nccer.org** or contact us at **1.888.622.3720**. Our website has information on the latest product releases and training, as well as online versions of our *Cornerstone* newsletter and Pearson's product catalog.

Your feedback is welcome. You may email your comments to **curriculum@nccer.org,** send general comments and inquiries to **info@nccer.org**, or fill in the User Update form at the back of this module.

V.1 12/11

80204-11
Underground Residential Distribution (URD) Systems

Objectives

When you have completed this module, you will be able to do the following:

1. Describe the history and applications of URD systems.
2. Describe trenching and backfill methods used for URD systems, including common-trench applications.
3. Identify and describe common types of cable conductors and termination methods used in URD installations.
4. Describe common types of lightning protection and fault-indicating devices used in URD systems.
5. Install lightning protection and fault-indicating devices in URD systems.
6. Identify and describe pad-mounted switchgear and transformers used in URD systems.
7. Select the proper types of conductors and termination methods for specific URD applications.

Performance Tasks

Under the supervision of your instructor, you should be able to do the following:

1. Install lightning protection and fault-indicating devices in URD systems.
2. Identify and describe pad-mounted switchgear and transformers used in URD systems.
3. Select the proper types of conductors and termination methods for specific URD applications.

Trade Terms

Ampacity
Common trench (joint trench)
Dead front
Double feed
Dual-rated
Lightning arrester
Live front
Loop feed
Metal oxide varistor (MOV)
Radial feed

Industry Recognized Credentials

If you're training through an NCCER-accredited sponsor you may be eligible for credentials from NCCER's Registry. The ID number for this module is 80204-11. Note that this module may have been used in other NCCER curricula and may apply to other level completions. Contact NCCER's Registry at 888.622.3720 or go to nccer.org for more information.

Contents

Topics to be presented in this module include:

1.0.0 Introduction 1
2.0.0 URD System Overview 1
2.1.0 History and Significance of URD Systems 1
2.2.0 Types of URD Systems 3
2.2.1 Radial Feed 3
2.2.2 Loop Feed 4
2.2.3 Double Feed 4
2.3.0 URD Trenching and Backfilling Methods 5
3.0.0 Cable Types and Termination Methods 7
3.1.0 Conductors 7
3.2.0 Termination Methods 9
3.2.1 Copper to Aluminum 10
3.2.2 Splices 11
3.2.3 Compression Terminations 12
4.0.0 Lightning Protection and Fault-Indicating Devices 13
4.1.0 Lightning Protection Devices 13
4.2.0 Fault-Indicating Devices 15
4.3.0 Lightning Protection and Fault-Indicating Device Installation 16
5.0.0 Pad-Mounted Switchgear and Transformers 18
5.1.0 Switchgear 18
5.2.0 Transformers 19
5.2.1 Pad-Mounted Transformer Installation 21
6.0.0 Transformer Connections 22
6.1.0 Primary Connections 23
6.1.1 Dead-Front Transformer 23
6.1.2 Live-Front Transformer 24
6.2.0 Secondary Connections 25

Figures and Tables

Figure 1 Overhead utility lines, circa 1890 1
Figure 2 Underground electrical distribution system vault and conduit 2
Figure 3 Single-phase overhead service drop 2
Figure 4 Primary feeder connection to a URD pad-mounted transformer 3
Figure 5 Radial-feed URD system 4
Figure 6 Radial-feed system with fault 4
Figure 7 Loop-feed URD system 5
Figure 8 Double-feed URD system 5
Figure 9 Trencher digging a URD cable trench 6
Figure 10 Backhoe and excavator 6
Figure 11 Boring equipment for URD installation 7
Figure 12 Primary URD cable layers 8
Figure 13 Secondary URD cable 9
Figure 14 Manual and battery-powered cable cutters 9
Figure 15 Primary cable-stripping tools 10
Figure 16 Dual-rated connector with ALCU marking 11
Figure 17 Compression terminal 11
Figure 18 Line worker attaching compression connector 11
Figure 19 Split-bolt mechanical connector 12
Figure 20 Primary URD cable layers 12
Figure 21 Prepackaged splice kit for primary URD cable 12
Figure 22 Burndy® MD6 mechanical crimping tool 12
Figure 23 Burndy® Y35 hand-operated hydraulic tool 13
Figure 24 Using a compression tool to install a terminal 13
Figure 25 Surge (lightning) arrester 13
Figure 26 Surge arrester cutaway showing MOV disk 14
Figure 27 Surge arrester on riser pole 14
Figure 28 Dead front surge arrester variations 15
Figure 29 Test point fault indicator 16
Figure 30 Cable-mounted (clamp-type) fault indicator 16
Figure 31 Fault location based on fault indicator displays 17
Figure 32 Elbow surge arrester installed at end of radial-feed system 17
Figure 33 Test point fault indicator and elbow 18
Figure 34 Recommended installation of cable-mounted fault indicator for exposed concentric neutral 18
Figure 35 Pad-mounted switchgear 19
Figure 36 Gang-controlled switches in pad-mounted switchgear 19
Figure 37 Pad-mounted transformer locations in loop-feed URD system 20
Figure 38 Single-phase pad-mounted transformer connections 20
Figure 39 Interior of an oil-filled, three-phase pad-mount transformer 21
Figure 40 Primary side 21
Figure 41 Bayonet fuse holder 22
Figure 42 Secondary side 22
Figure 43 Oil drain valve 22

Figures and Tables (*continued*)

Figure 44 Mounting pad features 22
Figure 45 Aerial feeder 22
Figure 46 Primary cable installation 23
Figure 47 Transformer installation 23
Figure 48 Primary connections in a dead-front URD transformer 24
Figure 49 Load break elbows 24
Figure 50 Primary terminals in a live-front URD transformer 24
Figure 51 Secondary terminal connections in a pad-mounted transformer 25
Figure 52 Example of a connection bolted to a terminal lug 26

Table 1 Typical URD Cable Measurements 8

1.0.0 INTRODUCTION

At one time, overhead power lines were the only way to get electricity to residential customers. In rural areas, many customers still get their power that way. But in the past few decades, utilities have increasingly used underground residential distribution (URD) systems to supply electricity to residential neighborhoods.

In a URD system, the electrical cables used to deliver power are designed to be buried underground without the need for protective conduit or extra sheathing. For that reason, URD systems are often called direct-buried systems. Transformers and switchgear used in these systems are commonly mounted above ground on polymer, fiberglass, or concrete pads.

URD systems cost more to install than overhead lines, but the benefits of using buried lines tend to outweigh the extra costs. Most utilities use URD systems for any new service in residential neighborhoods. Some areas require URD systems to be used.

As a power line worker, you need to be familiar with the layout and operation of URD systems and the components used in these systems. This module describes different types of URD systems and the cables, terminations, and other equipment used in them. It also explains how to select and install the proper components.

2.0.0 URD SYSTEM OVERVIEW

The most common use of URD systems is to provide single-phase service to residential customers. URD systems are also used to distribute single-phase and three-phase power to commercial and industrial customers. In many cases, power reaches a URD system by way of overhead primary distribution lines. The primary distribution voltage can vary. Some common voltages are 7.2 kilovolts (kV), 7.6kV, 14.4kV, 19.9kV, and 35kV.

A primary feeder line connects one of the overhead distribution primaries to a pad-mounted transformer in the URD system. Underground primary feeder cable continues from that transformer to all of the other pad-mounted transformers in the system. Each transformer reduces the primary voltage to a level that is appropriate for the customers being fed from that transformer.

The stepped-down secondary voltage is carried through buried conductors to the individual customers in the system. A typical system uses a three-wire secondary service. Two of the wires carry 120 volts each. The third wire is the neutral wire. This arrangement provides 120V power for lighting, outlets, and small appliances, as well as 240V power for larger equipment like electric clothes dryers, air conditioners, and electric stoves.

The layout of URD systems and the components used in them can vary, depending on the requirements of the system. To better understand how URD systems work, it helps to have some background knowledge about them and be able to identify common types of URD layouts and how they are installed.

2.1.0 History and Significance of URD Systems

When electrical distribution first began in the late 1800s, overhead power lines were used to deliver electricity from substations to customers. Before long, though, the sheer number of overhead lines in heavily populated areas became an eyesore and a safety concern. *Figure 1* shows a drawing of how crowded the skies became with overhead utility lines in the late 1800s.

By the early 1900s, utilities started using underground distribution systems for industrial and commercial customers in cities. These systems use manholes, underground vaults with high-voltage equipment, jacketed cables, and conduit encased in concrete. Such a system is cost effective and beneficial for areas with high electrical demand, but far too expensive for most residential areas where demand is much lower (*Figure 2*).

80204-11_F01.EPS

Figure 1 Overhead utility lines, circa 1890.

80204-11_F02.EPS

Figure 2 Underground electrical distribution system vault and conduit.

Prior to the 1960s, basically all residential customers received their power from overhead lines. In a common overhead service, such as the one shown in *Figure 3*, power arrives to the area from a substation by way of overhead primary distribution lines. A primary feeder line runs from an overhead distribution primary through a fused disconnect switch and a **lightning arrester** to a step-down transformer on the utility pole. The disconnect switch enables the system to be energized and de-energized. The fuse helps protect the equipment in the system by opening if there is a fault or an overcurrent condition. The lightning arrester protects the system from overvoltage conditions caused by lightning strikes. The transformer reduces the primary voltage to a secondary voltage level that is appropriate for the customer or customers serviced by that transformer. A three-wire secondary service drop runs from the transformer to the electric meter on the customer's home.

During the 1960s and 1970s, the suburban population increased dramatically. To avoid the use of so many overhead lines in neighborhoods, utilities started to use URD systems to supply electricity to customers.

A simplified illustration of part of a URD system is shown in *Figure 4*. As with the overhead service, power arrives to the area by way of overhead

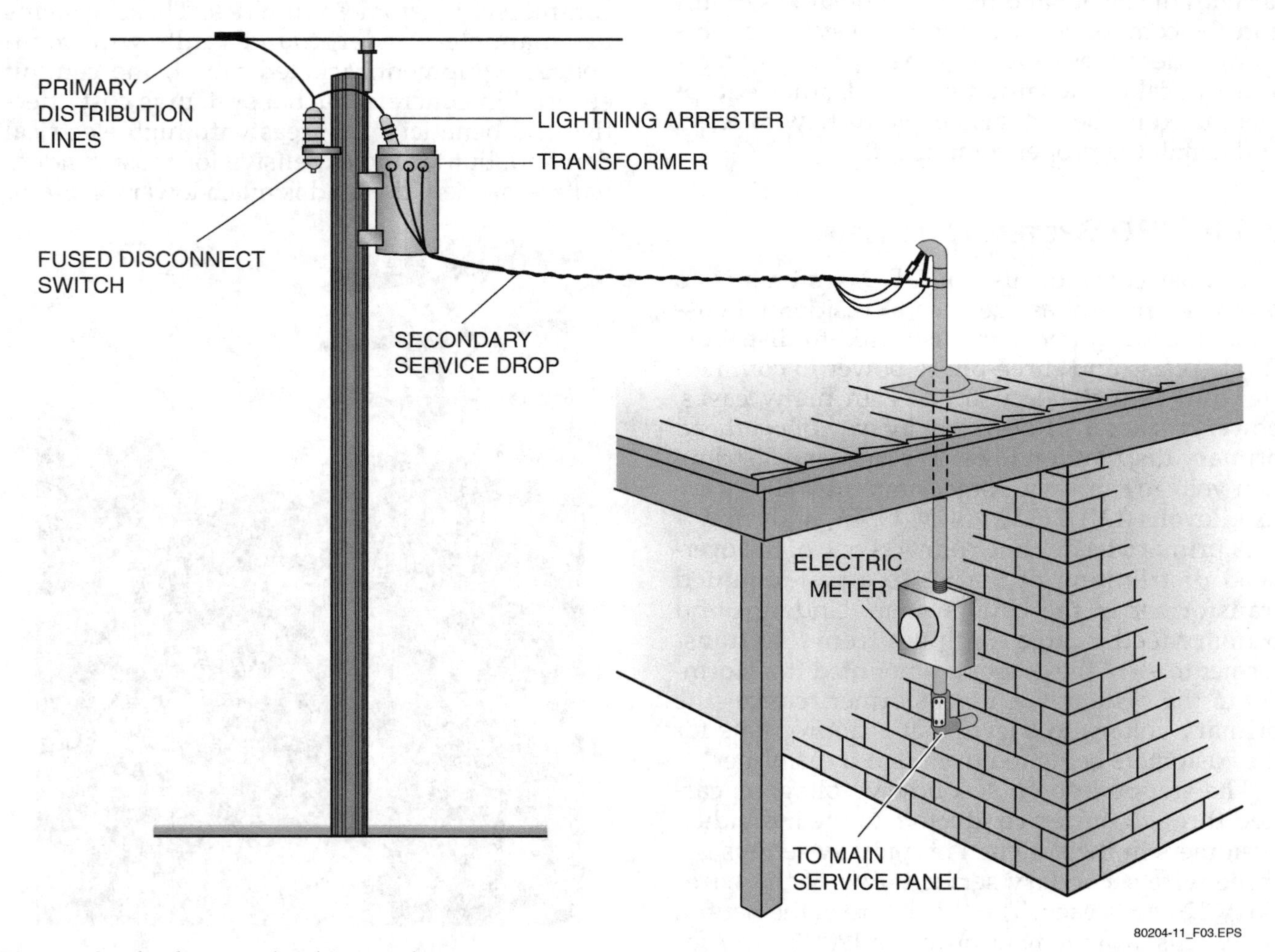

80204-11_F03.EPS

Figure 3 Single-phase overhead service drop.

primary distribution lines. A primary feeder cable runs from an overhead primary through a fused disconnect switch and a lightning arrester. The feeder cable continues down the riser pole and then goes underground to the first pad-mounted transformer in the URD system. The transformer reduces the primary voltage, and underground secondary lines carry the power to the customer or customers served by that transformer. The primary feeder line continues out from that transformer to other pad-mounted transformers in the system.

The cost to install early URD systems was much higher than that of overhead lines. Plus, these systems were more time consuming and costly to troubleshoot and maintain. However, there were, and continue to be, the following distinct advantages to using direct-buried systems compared to overhead lines:

- Visual clutter of exterior power lines is eliminated.
- Weather, downed trees, and accidents have fewer effects.
- There are fewer faults and generally more reliability.
- Less periodic maintenance is required.
- Less maintenance is required for related activities like tree trimming.
- Safety is improved, since fewer components are exposed to the public.

Advances in equipment and installation procedures in recent years have made URD systems more affordable, reliable, and less time consuming to install and maintain, so the use of URD systems continues to grow.

2.2.0 Types of URD Systems

The exact way that the primary and secondary lines run through a URD system depends on the type of URD design being used. Three of the most common URD system designs are radial feed, loop feed, and double feed systems. The type that is used depends on customer needs. The descriptions that follow include general information about each design. Additional system components, such as transformers, lightning protection devices, and fault-indicating devices, are covered in detail later in this module.

2.2.1 Radial Feed

A radial-feed design, like the one shown in *Figure 5*, has only one connection to its source of power. It is connected to an overhead primary distribution line at one end. The other end of the system is terminated and does not return, or reconnect, to the overhead line. This type of design gets its name from the fact that the URD system radiates out from the overhead line, with the primary feeder line extending from transformer to transformer.

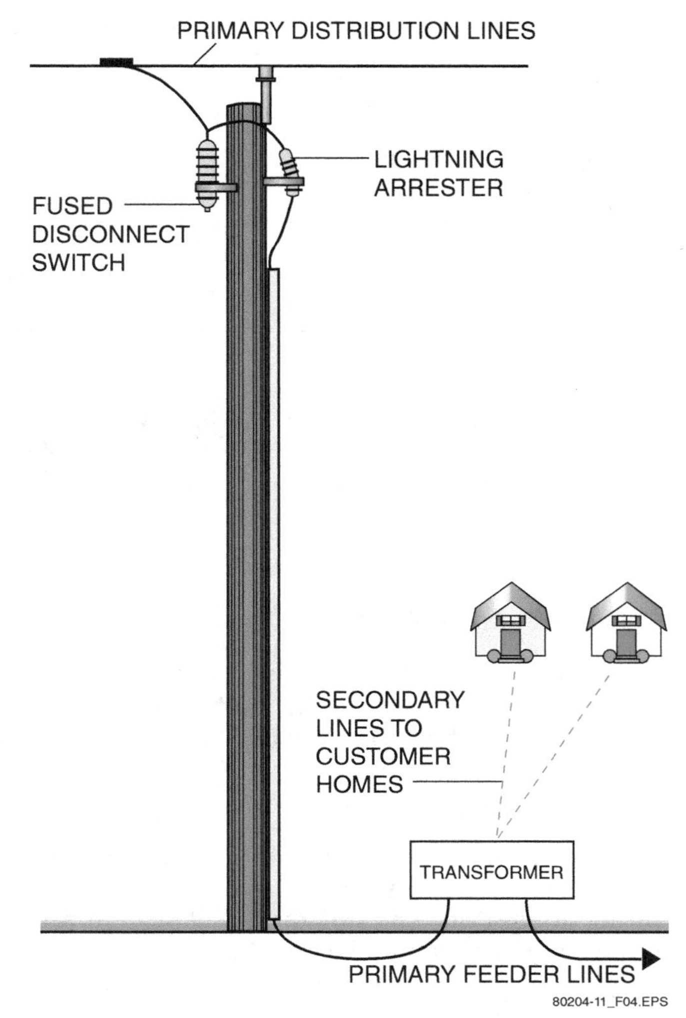

Figure 4 Primary feeder connection to a URD pad-mounted transformer.

In this radial-feed system, a primary feeder cable runs from the overhead primary line through a fused disconnect switch and a lightning arrester, down the riser pole, and underground to a pad-mounted transformer. The primary feeder continues from the first transformer to another transformer and so on, until it terminates at the final transformer in the URD system. Secondary lines run from each transformer to the individual customer homes.

A radial-feed design is one of the simplest types of URD systems; it is powered from only one side. This creates a distinct disadvantage over other designs. As *Figure 6* shows, if a fault

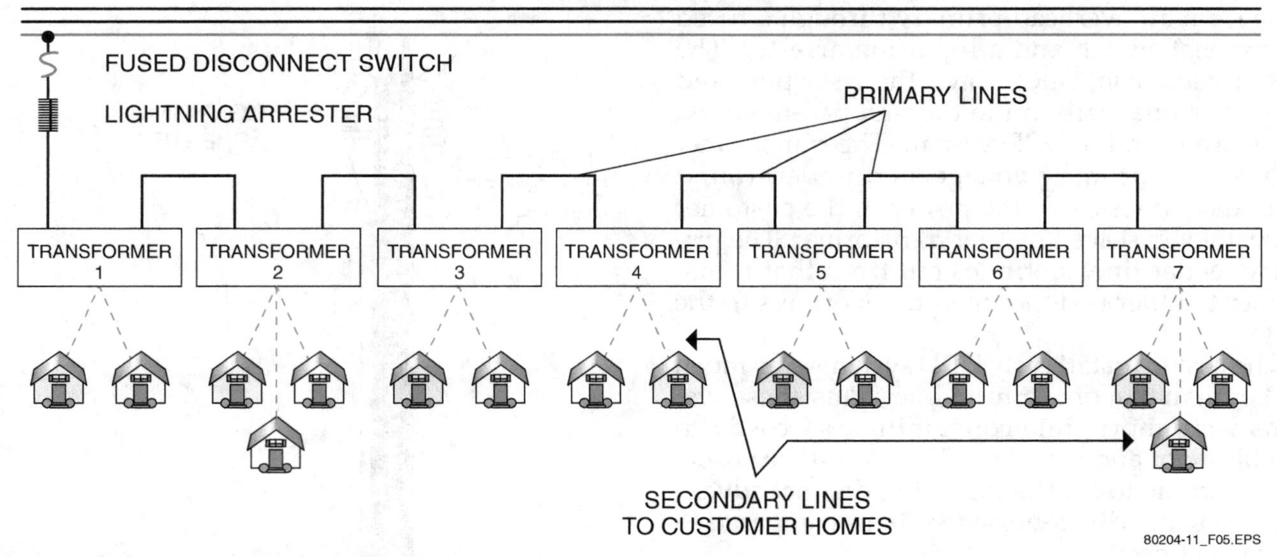

Figure 5 Radial-feed URD system.

occurs along a radial-feed system, any customers located downstream from the fault cannot be re-energized until the fault is located and corrected.

2.2.2 Loop Feed

A loop-feed design is one of the most common types of URD systems. It has two power source connections. Both ends of the URD system are connected to the same phase of the overhead primary. Half of the customers on the loop are fed from one energized feeder, and half are fed from the other energized feeder.

In many loop-feed systems, such as the one in *Figure 7*, there is a normally open connection in the loop. It is often at a transformer located at or near the middle of the loop. This design is called an open loop system. The design helps balance the load on the system. Also, the open connection is used so that if a fault occurs in one side of the loop, the customers who are fed from the other side of the loop can continue to receive power. This reduces the number of customers affected by problems in the loop. It also reduces the amount of troubleshooting needed to locate problems.

2.2.3 Double Feed

A double-feed URD system is used for customers who cannot tolerate being without electric power. An example of such a customer might be a hospital or a manufacturing facility where a constant supply of electric power is absolutely critical.

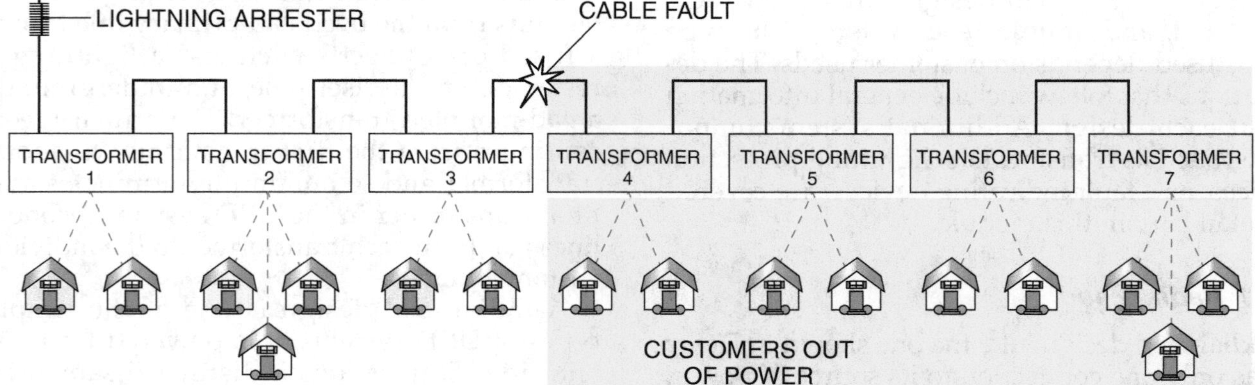

Figure 6 Radial-feed system with fault.

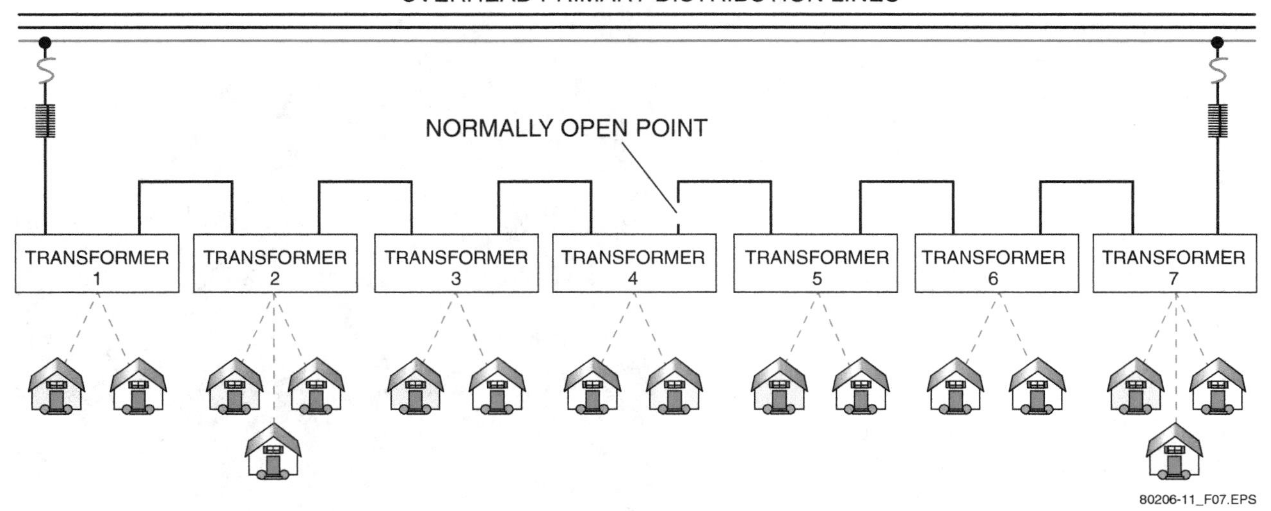

Figure 7 Loop-feed URD system.

In a double-feed design (*Figure 8*), two separate URD systems are connected to the customer. Each URD system is energized from a different primary feeder. A pad-mounted switchgear is located between the two systems so that the customer can be supplied by either one of the systems. If a problem occurs in one system, the switchgear automatically transfers the source of power from the failed system to the functioning system.

2.3.0 URD Trenching and Backfilling Methods

In order for any URD system to function correctly, the lines and other equipment used in the system must be properly installed. Otherwise, components could be damaged or fail prematurely, causing unexpected power outages to occur. A big part of any successful installation job has to do with planning, as well as the trenching and backfilling methods used to bury URD cable.

Before any installation occurs, a plan should be developed to identify the exact types and locations of cables, pad-mounted equipment, and any other components used in the system. It is very time consuming and costly to reopen a trench to make repairs or replace cable. Following basic installation guidelines is critical.

The equipment used to bury cables can depend on factors such as the following:

- Access to the area where the lines are being installed
- The depth at which the lines must be buried
- The width of the trench if other utilities are being buried with the electric lines
- The type of service being installed
- The presence of other utilities already installed in the area
- Local codes and company guidelines

Another consideration for URD trenching is whether a **common trench (joint trench)** ap-

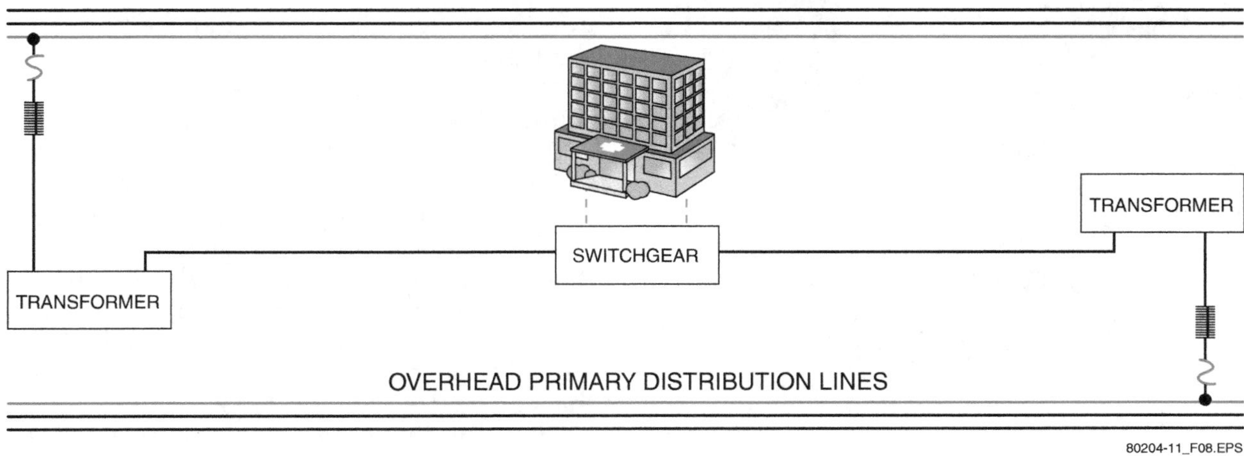

Figure 8 Double-feed URD system.

proach is being taken. In a common-trench installation, multiple utilities such as electric, gas, cable TV, and telephone lines are buried in a single trench. This approach reduces cost, since utilities agree to share in the expense. It also requires less time, fewer on-site workers, and makes it easier to locate utilities later on.

> **NOTE**
> There is a minimum separation required between gas and power lines.

Trenches for URD cable are commonly dug using a trencher, a tractor-mounted backhoe, or a tractor-mounted plow. *Figure 9* shows a typical trencher being used to dig a URD cable trench. Trenchers like this are small enough to maneuver in tight areas where other utilities might already be in place. If only the URD cable is being buried, a trencher might only dig a trench that is just eight inches wide.

A backhoe or excavator (*Figure 10*) might be used in a common-trench installation, where the trench needs to be wide enough to install other utilities with the URD cable. For instance, if telephone, cable TV, gas, water, or sewer lines are being placed in the same trench with URD cable, the required trench width might be 24 inches or more. An excavator or backhoe usually needs more room to maneuver than a trencher.

In rural areas where there are no other underground utilities to interfere with the URD installation, a tractor-mounted plow may be used for trenching and cable installation. This type of equipment has a reel of cable mounted on it. As the plow digs the trench, cable spools off the reel into the trench. Blades behind the plow backfill the trench as the equipment moves along. This way, the trenching, cable installation, and backfilling all occur at the same time.

80204-11_F09.EPS

Figure 9 Trencher digging a URD cable trench.

BACKHOE

EXCAVATOR

80204-11_F10.EPS

Figure 10 Backhoe and excavator.

Another method that is sometimes used to bury URD cable is boring with horizontal directional drilling (HDD) equipment. HDD equipment enables utilities to drill underground tunnels and then pull the URD cable through the tunnel. This method eliminates the need to dig up yards, sidewalks, roads, and driveways during installation. *Figure 11* shows one type of boring equipment that can be used for URD installation.

The equipment that is used to bury URD cable must be able to trench or bore in a way that satis-

80204-11_F11.EPS

Figure 11 Boring equipment for URD installation.

fies several vital requirements. For instance, the equipment must be able to trench or bore to the proper depth. Local codes and company guidelines often dictate the depth at which URD cable must be buried. For example, secondary cables might need to be buried a minimum of 30 inches, while primary cables might need to be 48 inches deep.

Trenching or boring must be done in as straight a line as possible. Sharp turns should be avoided to minimize the amount of bending that could damage the URD cable.

The bottom of a trench should be as smooth and flat as possible. Rocks and sharp objects can damage URD cable and lead to premature failure. Also, different types of soil can affect URD cable differently. The cable selected for burial must be suitable for the ground conditions.

The way a trench is backfilled is also important. Some utilities require sand bedding under and above the URD cables to help protect them. The excavated material from the trench can often be used for backfill, as long as it is free of rocks and other foreign material. Avoid using soil that could damage the cable. The trench should be backfilled in layers, and each layer should be carefully tamped down so that no voids are left around the cable. Some excess soil should be placed on top of the trench and tamped down to allow for settling.

3.0.0 Cable Types and Termination Methods

The primary and secondary cables used in URD systems are designed specifically to be used underground. These cables vary in size and design, depending on the voltages being conducted and other requirements of the system. The methods used to terminate those cables to system equipment or to connect different cables to one another can also vary. This section describes common types of primary and secondary URD cable and examines various termination and connection methods.

GOING GREEN

Erosion Control

When bare earth is exposed to rain, a great deal of soil can run off the site with the rains. If too much soil runs off, then you may need to go back over the grade and build it up. This takes time and costs money. A silt fence may be used to prevent this. A silt fence is a special woven cloth that is used to catch soil runoff from exposed soil.

80204-11_SA01.EPS

WARNING!

There are many electrical hazards associated with URD equipment. Failure to follow safety guidelines and wear the proper PPE can result in serious injury or death. Companies have specific rules and procedures about energizing and de-energizing equipment in URD systems. Always follow your company's guidelines when working in or around URD systems.

3.1.0 Conductors

URD cable is different from the cable used in overhead installations and, in most cases, underground systems with vaults and conduit. Overhead lines must be strong enough to support their own weight. URD cable is made up of stranded conductors. Stranded conductors allow the cable to be flexible enough to bend and follow curved trenches. However, the conductors must still be strong enough to withstand pulling and other physical demands encountered during installation and use.

The size and type of cable selected for a URD application is based on several factors. One factor is **ampacity**. Ampacity is the maximum amount of current that a cable can carry before it begins to deteriorate. This term is sometimes called *current rating* or *current-carrying capacity*. A cable's ampacity is based on factors such as the temperature rating of the insulation, the electrical properties of the conductor material, and the temperature surrounding the cable.

The conductor material in a URD cable is usually aluminum or copper. The cable has layers of electrical insulating material and sheathing designed to resist corrosion and other problems that might occur in underground applications. Like overhead cable, URD cable size is specified in a gauge size or circular mils, based on the American Wire Gauge (AWG) standard. *Table 1* shows some typical measurements for a 15kV primary URD cable.

One common type of primary URD cable is shown in *Figure 12*. As the lettering on the outer jacket indicates, this is a size 2/0 cable. It is rated for up to 35kV. The shielding and insulation have been stripped back to reveal the various layers of the cable.

This type of cable consists of a stranded aluminum conductor that is surrounded by a layer of semiconductor material. The semiconductor material is often called a shield or screen. Surrounding the semiconductor material is a layer of insulating material that keeps the conductor isolated from any other conductors and ground. This insulator is ethylene propylene rubber (EPR), but many primary cables use cross-linked polyethylene (XLPE). Surrounding the insulation is another layer of semiconductor material that serves as an insulation shield. A metallic shield of concentrically wound copper or aluminum wires is wrapped around the insulation shield to help protect the cable and provide a grounded neutral. The entire cable is encased in a linear low-density polyethylene (LLDPE) jacket, or sheath, that helps protect the cable from underground conditions such as moisture and corrosion.

Not all primary URD cables are jacketed. Some have the metallic shield as the outermost layer. Other primary cables have an exterior conduit already in place to further protect the cable in harsh environments. Line workers must be familiar with the types of primary cable used by their company and follow the installation and maintenance procedures that relate to them.

Secondary URD cables carry lower voltages from pad-mounted transformers to customer homes. The lower voltage requirements mean that secondary cables have fewer layers than primary cables. Typically, secondary URD cable consists of a stranded conductor that is covered in protective insulating sheath. However, some secondary cables do have a layer of concentric wiring to serve as the neutral connection.

Like primary URD cable, there are different sizes and types of secondary cable, depending on the system requirements. *Figure 13A* shows a single-conductor cable that consists of a stranded aluminum conductor that is encased in an XLPE insulation sheath. This is a size 1/0 cable that is rated for a maximum of 600 volts.

Two other common types of secondary URD cable are shown in *Figure 13B*. One is a two-con-

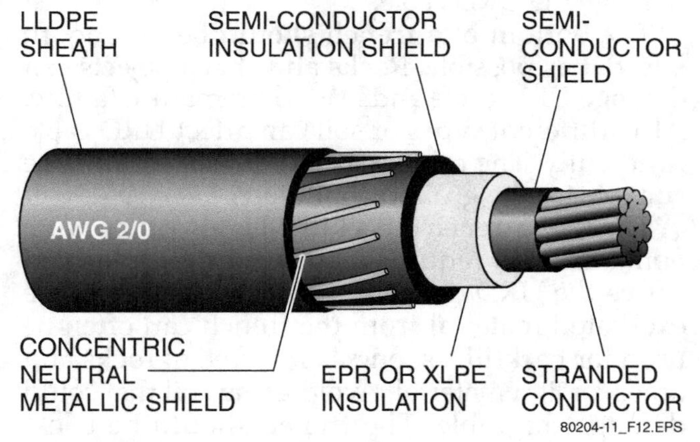

Figure 12 Primary URD cable layers.

Table 1 Typical URD Cable Measurements

AWG Size (Gauge)	Cable Diameter Circular Mils	Number of Aluminum Conductor Wires	Number of Copper Neutral Wires	Neutral Wire Size (AWG)
2	886	7	10	14
1	916	19	13	14
1/0	966	19	16	14
2/0	1,008	19	20	14
3/0	1,058	19	25	14
4/0	1,149	19	20	12

ductor, or duplex, cable and the other is a three-conductor, or triplex, cable. These cables are also rated for a maximum of 600 volts.

In most cases, two insulated cables are run from the transformer to the customer's home. Each of the cables carries 120 volts. A third cable, often a non-insulated cable, would be included as the neutral.

3.2.0 Termination Methods

Terminations and connectors come in many forms, and different procedures are used to terminate and connect cables. A few of the more common methods are covered in this section.

Regardless of the equipment and procedures used, it is very important to properly prepare the cable and install the components correctly. When a cable is cut to be terminated to equipment or connected to another cable, the layers of sheathing and insulation must be carefully removed to expose the conductor wire. Special tools are available for properly cutting cables (*Figure 14*). In addition to those shown, electric and hydraulic cable cutters are also available. Once the cable is cut, the insulation must be removed to expose the conductors. It is essential to use the proper tools for preparing the cable ends. *Figure 15A* shows a cable-stripping tool used to remove the outer cable jacket on primary conductors. It can also be used to strip off the insulation. *Figure 15B* is a semiconductor-scoring tool that can be adjusted for cutting depth and angle. Once it is set to the proper depth, a ring cut is made around the cable at the desired point. After the ring cut is made, the blade is angled so that each rotation around the cable results in a spiral cut. *Figure 15C* is a tool used to score the semiconductor material.

When the conductor has been properly cleaned, the appropriate termination or connector can be installed. In some cases, soldering and/or taping is done to properly seal the termination or splice. The care and expertise of the installer are vital parts of any successful termination or splice.

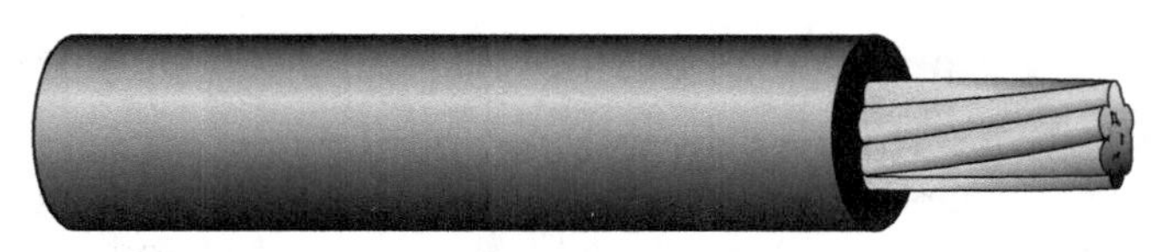

A. SECONDARY CABLE

B. DUPLEX AND TRIPLEX SECONDARY CABLE

80204-11_F13.EPS

Figure 13 Secondary URD cable.

80204-11_F14.EPS

Figure 14 Manual and battery-powered cable cutters.

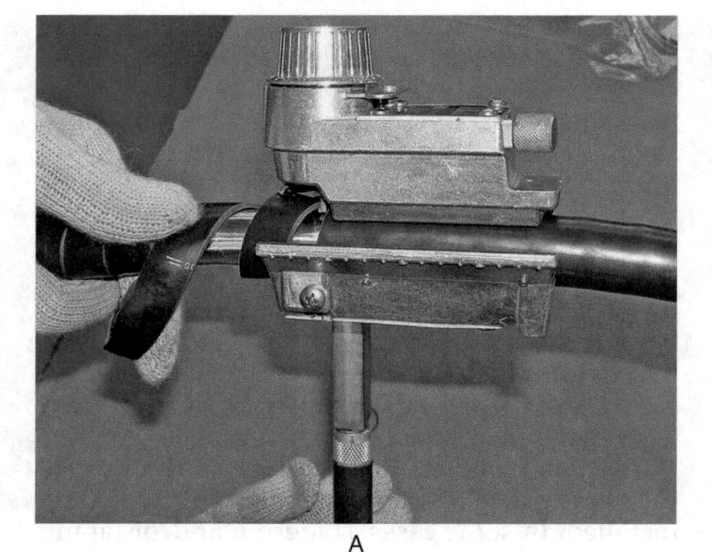
A

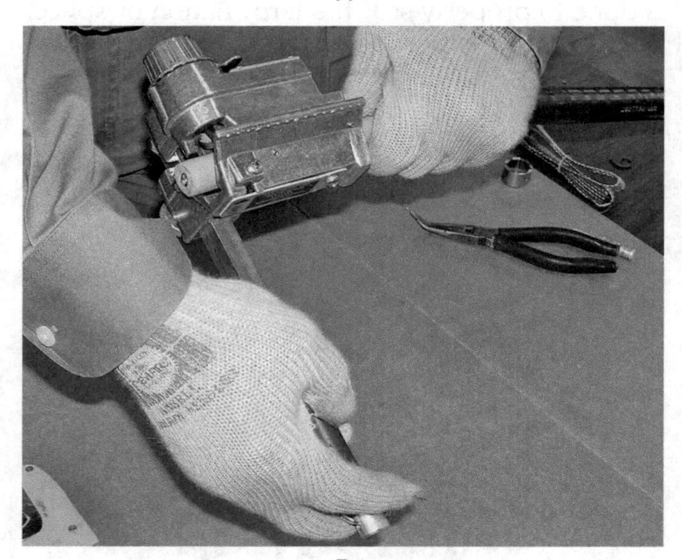
B

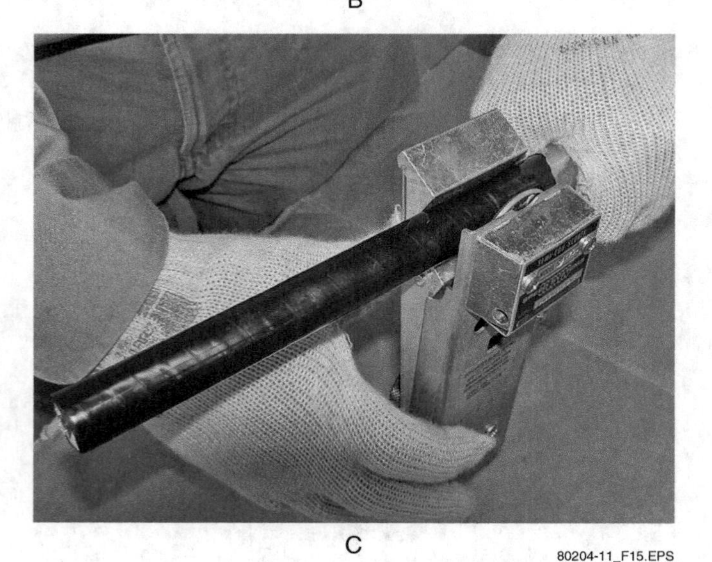
C

80204-11_F15.EPS

Figure 15 Primary cable-stripping tools.

On Site

Cutting Power Cables

Open-blade (knife) stripping can damage the conductors and is prohibited on many job sites. Adjustable strippers such as the tool shown here are used instead for preparing secondary cable ends.

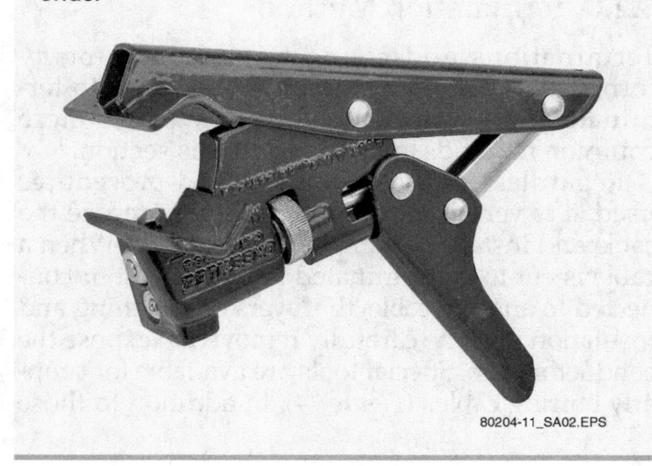

80204-11_SA02.EPS

3.2.1 Copper to Aluminum

URD conductors are made of either aluminum or copper. When a terminal or connector is used on URD cable, it must be the appropriate type for the conductor material. Some terminals and connectors are designed for copper conductors only, while others are designed for aluminum conductors only. Some terminals and connectors are **dual-rated**. Dual-rated connectors are made of aluminum, but they are designed to be used for both aluminum and copper. They can be used in situations where aluminum and copper conductors must be joined. Dual-rated connectors, such as the H-type connector shown in *Figure 16*, usually have some sort of ALCU marking on them. AL is the abbreviation for aluminum; CU is the abbreviation for copper.

Some of the terminals marketed under the name HYLUG™ by Burndy® LLC are dual-rated. These terminals are appropriate for either copper or aluminum. The HYLINK™ brand of Burndy connectors, as well as dual-rated connectors produced by other manufacturers, can be used to splice together aluminum and copper conductors.

When a compression connector is used to join aluminum and copper conductors, a compression tool and die are used to crimp the connector onto the conductors. Many connectors are stamped or

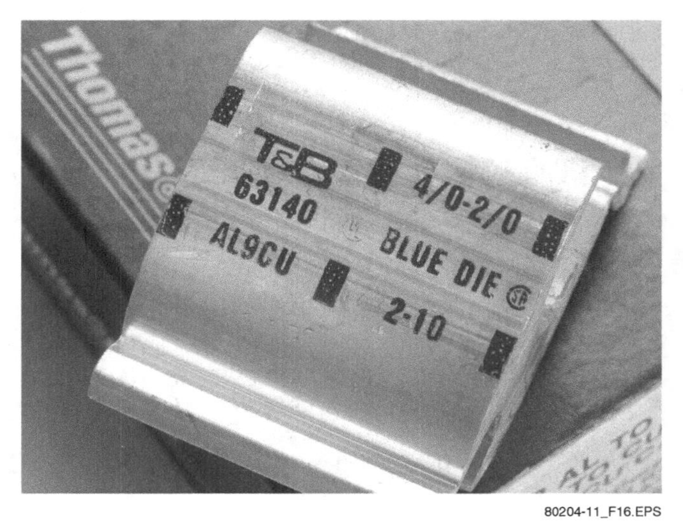

80204-11_F16.EPS

Figure 16 Dual-rated connector with ALCU marking.

color-coded to identify the proper die to use. The number of crimps needed to attach the connector to the cable can vary, but many connectors are marked with crimp locations (*Figure 17*). The basic procedure is to place the connector onto the ends of the stripped conductors and then start crimping in the center and work outward (*Figure 18*). Once the connector has been tightly crimped onto the two conductors, the splice can be sealed with tape or some other sealing material such as heat shrink tubing.

3.2.2 Splices

Splices can be used to connect secondary and primary cables in a URD system, but the way the splices are made can be very different. Secondary cables often consist of a single stranded conductor encased in an XLPE insulation sheath. Primary cables, on the other hand, typically have several layers of shielding and insulation along with concentric neutral wiring. So there are lot more layers and materials to deal with.

Figure 19 shows one type of mechanical connector that is suitable for secondary cable splicing. This is a split-bolt mechanical connector. With this type of connector, the two bolts are loosened so that the connector jaws can spread apart. Then, each conductor is placed into the jaws and the bolts are tightened.

When a split-bolt connector is used, the preliminary steps are basically the same as those used for other types of splices. The insulation shield of the secondary URD cable should be wiped clean with a nonabrasive cloth. Then, the sheath on each cable end should be cut and removed far enough to allow the connector to be installed.

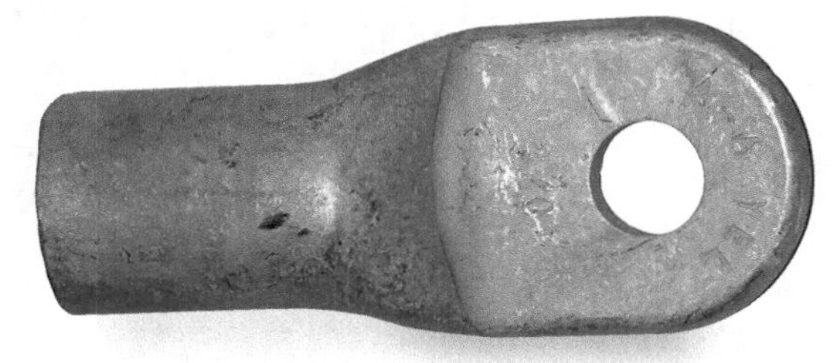
80204-11_F17.EPS

Figure 17 Compression terminal.

80204-11_F18.EPS

Figure 18 Line worker attaching compression connector.

On Site

Rejacketing

This is an example of a heat-shrinkable wraparound jacket repair sleeve used to protect splices in cables subjected to mechanically abusive environments. It can also be used to provide a moisture barrier for repairing and rejacketing lead-covered shield and moisture-impervious cables.

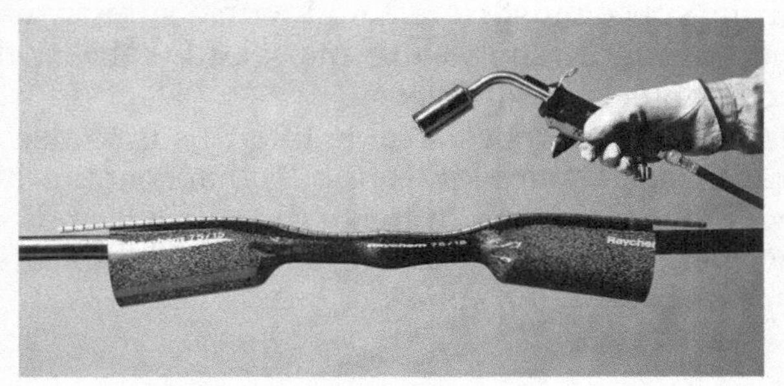
80204-11_SA03.EPS

80204-11_F19.EPS

Figure 19 Split-bolt mechanical connector.

Next, the bolts of the connector are loosened so that the connector jaws can slide open. Then, one cable end is slid into the upper part of the opening above the spacer, and the other cable end is slid into the opening below the spacer. The bolts are then tightened to secure the cables.

Once the cables are secure in the split-bolt connector, the splice should be taped to replace the insulating sheath that was removed. This taping procedure is basically the same as the one used for the solder connection. Rubber tape should be applied in a half-lap manner, beginning with the space between the two ends of the insulation. The taping should continue and extend onto the insulating sheath on both sides of the splice. Once the tape is about 75 percent thicker than the existing insulating sheath, the splice is complete.

A primary cable consists of several layers of shielding, insulation, and wiring, so a primary cable splice is considerably more complicated than a typical secondary cable splice. Each layer of material that is removed from the cable to make room for the splice must be restored with a similar material. *Figure 20* shows a primary URD cable with the various layers labeled.

One way to splice a cable like this is to use a prepackaged splice kit (*Figure 21*) that contains all the material needed to make the splice. Detailed instructions are included in the kit for the specific type of cable being spliced.

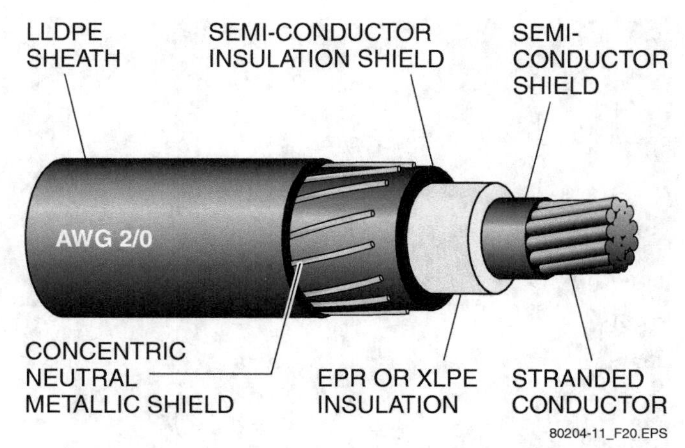

80204-11_F20.EPS

Figure 20 Primary URD cable layers.

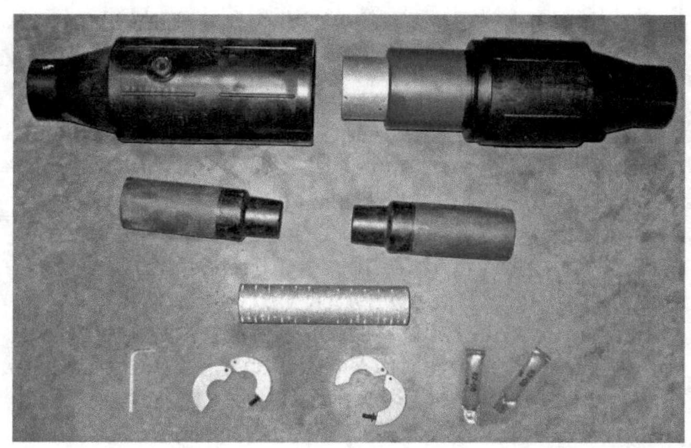

80204-11_F21.EPS

Figure 21 Prepackaged splice kit for primary URD cable.

3.2.3 Compression Terminations

Compression terminals and connectors are among the most common types of connectors used for URD cable. They are relatively easy to install when the proper tool is used. Most tool and connector manufacturers use numbering or color coding systems to make selection and installation easier.

Figure 22 shows one type of compression tool that is used to install compression terminals and connectors. This particular tool is an MD6 mechanical crimper made by Burndy®. It is a hand-

80204-11_F22.EPS

Figure 22 Burndy® MD6 mechanical crimping tool.

operated tool that is capable of producing 9,000 pounds of crimping force.

The MD6 can be used to crimp many varieties of compression terminals and connectors. Some of the more common terminations can be done without using extra dies. But extra dies are available to expand the tool's capabilities.

Another type of compression tool used to install compression terminals and connectors is shown in *Figure 23*. This is a Burndy® Y35 hydraulic crimper. It is a hand-operated hydraulic tool. It is capable of developing 12 tons of crimping force. Also, with the Y35, only one handle needs to be operated to build the pressure needed to make a crimp. The Y35 tool uses numerous dies to crimp a wide variety of compression terminals and connectors.

Figure 24 shows a compression tool used to crimp a compression connector onto a cable end. The number of crimps needed and the locations of the crimps are indicated in the manufacturer's literature and, in many cases, on the connector itself.

4.0.0 Lightning Protection and Fault-Indicating Devices

The fact that URD systems have buried cables does not mean they are immune to many of the same problems that affect overhead lines. Cables wear out and fail, faults occur because of lightning strikes, and system components can be damaged by accidents. When problems occur, it can take a long time to locate the problem, dig up the cable, make the necessary repairs, and bury the cable again. To help minimize the time and cost of maintaining URD systems, devices are used to help protect against faults and make troubleshooting easier. Two common examples are lightning protection devices and fault-indicating devices.

4.1.0 Lightning Protection Devices

Many of the problems related to URD systems occur because of cable faults caused by lightning strikes and voltage surges. Lightning protection for URD systems is typically provided in the form of surge arresters. A surge arrester (*Figure 25*) is a protective device that is installed between an electrical conductor and ground to limit the effect of overvoltage on equipment. The arrester does this by absorbing some of the energy and discharging the overvoltage.

Inside a surge arrester (*Figure 26*) is a **metal oxide varistor (MOV)** disk. The MOV disk is a semiconductor. During normal conditions, the disk is an insulator; but during a surge, the disk becomes a conductor to divert the overvoltage to ground. Since overvoltage in a URD system is typically caused by lightning strikes or voltage surges to ground, these devices are sometimes called lightning arresters.

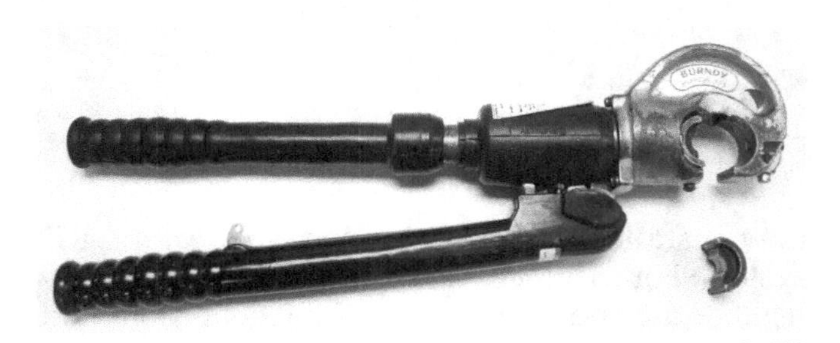

80204-11_F23.EPS

Figure 23 Burndy® Y35 hand-operated hydraulic tool.

80204-11_F24.EPS

Figure 24 Line worker using a compression tool to install a terminal.

80204-11_F25.EPS

Figure 25 Surge (lightning) arrester.

On Site

Compression Tools

There are many tools available for crimping connections. These are just a few examples.

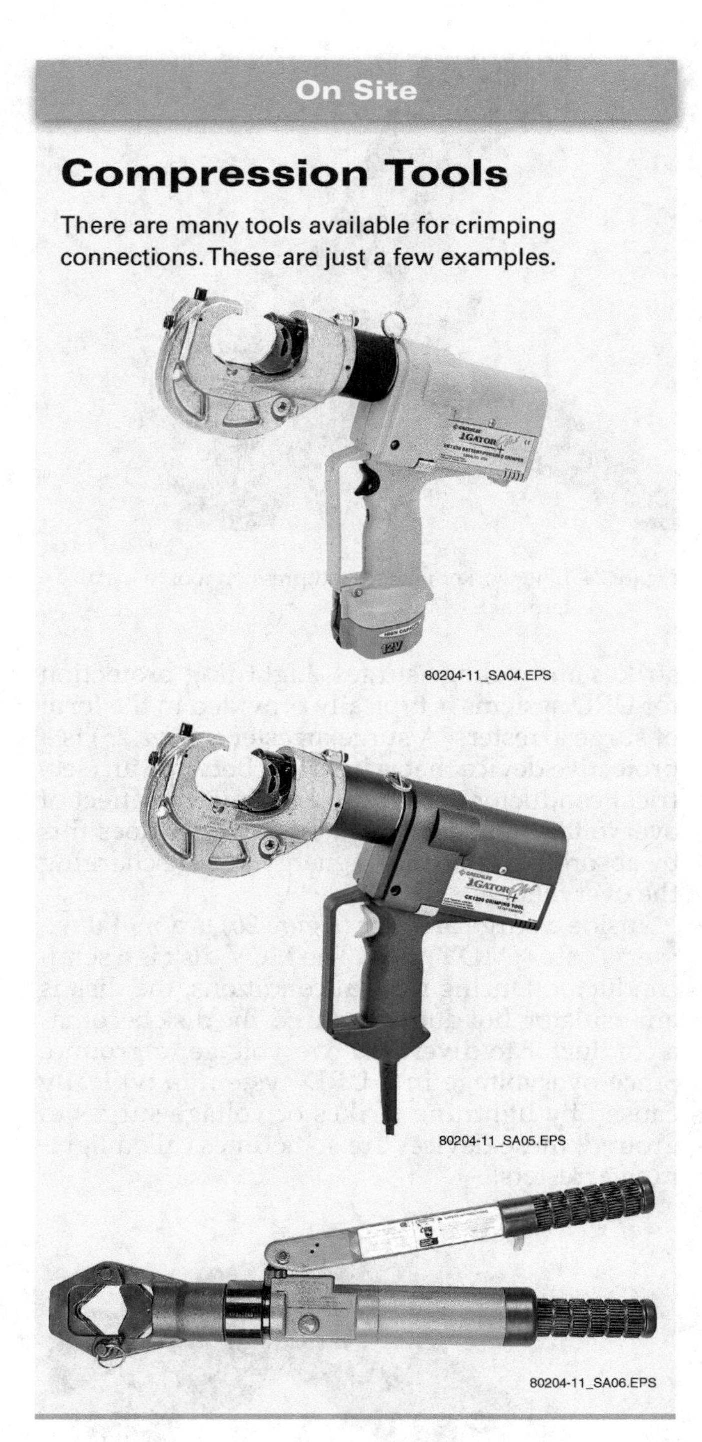

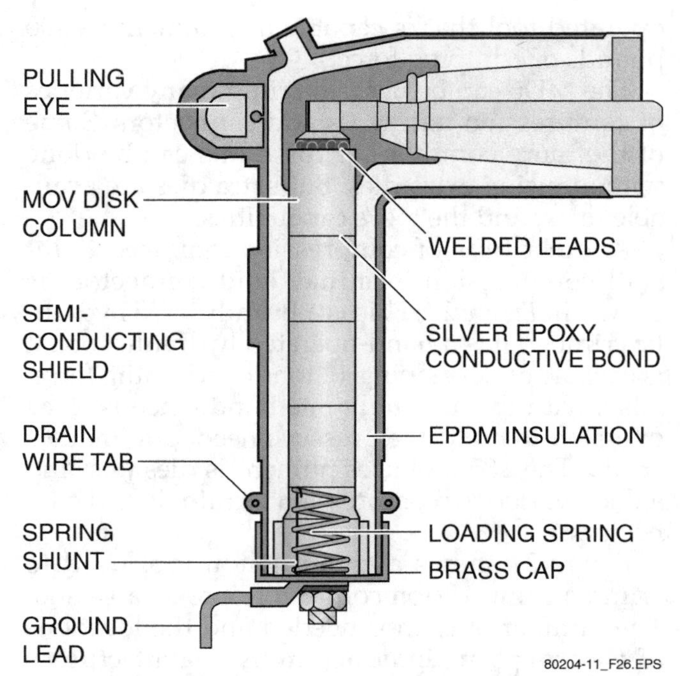

Figure 26 Surge arrester cutaway showing MOV disk.

Figure 27 Surge arrester on riser pole.

In a typical URD system that is fed from an overhead primary distribution line, a surge arrester is installed on the primary feeder at each riser pole. As *Figure 27* shows, the primary feeder passes through a fused disconnect switch, then through a surge arrester, down the riser pole, and then underground to the URD system.

When URD systems first appeared, it was believed that a riser pole surge arrester would provide enough protection for the URD system. Over the years, companies have come to realize that more protection is needed, so surge arresters are now installed at numerous locations in a URD system. For instance, surge arresters are typically used at the end point in a radial-feed system and at the normally open points in a loop-feed system.

Different types of surge arresters are available for URD systems. Most of them are dead-front surge arresters of one type or another. *Figure 28* shows three different variations of dead-front surge arresters.

The bushing surge arrester (*Figure 28A*) is designed to fit directly into a transformer bushing without the need for a bushing insert. This takes up less space in the transformer housing. It enables a source cable to be attached directly to it. The more conventional dead-front surge arrester (*Figure 28B*) requires a transformer bushing insert. Another configuration is the parking stand surge arrester (*Figure 28C*). A parking stand arrester fits onto the parking stand bracket in a transformer and provides surge protection for energized cables that are parked.

4.2.0 Fault-Indicating Devices

A fault indicator is a device that senses fault current, which is the unusually high current flow that occurs when there is a fault. It indicates the fault with a warning light, flag, or similar means. Fault indicators have been in use for many years and are reliable ways to reduce the time it takes troubleshooters to locate faulted equipment.

Various types of fault-indicating devices are available for URD systems. One type, shown in *Figure 29*, is a test point-mounted fault indicator.

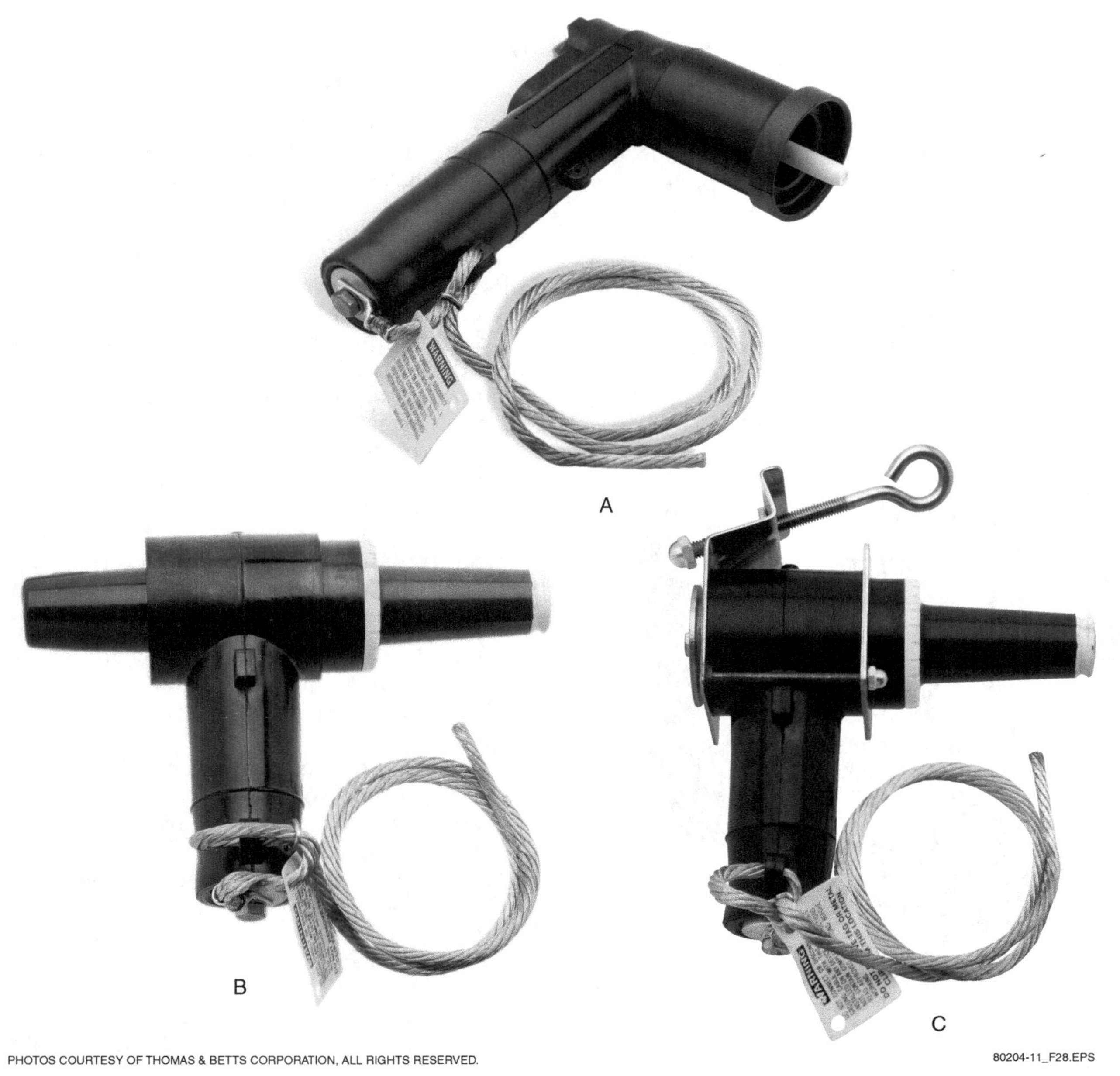

Figure 28 Dead-front surge arrester variations.

80204-11_F29.EPS

Figure 29 Test point fault indicator.

This type of fault indicator is designed to mount directly onto an elbow or cable component that has an *IEEE 386 Standard* capacitive test point.

Another type of fault indicator is a cable-mounted, or clamp-type, indicator (*Figure 30*). This type of fault indicator is designed to clamp directly onto a URD cable.

Regardless of the type of fault indicator used, they all work on the same basic principle. When a fault occurs in a circuit, a magnetic field related to the fault is produced. This field closes a reed switch in the fault indicator. The closed switch causes the indicator to display a light or flag to indicate a trip. Once the system is re-energized, the indicator automatically resets.

80204-11_F30.EPS

Figure 30 Cable-mounted (clamp-type) fault indicator.

Figure 31 shows a typical loop-feed URD system with fault indicators located at each pad-mounted transformer. This image shows how fault indicators can provide a great deal of help for troubleshooters. In this case, the fuse between the overhead distribution primary and the URD system has blown. Also, three of the fault indicators on that side of the loop are displaying tripped lights. The fault indicator at the fourth transformer on that side of the loop is not displaying a tripped light. The cable section between the last tripped fault indicator and the first non-tripped indicator is where the fault occurred.

4.3.0 Lightning Protection and Fault-Indicating Device Installation

Modern surge arresters and fault-indicating devices are relatively easy to install, but working around high-voltage equipment is always dangerous. Always follow your company's safety and installation guidelines when installing lightning protection and fault-indicating devices. Part of this involves wearing the proper protective clothing and using the proper tools.

Ideally, surge arresters work best when they are installed directly at the terminals of the equipment they are intended to protect. One example of surge arrester installation is adding an elbow surge arrester to a two-bushing transformer located at the end of a radial-feed URD system. This procedure involves installing the surge arrester in the unoccupied bushing of the transformer. The exact installation steps that are used can vary, depending on the system and the surge arrester being used. But the general procedure is as follows:

- Open the transformer housing to access the transformer bushings.
- Locate the transformer housing grounding point.
- Verify that the surge arrester is the proper size for the system voltage.
- Remove the surge arrester from its packaging and check it for damage.
- Use a hot stick to attach the arrester's ground lead to the system ground.
- Lubricate the arrester interface using the lubricant provided.
- Use the hot stick to remove the dead end cap from the transformer bushing.
- Firmly grasp the arrester's pulling eye with the hot stick.

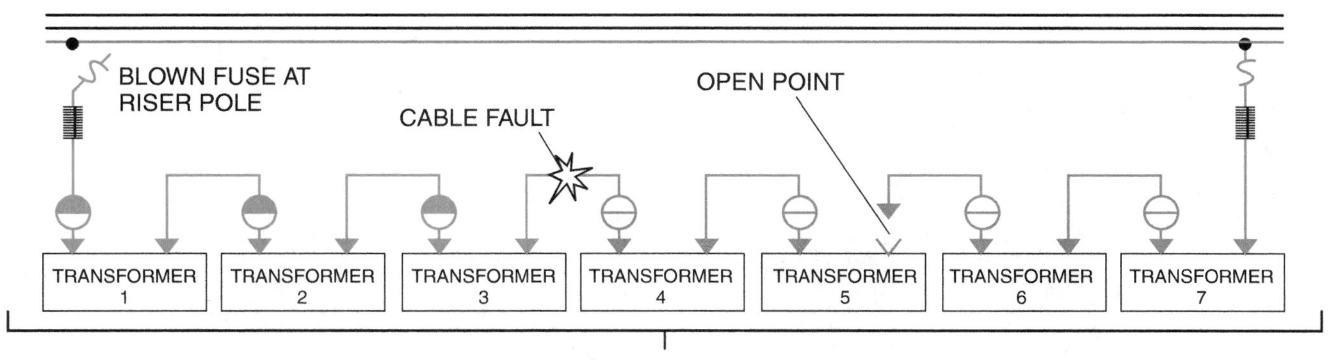

80206-11_F31.EPS

Figure 31 Fault location based on fault indicator displays.

- Position the arrester just into the bushing with the ground facing down.
- Firmly push the arrester onto the bushing.
- Remove the hot stick and close the transformer housing.

Figure 32 shows a single-phase URD transformer after an elbow surge arrester has been installed in the previously unoccupied (dead end) bushing.

Other types of surge arresters might require the use of bushing inserts and feed-through inserts. Again, it depends on the system and the type of surge arrester being installed.

The procedures for installing fault-indicating devices can also vary, but some general steps apply in most cases. Fault-indicating devices can be installed on energized or de-energized circuits. The following method can be used to install a test point fault indicator on an energized elbow:

- Open the transformer housing to access the cable elbows.
- Use a hot stick to remove the elbow test point cap.
- Verify that the area surrounding the test point is clean and dry.
- Lubricate the inside of the fault indicator boot with silicone lubricant.
- Attach the fault indicator's pulling eye to the hot stick.
- Use the hot stick to push the indicator boot over the test point while rotating the boot.
- Align the fault indicator's pulling eye with the elbow's pulling eye.
- Remove the hot stick and close the transformer housing.

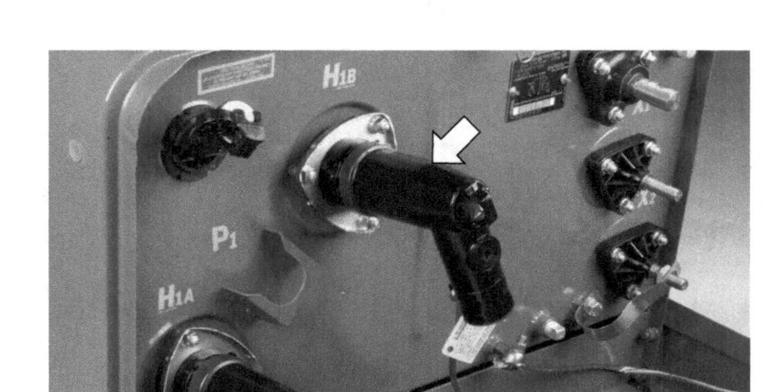

80204-11_F32.EPS

Figure 32 Elbow surge arrester installed at end of radial-feed system.

Figure 33 shows a test point fault indicator as it would be installed on a transformer elbow.

Installing a cable-mounted fault indicator onto a URD cable is no more difficult than installing a test point indicator, but it is important to *not* install the indicator over an exposed concentric neutral. This could affect the accuracy of the indicator. *Figure 34* shows correct and incorrect installation methods for cable that has an exposed concentric neutral.

The exact steps for installing a cable-mounted fault indicator can vary. But most cable-mounted indicators have a trigger mechanism that automatically clamps the indicator onto whatever size cable is being used. The basic procedure is as follows:

- Open the transformer housing to access the URD cables.

80204-11_F33.EPS

Figure 33 Test point fault indicator and elbow.

- Attach the fault indicator's pulling eye to a hot stick.
- Use the hot stick to push the indicator onto the cable.
- Make sure the trigger mechanism clamps the indicator to the cable.
- Remove the hot stick and close the transformer housing.

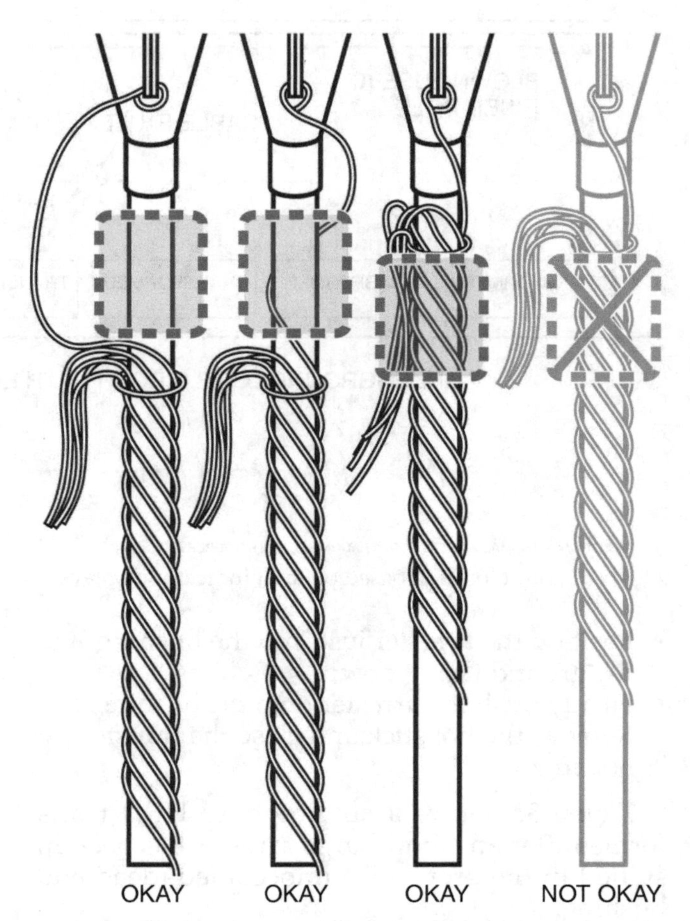

80204-11_F34.EPS

Figure 34 Recommended installation of cable-mounted fault indicator for exposed concentric neutral.

5.0.0 Pad-Mounted Switchgear and Transformers

A typical URD system consists of a lot more than just cables and terminations. Equipment such as switchgear is used to sectionalize and protect the system. Transformers are used to reduce the primary distribution voltage to a level that can be used by customers fed from the system.

Like other URD system components, pad-mounted switchgear and transformers can differ in their appearance and features. The size and capacity of the switchgear and transformers are based on system needs, so any time you work on this type of equipment you should always refer to the manufacturer's guidelines and your company's procedures.

5.1.0 Switchgear

Pad-mounted switchgear consists of electrical disconnects, along with fuses and/or circuit breakers that are used to sectionalize and protect a URD system. The basic purpose of this switchgear is to isolate electrical equipment. It can be used to de-energize equipment in specified sections of the system so that maintenance can be done. It can also be used to clear system faults. In the case of a double-feed URD system, switchgear can be used to isolate one power supply circuit so that a second power supply circuit can provide power to a critical customer.

Refer to *Figure 35*. In the left compartment, three underground primary feeder cables enter the switchgear. Three other primary cables exit the switchgear to another section of feeder. All six primary cables are terminated with deadbreak elbows.

WARNING!

Deadbreak elbows should not be switched under load. Doing so can cause serious damage to the equipment and pose a major safety risk to personnel.

In the right compartment, there are two primary cables that exit the switchgear and run to

80204-11_F35.EPS

Figure 35 Pad-mounted switchgear.

two URD systems. The terminations for these two primary cables are made with loadbreak elbows. Loadbreak elbows are designed to be operated under load.

This switchgear unit in *Figure 36* has three gang-controlled switches. One switch controls the primary feeder cables entering the switchgear. Another switch controls the primary feeder cables exiting the switchgear. The third switch controls the primary cables exiting the switchgear and running to the two URD systems.

If a URD system is fed directly from a substation, a pad-mounted switchgear is used as the entrance point to the URD system. In other words, the switchgear basically replaces the riser pole fuses and connections found in a URD system that is fed from an overhead primary distribution line.

80204-11_F36.EPS

Figure 36 Gang-controlled switches in pad-mounted switchgear.

Switchgear is available in many different sizes and configurations. Line workers working on or around switchgear need to be familiar with the specific types of switchgear used in their company's URD systems.

Switchgear is also categorized in different ways. Sometimes it is categorized according to the medium used to prevent or extinguish arcing that occurs when the switches are operated. A few common mediums that are used in switchgear are oil, air, SF6 gas, and vacuum. Switchgear can also be categorized as **live front** or **dead front**. Live front basically means that energized components are exposed and accessible when the housing is opened. Dead front means that energized components are shielded in some manner so that they are not exposed and accessible when the housing is opened.

5.2.0 Transformers

A pad-mounted transformer in a URD system performs the same function as a pole-mounted transformer in an overhead system. Its main purpose is to step down the primary voltage to a secondary voltage that is suitable for customers. In a URD system, the secondary voltage is typically 120/240V. *Figure 37* shows the locations of pad-mounted transformers in a loop-feed URD system.

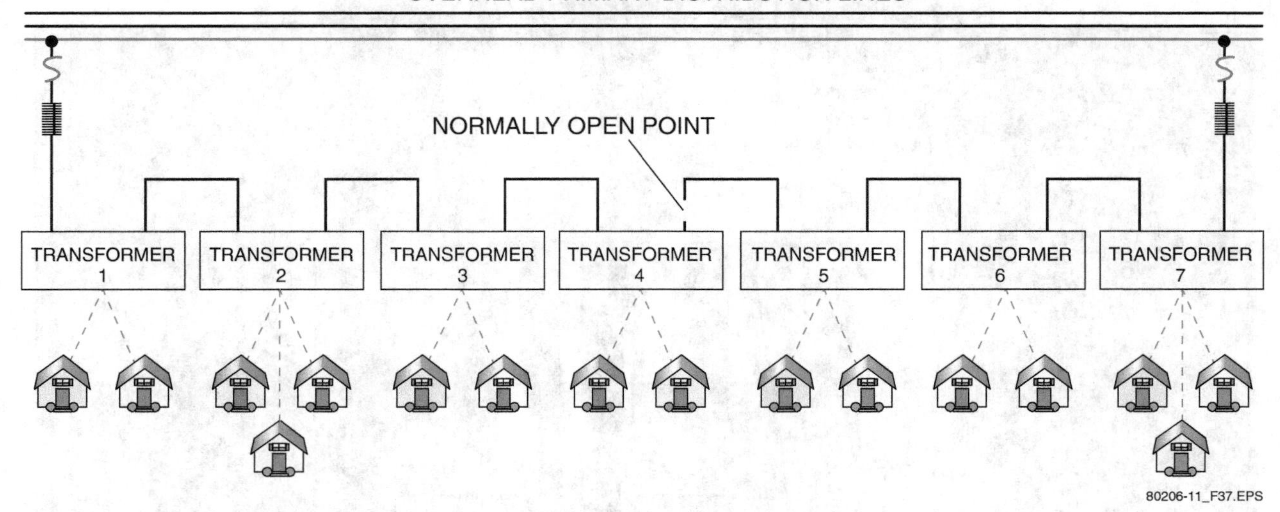

Figure 37 Pad-mounted transformer locations in loop-feed URD system.

There are many different sizes and styles of pad-mounted transformers. *Figure 38* shows a typical live-front single-phase transformer.

On the left side of the transformer are two primary connections. One is the incoming primary line that feeds the transformer. The other is the outgoing primary that exits this transformer and continues to the next transformer in the loop. Located between the two primary connections is a parking stand bracket. If either primary connector is disconnected from its bushing, it can be placed on the parking stand. A grounding strip is provided at the bottom of the transformer. It connects to a grounding rod. Ground lines from any of the transformer connections can be attached to the grounding strip.

SECONDARY TERMINALS

PRIMARY TERMINALS

80204-11_F38.EPS

Figure 38 Single-phase pad-mounted transformer connections.

On the right side of the transformer are three secondary connections. They exit the transformer to provide secondary voltage to the customers being fed by this transformer. Rubber covers are used to cover the secondary connections at the transformer to protect workers from the energized parts.

Modern pad-mounted transformers often have other protective devices built in. For instance, they may have surge arresters on the primary side to prevent overvoltage conditions. They might also have a surge arrester on the secondary side to prevent overvoltage.

Figure 39 shows the interior of an oil-filled, three-phase pad-mount transformer. The primary side is on the left and the secondary terminals are on the right. *Figure 40* is a close-up view of the primary side. The three terminals on the left are the inputs to the transformer primary. The terminals on the right are used to supply primary voltage to a downstream transformer. The switch in the center is a load break sectionalizer that allows users to select different primary taps. The three phases are protected by bayonet fuses (*Figure 41*), which are immersed in oil. The fuse holder is equipped with a pulling eye and is removed with a hot stick. Note the Caution label next to F2. It draws attention to the fact that oil pressure must be relieved before pulling a fuse. This transformer is also equipped with a tap changer.

The secondary side of the transformer (*Figure 42*) contains the secondary spade lug terminals, as well as meters to monitor the pressure, temperature, and condition of the oil. The oil pressure-relief valve is located next to the upper gauge. The

oil drain valve (*Figure 43*) is located at the bottom of the secondary compartment.

5.2.1 Pad-Mounted Transformer Installation

The transformer is mounted on a pad. Composite pads are commonly used for single-phase transformers; concrete pads are more common for three-phase transformers. However, pads and pedestals made of plastic and other materials may be used if allowed by local codes. The pad must have an opening that aligns with an opening in the bottom of the transformer. Primary and secondary cables enter and leave the transformer by way of this opening. Sections of conduit with long-radius elbows protect the cables as they enter and leave the transformer. *Figure 44* shows features of a typical mounting pad.

80204-11_F39.EPS

Figure 39 Interior of an oil-filled, three-phase pad-mount transformer.

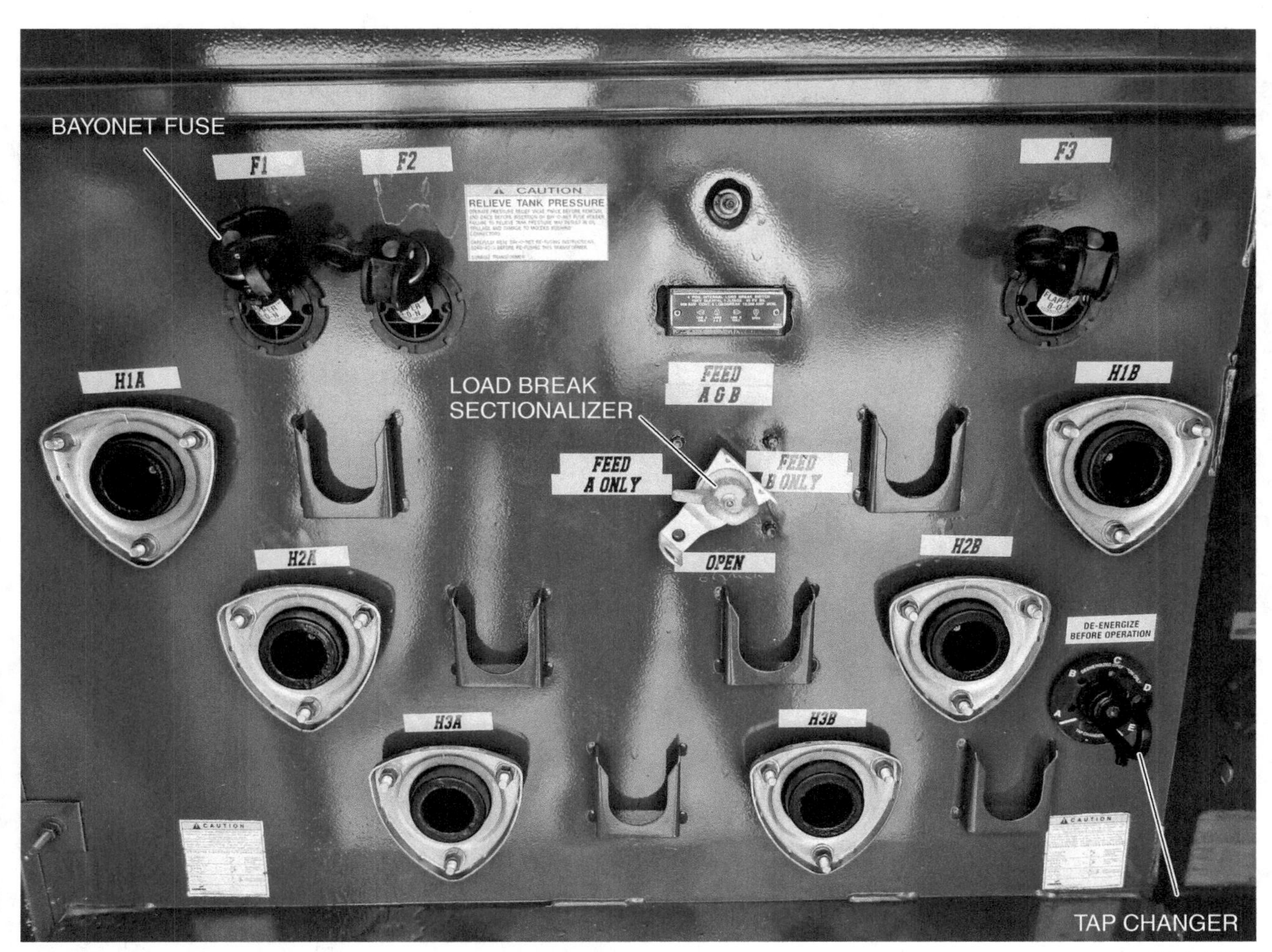

80204-11_F40.EPS

Figure 40 Primary side.

80204-11_F41.EPS

Figure 41 Bayonet fuse holder.

80204-11_F42.EPS

Figure 42 Secondary side.

> **NOTE**
> Pad-mounted transformers located in areas where there is vehicle traffic may require protective posts (bollards) set in concrete.

Primary cables are fed from an aerial feeder (*Figure 45*), through conduit (*Figure 46*), to the transformer pad. The transformer is then lowered into place and secured to the pad (*Figure 47*).

80204-11_F43.EPS

Figure 43 Oil drain valve.

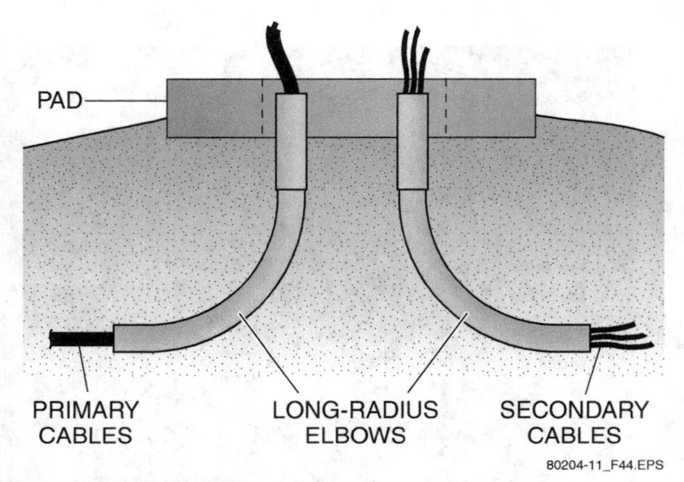

80204-11_F44.EPS

Figure 44 Mounting pad features.

80204-11_F45.EPS

Figure 45 Aerial feeder.

80204-11_F46.EPS

Figure 46 Primary cable installation.

80204-11_F47.EPS

Figure 47 Transformer installation.

6.0.0 Transformer Connections

In order for any URD system to function properly, the correct cables and terminations must be used to connect the various types of equipment. Often, the sizes and types of conductors used will be predetermined during system design. In other cases, such as when cable is being repaired or replaced, line workers might select the cable based on what was in use before. In any case, it is important to be familiar with the types of primary and secondary cable used by your company and know how to select the proper cable for a specific application.

The methods used to terminate URD cable can depend on the equipment being connected and company guidelines. Some companies might prefer a particular termination method or use a particular type of connector or tool for a specific application. Some general rules typically apply to some of the more common terminations.

6.1.0 Primary Connections

The primary connections used at the pad-mounted transformers in a URD system are designed to handle the incoming, or primary, voltage of the system. The connections can differ, depending on the type of transformer, the voltage level, and other factors. These connections can also depend on whether the transformer is a dead-front transformer or a live-front transformer.

> **NOTE**
>
> Ninety-degree elbows are used to connect primary cables in pad-mounted transformers. Here are reasons why they are used.
> - Cables come out of and into the ground vertically. Elbows enable cables to connect to horizontal bushings in a tight and confined space.
> - The pulling eye on the elbow enables a worker to pull the elbow off the bushing horizontally and safely with a hot stick.

6.1.1 Dead-Front Transformer

Figure 48 shows the connections in a typical dead-front three-phase URD transformer. Here, as in most cases, the energized parts are insulated with molded rubber. The terminations for the two primary cables are made with load break elbows (*Figure 49*). Load break elbows are designed to be operated under load. Some load break elbows are made to fit directly into the bushing well of a transformer. Other load break elbows require an insert for the bushing well. Either way, line workers can make or break a primary connection by removing the load break elbow from the bushing using a hot stick.

80204-11_F48.EPS

Figure 48 Primary connections in a dead-front URD transformer.

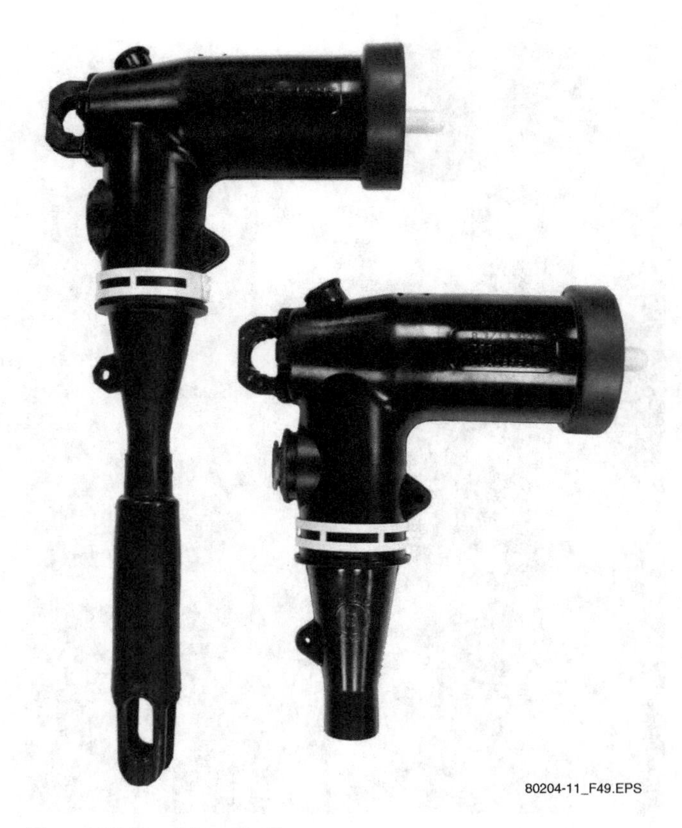

80204-11_F49.EPS

Figure 49 Load break elbows.

When load break elbows are added to the primary cable ends, line workers must select the proper elbow terminals or kits based on the type of primary cable being used in the system. Once the connections are made, they need to be sealed using self-fusing tape or whatever appropriate sealing material is supplied in the kit.

6.1.2 Live-Front Transformer

A live-front URD transformer (*Figure 50*) typically has porcelain bushings with some type of exposed terminal. In most cases, the terminals are either eyebolt or spade-type terminals. The primary cable terminations for live-front transformers are often compression lugs with holes for bolting the cable to the terminal spade, similar to the way secondary connections are made.

The size and type of primary cable being used again dictates the type of termination that will work. The terminations are typically color coded and/or stamped with the cable size to help line workers select the proper connector. Once the

80204-11_F50.EPS

Figure 50 Primary terminals in a live-front URD transformer.

primary cable is stripped, the conductor should be prepared. If the conductor is aluminum, it should be wire brushed and coated with antioxidant. Then the terminal lug can be crimped onto the conductor. The termination can then be completed by following instructions in the termination kit or company guidelines.

The aluminum lugs on the cable and the transformer should be wire brushed and coated with antioxidant before the connection is made. Then, the primary terminal lug can be bolted to the transformer terminal spade to complete the connection.

6.2.0 Secondary Connections

The secondary cables and terminations used in a typical residential URD system must be able to provide single-phase, three-wire service to customers. In this type of system, two of the three secondary cables carry 120 volts each, and the third cable is the neutral. The cable that is used for a system like this is typically rated for 600 volts. It can be separate pieces of single-conductor cable or one piece of two-conductor or three-conductor cable. Either way, each cable section typically consists of a stranded aluminum or copper conductor that is encased in an XLPE insulation sheath. A common size for secondary URD cable is 1/0.

Secondary cables are used to connect pad-mounted transformers to the line side of electric meters on customer homes. They are also commonly used to provide power for street lighting in many residential neighborhoods. So when a URD system is installed, line workers have to be able to select and install terminations at the transformer, the meter, and any street lighting that might be present.

Pad-mounted transformers usually come with some type of secondary terminal studs or spades already in place. It is the line worker's job to follow company guidelines or determine what type of termination to use to connect the secondary cables to those secondary terminals.

Figure 51 shows two types of secondary terminals on single-phase pad-mounted transformers. In one case, the secondary terminals each have a four-hole spade onto which the secondary cables must be connected using nuts and bolts. On the other transformer, the secondary cable terminal lug will be bolted onto the short bus bar attached to the X1 terminal. The type of terminal is se-

80204-11_F51.EPS

Figure 51 Secondary terminal connections in a pad-mounted transformer.

lected based on the size of the secondary cable, the type of conductor material in the cable (aluminum or copper), and the size of the holes in the lug.

If the secondary cable is aluminum, it should be prepared before the termination is made. The conductor should be wire brushed to clean off any oxidation. Then, if the connector being used is not already prefilled with antioxidant, an antioxidant should be applied to the cable end.

Once the cable has been prepared and the appropriate terminal lug has been selected, the proper crimping tool must be chosen. These tools vary with the manufacturer, but they all perform the same basic function. Most crimping tool dies and terminals are color coded or stamped for easy matching. When the crimping tool die is placed into the jaws of the tool, the tool is operated to compress the terminal onto the cable. Terminal lugs are often marked with specific points where they should be crimped. The termination can then be sealed with tape, heat-shrink material, or some similar method.

Before bolting the secondary cable to the transformer terminals, the aluminum lugs on the cable and the transformer should be wire brushed and coated with antioxidant to prepare them. Then, the secondary terminal lug can be bolted to the transformer terminal spade to complete the connection. *Figure 52* shows an example of a bolted connection on a secondary terminal lug of a three-phase transformer.

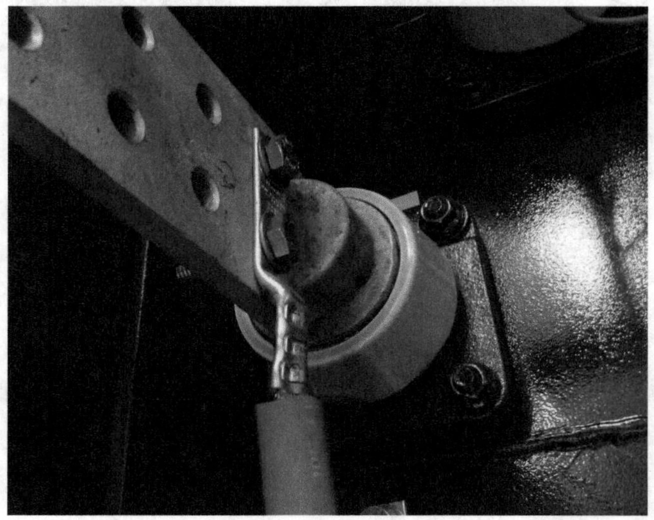

80204-11_F52.EPS

Figure 52 Example of a connection bolted to a terminal lug.

SUMMARY

The number of URD systems being used to supply power to residential customers continues to grow. While single-phase residential customers are the largest segment of URD customers, commercial and industrial customers are also supplied single- and three-phase voltage from URD systems. Radial-feed, loop-feed, and double-feed systems are the main types of URD systems. The loop-feed system is the most common. The double-feed system is used for critical customers such as hospitals. As a power line worker, it is important for you to be familiar with the layout and operation of URD systems and the components used in those systems. This includes being familiar with the different types of URD systems, as well as the cables, terminations, and other equipment used in those systems. It also involves knowing how to select and install the proper components. By following safety rules and applying your knowledge of URD systems, you can help ensure that URD customers have a reliable supply of power.

Review Questions

1. A typical URD system provides _____.
   a. three-phase, three-wire service to commercial customers
   b. two-phase service to industrial and residential businesses
   c. single-phase, three-wire secondary service to residential customers
   d. primary distribution service to rural substations

2. The use of URD systems began when residential populations increased dramatically during the _____.
   a. early 1900s
   b. 1920s and 1930s
   c. 1940s and 1950s
   d. 1960s and 1970s

3. The URD system represented in *Figure 1* can be best described as a _____.
   a. radial-feed system
   b. loop-feed system
   c. double feed system
   d. lateral helix system

4. A type of URD system in which both ends of the system are connected to the same phase of an overhead primary is called a _____.
   a. radial-feed system
   b. loop-feed system
   c. double feed system
   d. cross feed system

5. A URD installation in which multiple utilities such as electric, gas, cable TV, and telephone lines are buried in the same trench is known as a _____.
   a. shallow, or bored, trench
   b. common, or joint, trench
   c. dependent, or narrow, trench
   d. radial, or plowed, trench

6. To help protect URD cables during backfilling, some utilities fill the areas under and above the cables with _____.
   a. coarse gravel
   b. concrete
   c. insulating foam
   d. sand

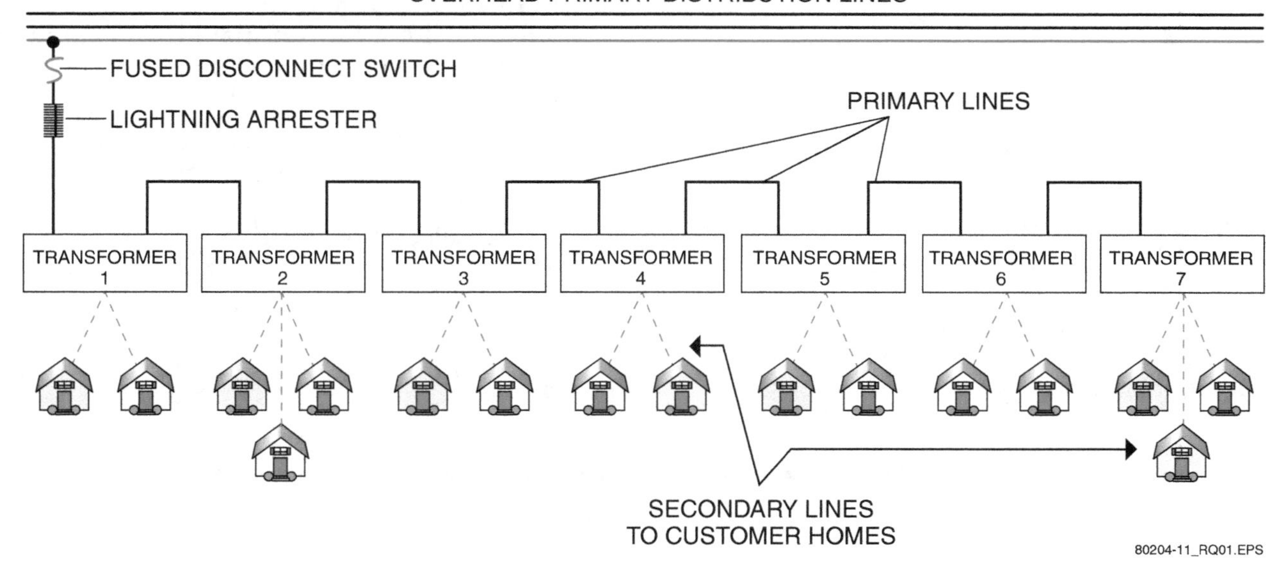

Figure 1

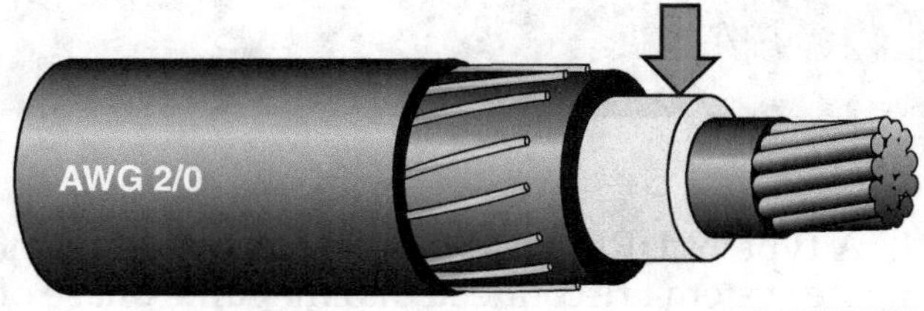

Figure 2

7. In the URD cable cutaway shown in *Figure 2*, the arrow is pointing to a layer of material called the _____.
   a. LLDPE sheath
   b. concentric metallic shield
   c. EPR or XLPE insulation
   d. semiconductor screen

Figure 3

8. The cable that is shown in *Figure 3* can be best described as _____.
   a. concentric metallic primary cable
   b. triplex secondary cable
   c. non-stranded primary cable
   d. duplex secondary cable

9. A dual-rated cable connector is most likely to be stamped or marked with _____.
   a. ALCU
   b. TWO2
   c. ACLU
   d. ALUM

10. Lightning protection for a URD system is most likely to come in the form of _____.
    a. lightning rods
    b. capacitance discharge units
    c. bayonet fuses
    d. surge arresters

11. The basic purpose of a fault indicator in a URD system is to sense fault current and display some type of indication that _____.
    a. de-energizes the entire loop
    b. pulses the system with high voltage
    c. reduces troubleshooting time and effort
    d. re-energizes the system without the need for repairs

12. A test point-mounted fault indicator is designed to mount directly onto an elbow or cable component that has a _____.
    a. high-potential testing outlet
    b. bleeder resistor
    c. surge arrester driver
    d. capacitive test point

13. If an elbow surge arrester is being added to a two-bushing transformer located at the end of a radial-feed URD system, the best place to install the surge arrester is in the _____.
    a. elbow connection at either secondary
    b. unoccupied bushing of the transformer
    c. bayonet fuse holder
    d. tap selector opening

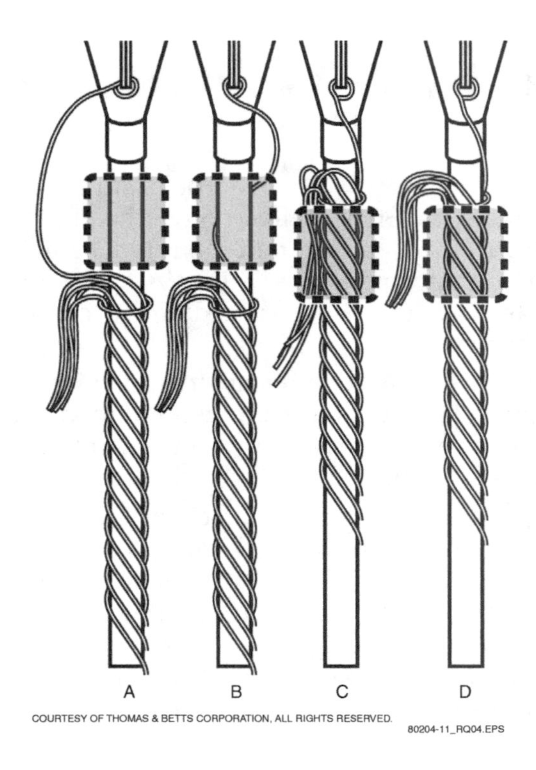

80204-11_RQ04.EPS

Figure 4

14. Which of those methods shown in *Figure 4* is the *incorrect* way of attaching a fault indicator to a cable with an exposed concentric neutral?
    a. A
    b. B
    c. C
    d. D

15. In a double-feed URD system, one power supply circuit can be isolated so that a second power supply circuit can feed a critical customer by using a _____.
    a. switchgear
    b. loop-feed connector
    c. jumper wire
    d. radial breaker

16. The equipment shown in *Figure 5* can be best described as a _____.
    a. live-front switchgear
    b. dead-front transformer
    c. live-front transformer
    d. dead-front switchgear

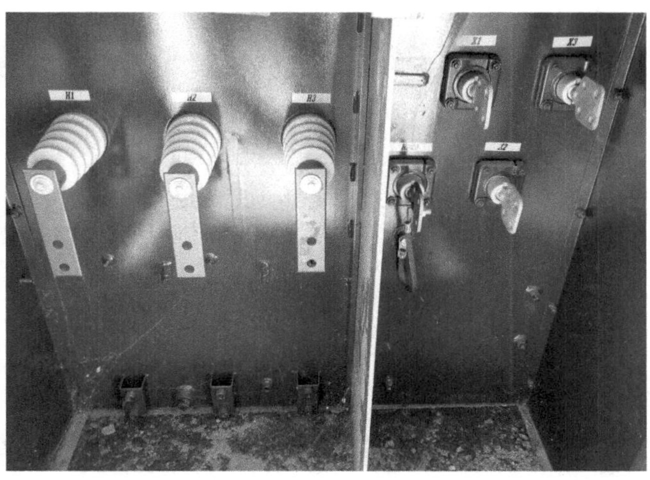

80204-11_RQ05.EPS

Figure 5

17. Before a termination is applied to the end of a secondary URD cable that has an aluminum conductor, the conductor should be wire brushed and coated with _____.
    a. rubberized tape
    b. an insulating gel
    c. an antioxidant
    d. heat shrink

18. Porcelain bushings with some type of exposed eyebolt or spade terminal are typically found on a _____.
    a. load-side electric meter
    b. live-front URD transformer
    c. capacitor discharge unit
    d. dead-front URD transformer

19. In a typical electric meter, two secondary cables from a pad-mounted transformer enter the meter and are terminated on the _____.
    a. live front
    b. load side
    c. dead front
    d. line side

20. Secondary cables are usually attached to spade lugs on a pad-mounted transformer by _____.
    a. welding them
    b. soldering them
    c. using nuts and bolts
    d. crimping them

Trade Terms Introduced in This Module

Ampacity: The maximum amount of current that a cable can carry before it begins to deteriorate. Sometimes called current rating or current-carrying capacity.

Common trench (joint trench): A trench in which multiple utilities, such as electrical cable, gas, communications, and cable TV, lines are installed.

Dead front: A switchgear or transformer configuration in which energized components are shielded in some manner so that they are not exposed and accessible when users open access doors or panels.

Double feed: A type of URD system design in which two separate URD systems are connected via switchgear to a single customer who cannot tolerate being without electric power. If a problem occurs in one URD system, the switchgear automatically transfers the source of power from the failed system to the other system.

Dual-rated: A type of electrical terminal or connector that is designed to be used on both aluminum and copper conductors.

Lightning arrester: A type of surge arrester that is typically installed on overhead primary feeder lines and at pad-mounted switchgear and transformers in a URD system to protect system components from damage caused by lightning strikes.

Live front: A switchgear or transformer configuration in which energized components are exposed and accessible when users open access doors or panels.

Loop feed: A type of URD system design in which both ends of the primary feeder line are connected to the same phase of the overhead primary, thereby forming a loop.

Metal oxide varistor (MOV): An electronic component in which the resistance to current flow drops as the voltage applied to it increases.

Radial feed: A type of URD system design in which the primary feeder line is connected to an overhead primary distribution line at only one end. The other end of the system is terminated.

Additional Resources

This module presents thorough resources for task training. The following resource material is suggested for further study.

Electric Power Distribution Handbook. T. A. Short. Boca Raton, FL: CRC Press LLC.

Figure Credits

Topaz Publications, Inc., Module opener, Figures 16–18, 27, 38–43, 48, and 50–52

www.sandman.com, Figure 1

C.H. Nickerson & Company, Inc., Figure 2

The Charles Machine Works, Inc., Manufacturer of Ditch Witch products, Figures 9 and 11

Caterpillar, Inc., Figure 10

Devon Distributing Corporation, SA01

Greenlee Textron, Inc., A subsidiary of Textron Inc., Figures 14, SA02, and SA04–SA06

Jim Mitchem, Figure 15

Tyco Electronics Energy Division, SA03

Photo courtesy of Thomas & Betts Corporation, all rights reserved., Figures 19, 25, and 28–36

Harry Wood, Figures 22–24

Courtesy of Cooper Power Systems, Figures 26 and 49

Tony Vazquez, Figures 21 and 45–47

NCCER CURRICULA — USER UPDATE

NCCER makes every effort to keep its textbooks up-to-date and free of technical errors. We appreciate your help in this process. If you find an error, a typographical mistake, or an inaccuracy in NCCER's curricula, please fill out this form (or a photocopy), or complete the online form at **www.nccer.org/olf**. Be sure to include the exact module ID number, page number, a detailed description, and your recommended correction. Your input will be brought to the attention of the Authoring Team. Thank you for your assistance.

Instructors – If you have an idea for improving this textbook, or have found that additional materials were necessary to teach this module effectively, please let us know so that we may present your suggestions to the Authoring Team.

NCCER Product Development and Revision
13614 Progress Blvd., Alachua, FL 32615

Email: curriculum@nccer.org
Online: www.nccer.org/olf

❑ Trainee Guide ❑ AIG ❑ Exam ❑ PowerPoints Other ______________________

Craft / Level: Copyright Date:

Module ID Number / Title:

Section Number(s):

Description:

____

____

____

____

Recommended Correction:

____

____

____

____

Your Name:

Address:

____

Email: Phone:

80205-11

Overhead and URD Service Installations

Module Five

Trainees with successful module completions may be eligible for credentialing through NCCER's National Registry. To learn more, go to **www.nccer.org** or contact us at **1.888.622.3720**. Our website has information on the latest product releases and training, as well as online versions of our *Cornerstone* newsletter and Pearson's product catalog.

Your feedback is welcome. You may email your comments to **curriculum@nccer.org**, send general comments and inquiries to **info@nccer.org**, or fill in the User Update form at the back of this module.

V.1 12/11

80205-11
Overhead and URD Service Installations

Objectives

When you have completed this module, you will be able to do the following:

1. Describe the methods and equipment used in the installation of residential and commercial electrical services.
2. Install single- and three-phase overhead loops.
3. Install risers on poles (conduit or U-guard).
4. Terminate single- and three-phase underground secondary services.

Performance Tasks

Under the supervision of your instructor, you should be able to do the following:

1. Install single- and three-phase overhead loops.
2. Install risers on poles (conduit or U-guard).
3. Terminate a single- and/or three-phase underground secondary service.

Trade Terms

Drip loop
Loop-fed distribution system
Meter loop
Quadruplex
Radial distribution system
Rigid metal conduit (RMC)
Riser
Secondary pedestal
U-guard

Industry Recognized Credentials

If you're training through an NCCER-accredited sponsor you may be eligible for credentials from NCCER's Registry. The ID number for this module is 80205-11. Note that this module may have been used in other NCCER curricula and may apply to other level completions. Contact NCCER's Registry at 888.622.3720 or go to nccer.org for more information.

Contents

Topics to be presented in this module include:

1.0.0 Introduction ... 1
2.0.0 Safety ... 1
3.0.0 Types of Distribution Systems ... 2
3.1.0 Overhead Distribution Systems ... 2
3.1.1 Three-Phase Distribution ... 3
3.2.0 Underground Distribution Systems ... 5
4.0.0 Installing Overhead Services ... 7
4.1.0 Types of Overhead Meter Loops ... 7
4.1.1 Meter Loop on Building ... 7
4.1.2 Meter Loop with Mast ... 10
4.1.3 Overhead Meter Pole ... 11
4.2.0 Installing an Overhead Service Drop ... 12
4.3.0 Commercial Overhead Service Installation ... 13
4.4.0 Watt-Hour Meters ... 14
4.4.1 Meter Installation ... 14
5.0.0 Installing and Terminating Underground Secondary Services ... 15
5.1.0 Underground Conductors ... 15
5.2.0 Pad-Mounted Transformers ... 16
5.3.0 Underground Secondary Distribution Systems ... 17
5.4.0 Installing Underground Meter Loops ... 18
5.4.1 Underground Meter Loop On Building ... 18
5.4.2 Underground Meter Pole ... 19
5.5.0 Street Lighting ... 19
5.6.0 Commercial Underground Service ... 20

Figures and Tables

Figure 1 Loop-fed distribution system ... 2
Figure 2 Radial distribution system ... 3
Figure 3 Rural radial distribution system ... 3
Figure 4 Transformer connected to aerial cable ... 4
Figure 5 Triplex cable ... 4
Figure 6 Service drop tapped into overhead cable ... 4
Figure 7 Three-phase service drop attached to a pole ... 5
Figure 8 Example of a three-phase transformer wiring configuration ... 5
Figure 9 Underground distribution system in a new housing tract ... 6
Figure 10 Riser with U-guard ... 6
Figure 11 Riser primary conductors with fused cutouts ... 6
Figure 12 Underground service drop from an aerial transformer ... 7
Figure 13 Meter loop ... 7
Figure 14 Three-phase overhead meter loop on a building ... 8
Figure 15 Single-phase meter base connections ... 9
Figure 16 *NEC®* requirements for meter loop on building ... 10
Figure 17 Ground clearance dimensions ... 11
Figure 18 Meter loop with mast ... 11
Figure 19 Drip loop ... 11
Figure 20 Overhead meter pole ... 13
Figure 21 Service drop connections ... 13
Figure 22 Three-phase meter loop with mast ... 14
Figure 23 Single-phase watt-hour meters ... 14
Figure 24 Reading an analog meter ... 15
Figure 25 Three-phase watt-hour meter ... 16
Figure 26 Meter base ... 16
Figure 27 Meter installation ... 17
Figure 28 Underground triplex cable ... 17
Figure 29 Pad-mounted transformer ... 17
Figure 30 Residential underground distribution system ... 18
Figure 31 Secondary pedestal ... 18
Figure 32 Underground meter loop on a building ... 19
Figure 33 Underground meter pole ... 19
Figure 34 Junction box for a street lighting circuit ... 19
Figure 35 Examples of commercial underground services ... 20
Figure 36 Typical CT cabinet arrangement ... 21

Table 1 Underground Cable Burial Depths ... 18

1.0.0 INTRODUCTION

Systems that deliver power are designed so that the fewest number of customers are affected in the event of a power outage. A loop-fed distribution system or a radial distribution system is used. Power supplied to customers is delivered by way of overhead or underground cables to the meter loop that is part of the customer's service entrance. The meter loop is often installed by the customer and must meet the utility's requirements before power can be turned on. However, in some locations, the utility installs the meter loop. Power line workers are tasked to install and repair the cables and other hardware used to provide customer power. In this module, you will learn how to install and repair overhead and underground electrical services that supply residential and light commercial customers.

2.0.0 SAFETY

All power line workers are required to wear personal protective equipment (PPE) to prevent on-the-job injuries.

- *Clothing* – Clothing should be made of natural fibers, such as cotton or wool, and treated to be arc-resistant and fire-retardant. Avoid clothing made of certain synthetic fibers, such as rayon, nylon, and polyester, that could burn or melt and cling to skin. Pant legs should extend over the top of shoes or boots. Never wear shorts on the job. Shirts should be long-sleeved to protect the skin from the sun, to minimize injuries caused by wood splinters, and to provide protection from chemicals used to treat the wood against rot. On most job sites, workers are required to wear a bright-colored shirt or vest to increase their visibility.
- *Footwear* – Only safety-toe leather shoes or boots should be worn on the job. High-top boots have the added advantage of greater support for the ankles. Many companies require dielectrically rated boots or overshoes. Never wear sandals or canvas-type shoes on the job.
- *Eye protection* – Safety glasses with side shields in accordance with *ANSI Standard Z87.1* must be worn at all times while on the job. If a power tool such as a chain saw is used on the job, a full face-shield may be necessary in addition to the safety glasses.
- *Hearing protection* – If power tools or noisy machinery, such as a chain saw, trencher, or air compressor, are used on the job, approved earmuffs or earplugs are generally required.
- *Dust mask* – If a wood pole is cut or drilled, a dust mask must be worn. The preservatives in the resulting sawdust may be harmful if inhaled.
- *Gloves* – Working on wood poles used in aerial distribution systems can be rough on hands. Leather work gloves provide protection from rope burns, wood splinters, and other hazards.
- *Rubber gloves and sleeves* – Power line workers often work near energized lines or equipment. Rubber gloves and sleeves protect against electrical shock. Use an air and water test on the gloves and roll the sleeves in each direction to look for damage. Inspect all other PPE, and tools according to the manufacturer's recommendations to ensure they are safe to use.
- *Safety helmet/hard hat/hard cap* – An approved *ANSI Standard Z89.1 Class E* safety helmet made of a non-conducting plastic material, such as polypropylene or polycarbonate, must be worn at all times while on the job site. The helmet protects the head from injuries caused by bumps and falling objects, and it provides shade from the sun.

In addition to wearing the proper PPE, follow these safety precautions when connecting an electrical service:

- Avoid installing or repairing an electrical service in wet or bad weather except in emergencies.
- Install protective covers around conductors and other components when working near energized lines.
- Before starting any job, inspect rubber gloves and sleeves, PPE, and all tools to ensure they are safe to use.
- Only use tools and equipment that are approved for power line use.
- Use fall protection equipment when working at heights above six feet.
- If an insulated lift bucket is used on the job, test the insulating arm for excessive leakage current each day before it is used.
- Never dig in an area until all underground utilities in that area have been located and marked.
- Have a qualified observer at all job sites to ensure that minimum approach distances to energized equipment are observed, and correct PPE and tools are being used.

3.0.0 Types of Distribution Systems

Loop-fed and radial distribution systems are used by utilities to supply power to customers. In an open loop-fed system, power can be delivered from two directions (*Figure 1*). If a fault occurs in the loop, automatic switches can redirect power from another source so that the size of the outage is kept small. Loop-feed systems are widely used in URD applications. A radial system (*Figure 2*) can be compared to the spokes on a wheel, with a substation at the hub that feeds the various loads on each spoke. If the central power source is lost, all customers fed by that substation will have an outage. Radial systems are widely used in overhead systems in both urban and rural settings. Overhead radial systems are the most common type found in rural settings. They may have widely spaced substations with radial spokes that may run for miles (*Figure 3*).

> **NOTE**
> Radial systems are found within loop-fed systems. In an underground distribution system, all the transformers are loop-fed. However, the homes supplied by each transformer are connected in a radial pattern.

Power is supplied to customers through overhead or underground cables. Both methods have advantages and disadvantages. Major advantages of overhead systems include the following:

- Overhead systems are easier and faster to install.
- Overhead systems are easier and less costly to repair and maintain.

Major disadvantages of overhead systems include the following:

- Overhead systems are prone to damage from weather and accidents.
- Overhead systems have poor aesthetics.

Major advantages of underground systems include the following:

- Underground systems are not prone to damage from weather.
- Underground systems have good aesthetics.

Major disadvantages of underground systems include the following:

- Underground systems are harder to install.
- Underground systems are prone to corrosion and insulation breakdown.
- It is harder to find and repair damage in underground systems.

3.1.0 Overhead Distribution Systems

Conductors are used to supply power to customers in neighborhoods with overhead distribution systems. In housing tracts, a centrally located single-phase aerial transformer is sized to handle the electrical load of several homes. The total load a transformer must carry (in kilowatts) determines its size. Secondary voltage is typically 120/240V.

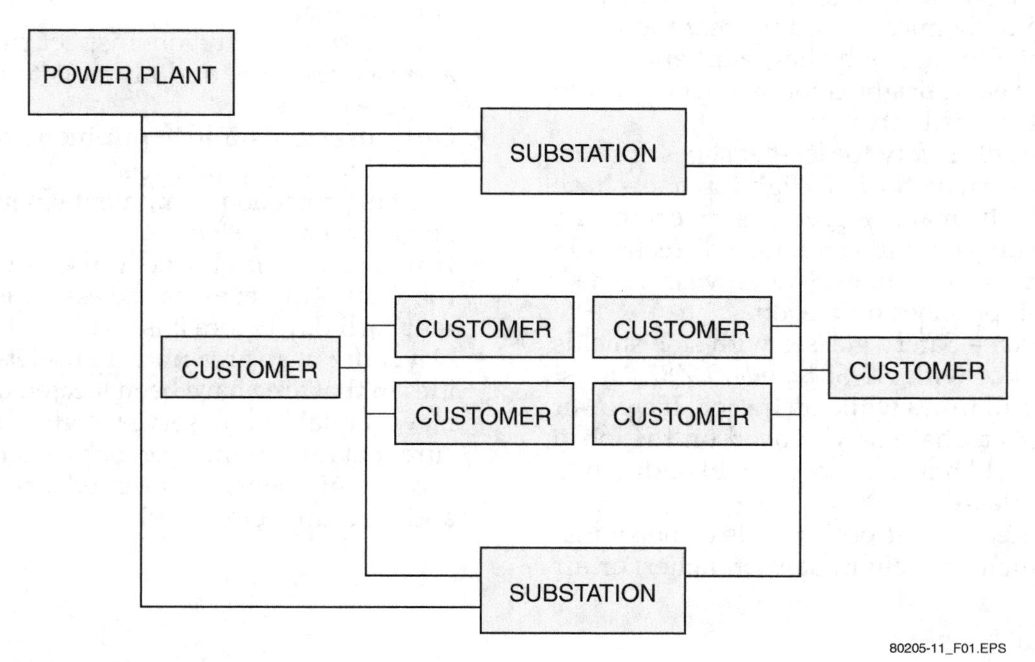

Figure 1 Loop-fed distribution system.

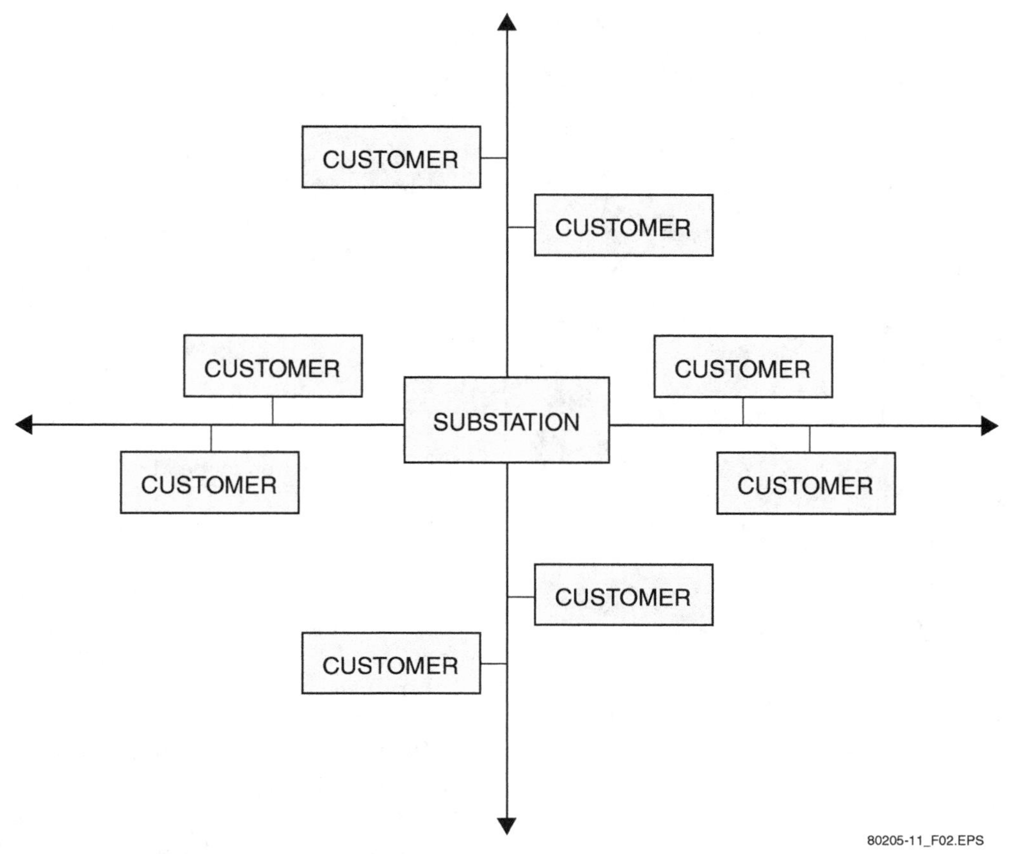

Figure 2 Radial distribution system.

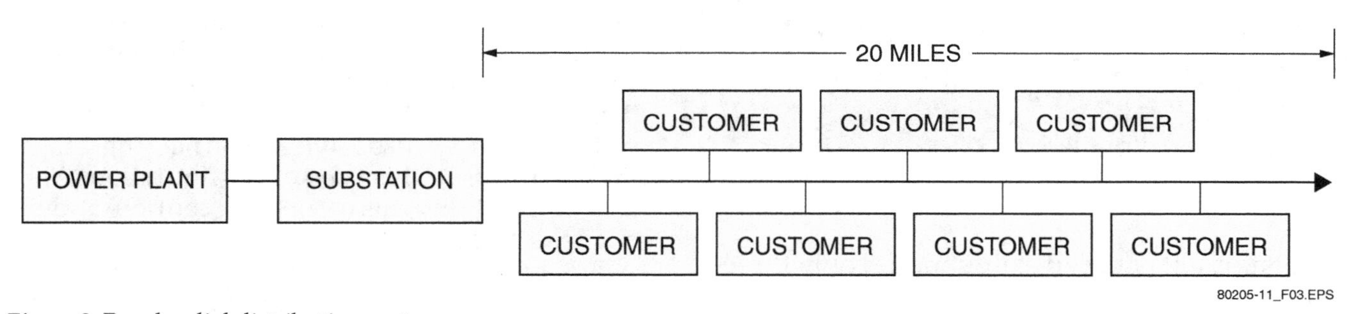

Figure 3 Rural radial distribution system.

> **NOTE**
> In rural areas with widely spaced homes and farms, it is common to have one transformer for each home or farm.

Insulated conductors connect the low-voltage terminals of the transformer to the conductors in the aerial cable used to supply the neighborhood (*Figure 4*). Triplex cable (*Figure 5*) is widely used. It consists of a bare neutral and two insulated conductors. The neutral acts as a messenger. The service drop to the customer is generally connected at a pole. In some special cases, it may be tapped directly into the cable between poles (*Figure 6*). An example would be providing power to a lighted billboard located between poles.

> **NOTE**
> Dedicated poles may carry secondary cable, or the poles that carry primary conductors can be used. If the cable shares a pole with primary conductors, the cable must be placed at least 3 feet below the lowest primary conductor.

3.1.1 Three-Phase Distribution

In three-phase applications, three-phase power is delivered to a business by way of aerial cables. Three single-phase transformers are typically mounted together on a single pole. They can be connected in a wye or delta configuration to supply three-phase power through four conductors. Three-phase power can only be delivered through three conductors that are 120 degrees out of phase

Figure 4 Transformer connected to aerial cable.

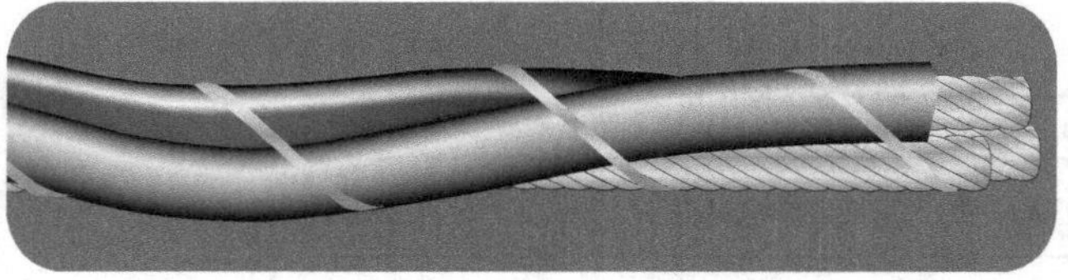

Figure 5 Triplex cable.

from one another. It typically supplies only three-phase devices, such as motors and industrial machinery. However, some electricians have been known to connect a single-phase 240V motor, such as an air compressor in an auto shop, to a 240V three-phase service. The motor works because the voltage is correct. However, the single-phase motor was actually designed to operate on a single system where the two 120V sources (120/240) fed from the single-phase load center are 180 degrees out of phase. Using only two of the three phases in a three-phase load center also unbalances the load on the three-phase transformers.

Unlike single-phase service, the service drop to the customer is almost always attached to the pole on which the transformers are mounted (*Figure 7*).

Wye/wye-connected and delta/wye-connected utility distribution transformer configurations typically supply 120/208V or 277/480V secondary voltages and typically use a four-wire bus or **quadruplex** service loop. Wye/delta-connected

Figure 6 Service drop tapped into overhead cable.

and delta/delta-connected utility transformers typically supply 240V or 480V secondary voltages and would typically use a three-wire bus or triplex service loop. Note that some utilities use what is called a high-leg delta system where one of the three-phase transformer secondary coils is center tapped to provide a 120/240V secondary configuration. This is not the same as a delta/wye four-wire secondary configuration. The secondary service then uses a quadruplex service loop, thus providing a 120/240V service voltage. However, as shown in *Figure 8*, if the phase A transformer is the transformer with the center-tapped secondary coil, then 120 volts only obtainable from phase A to neutral and phase B to neutral. Phase C to neutral produces a 208V potential, so caution must be used when connecting single phase services to this type of configuration. Also, keep in mind that this secondary configuration is still delta and not wye.

The number of customers a three-phase bank may supply depends on the application of the bank. For example, in a city setting, three-phase banks may be larger in size and supply a bus with many customers attached due to the close proximity of the buildings. In a rural area, one, two, or three customers per three-phase bank may be more typical.

GOING GREEN

Underground vs. Overhead

If you compare underground distribution systems to overhead systems, you will see that they are better for the environment. Here is why. Trees do not have to be cut down for poles, and those poles do not have to be treated with toxic chemicals to prevent rot. Underground systems are also safer for wildlife.

Figure 7 Three-phase service drop attached to a pole.

3.2.0 Underground Distribution Systems

Underground distribution systems are widely used in cities and newer residential subdivisions. Underground systems make sense in a crowded urban setting. The conductors are out of sight where they are less prone to damage. Newer housing tracts demand underground systems for both practical and aesthetic reasons. *Figure 9* shows a new housing tract with pad-mounted transformers used in an underground distribution system. Overhead primary power is carried on poles to the edge of the subdivision. From there it transitions from the pole to underground cables that feed pad-mounted transformers that serve several homes. The pole on which the transition takes place is called a **riser**. The cables on the riser are protected by conduit or a **U-guard**. Conduit (metal or PVC) is mounted on the pole with standoffs and extends below grade to protect the cables. U-guard (*Figure 10*) is shaped like the letter U. It is made of metal or PVC and is fastened

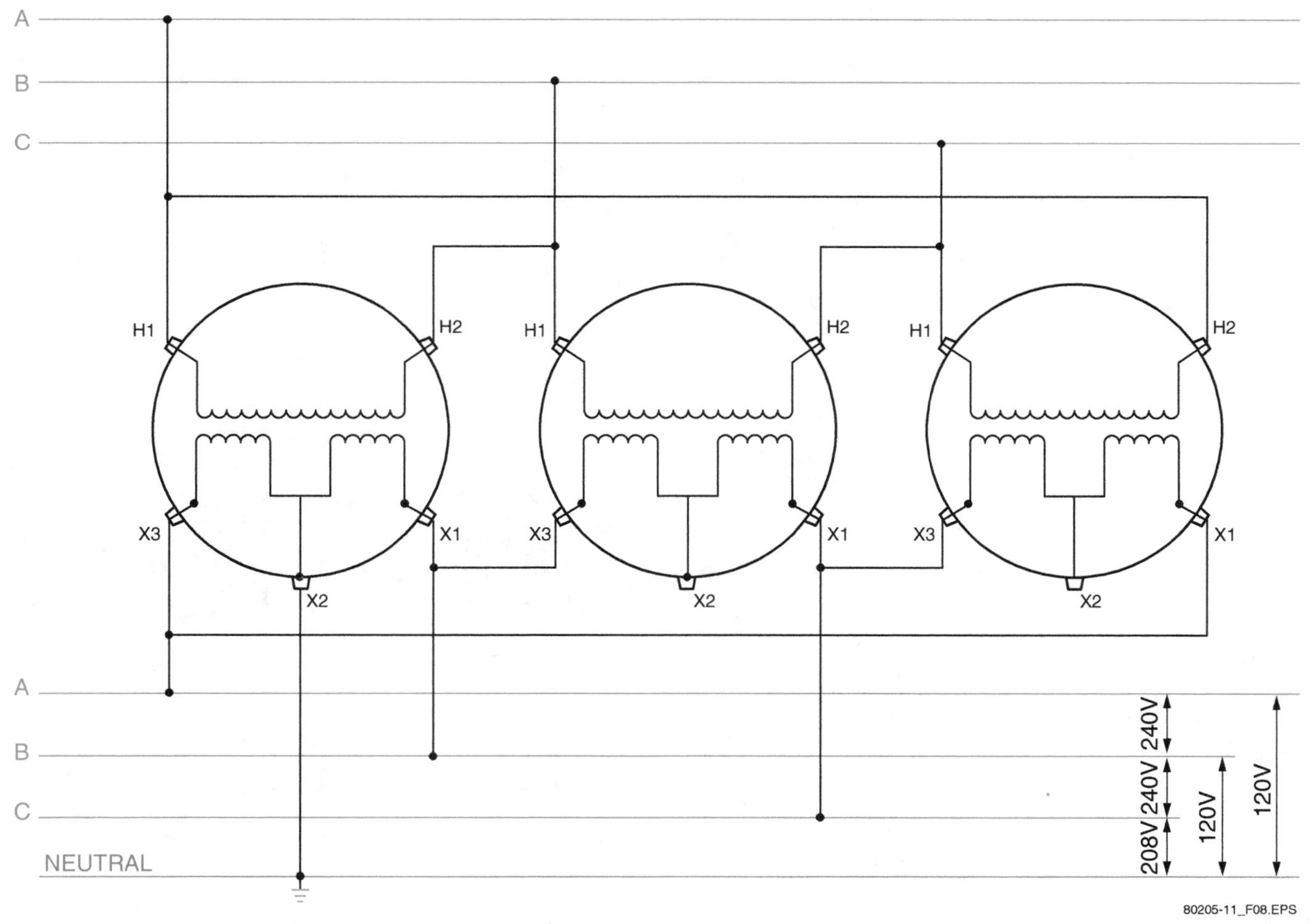

Figure 8 Example of a three-phase transformer wiring configuration.

directly to the pole. It is less costly to install than conduit. Primary conductors that transition from overhead to underground are equipped with fused cutouts (*Figure 11*). Clip-on fault indicators are often applied to the overhead primary conductors near the riser and before they transition to underground. The same type of fault indicator may be installed on a pad-mounted transformer. Fault indicators can be mounted flush with the door of the transformer or inside the transformer cabinet.

A riser is also used when a customer does not want an overhead service drop. In that case, power from an aerial transformer is fed by underground cables to the customer's meter loop. A short length of conduit protects the cable as it enters the ground. The cable is then buried in the soil. Conduit again protects the cable where it leaves the ground and connects to the meter loop or service entrance (*Figure 12*).

80205-11_F09.EPS

Figure 9 Underground distribution system in a new housing tract.

80205-11_F10.EPS

Figure 10 Riser with U-guard.

80205-11_F11.EPS

Figure 11 Riser primary conductors with fused cutouts.

On Site

U-Guard Ventilation

U-guards can be fitted with an adapter that allows air to travel up the U-guard by way of the chimney effect. The air cools the cable and prevents it from having to be re-rated for a higher temperature.

> **NOTE**
> A total underground distribution system carries power from the substation to the end user completely underground. There are no overhead power lines. This type of system is not widely used.

4.0.0 Installing Overhead Services

The procedures for installing single- and three-phase overhead service drops are similar. For that reason, this module focuses on installing a single-phase overhead service on a home. Different utilities in the United States have different policies that cover service installations. Some utilities require that the customer install the meter loop (*Figure 13*). Others require that the utility install the meter loop. This module assumes the meter loop will be installed by utility workers.

4.1.0 Types of Overhead Meter Loops

There are three basic types of overhead meter loops used on homes. They include the following:

- Meter loop on building
- Meter loop with mast
- Overhead meter pole

4.1.1 Meter Loop on Building

A meter loop on a building (*Figure 14*) is attached to the side of a building and does not extend above the roof. Triplex or quadruplex is used for the service drop. The messenger (bare neutral) is attached to the building using a clevis-type wire holder that is bolted through the wall. The messenger supports the service drop. Conduit equipped with a service head protects the conductors and is attached to the meter base. Lugs for connecting the hot leads and neutral are in the meter base (*Figure 15*). The size of the conduit and the conductors used is based on the size (in amperes) of the home's electrical service. A 200-amp service is common in modern homes.

The utility specifies various dimensions that must be maintained when installing the meter loop, including the following:

- Height of the service head above ground level
- Height of the meter base above ground level
- Distance between conduit clamps
- Distance between the meter base and service disconnect
- Minimum distance between the meter loop and any door or window on the building

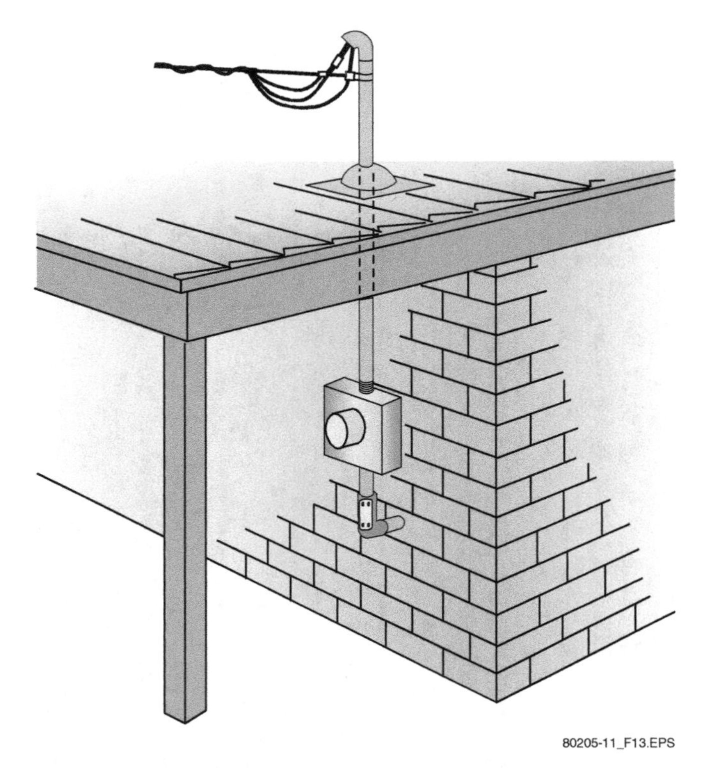

80205-11_F13.EPS

Figure 13 Meter loop.

80205-11_F12.EPS

Figure 12 Underground service drop from an aerial transformer.

> **On Site**
>
> **Prohibited Installations**
>
> Most utilities do not permit meter loops to be installed on mobile homes or portable buildings.

80205-11_F14.EPS

Figure 14 Three-phase overhead meter loop on a building.

Other types of utility installation specifications include the following:

- Materials allowed for conductors (copper or aluminum)
- Color of conductor insulation
- Materials allowed for conduit
- Type and length of ground rod and grounding conductor

All meter loop installations must conform to standards set forth in the *National Electrical Code*®(*NEC*®). *Figure 16* shows code requirements for a meter loop attached to the side of a building. The building's outside wall must be high enough to accommodate this type of installation. A high wall permits the service head to be placed high enough to allow the service drop conductors to maintain minimum ground clearances. Typical minimum ground clearances include the following:

- Service head to ground – 12 feet
- Service drop conductors to walkway – 12 feet
- High point below service drop to service drop – 12 feet
- Street or driveway to service drop – 16 feet

On Site

Early Distribution Systems

When American cities were first electrified, distribution was overhead. Photos taken in cities at that time show the mess that was created. Cities were forced to go underground to reduce that tangle of overhead wires.

80205-11_SA01.EPS

NOTE

Ground clearance dimensions may vary between utilities.

Figure 17 shows minimum clearances from the ground to the service drop conductors. The maximum span length of service drop cable from a pole to the building is 125 feet. A span longer than 125 feet requires the installation of another pole to support the cable. Ground clearance and span length dimensions of service drop cables are similar for all types of overhead meter loop installations.

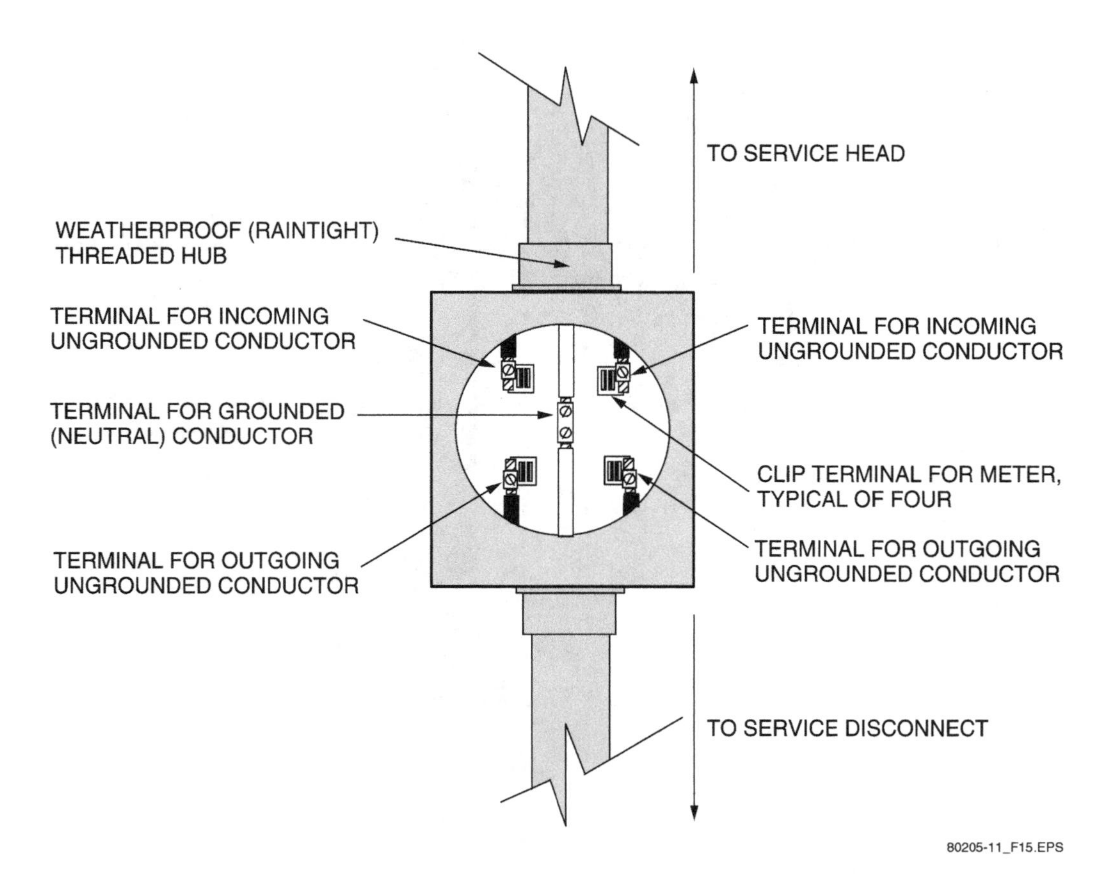

Figure 15 Single-phase meter base connections.

On Site

Smart Meters

There is a nationwide project to replace existing watt-hour meters with so-called smart meters like the one shown. These meters allow utilities to wirelessly collect information about power consumption by individual residences and businesses. Power companies are increasingly using time-of-day billing, in which the cost of power is the greatest during peak usage times. Given access to information about their own power consumption in relation to the time-of-day rates, consumers then have the opportunity to adjust their consumption to match low-rate times. On the utility side, smart meters eliminate the need for meter readers, provide instant reporting of power outages, and make it easier to balance power demand.

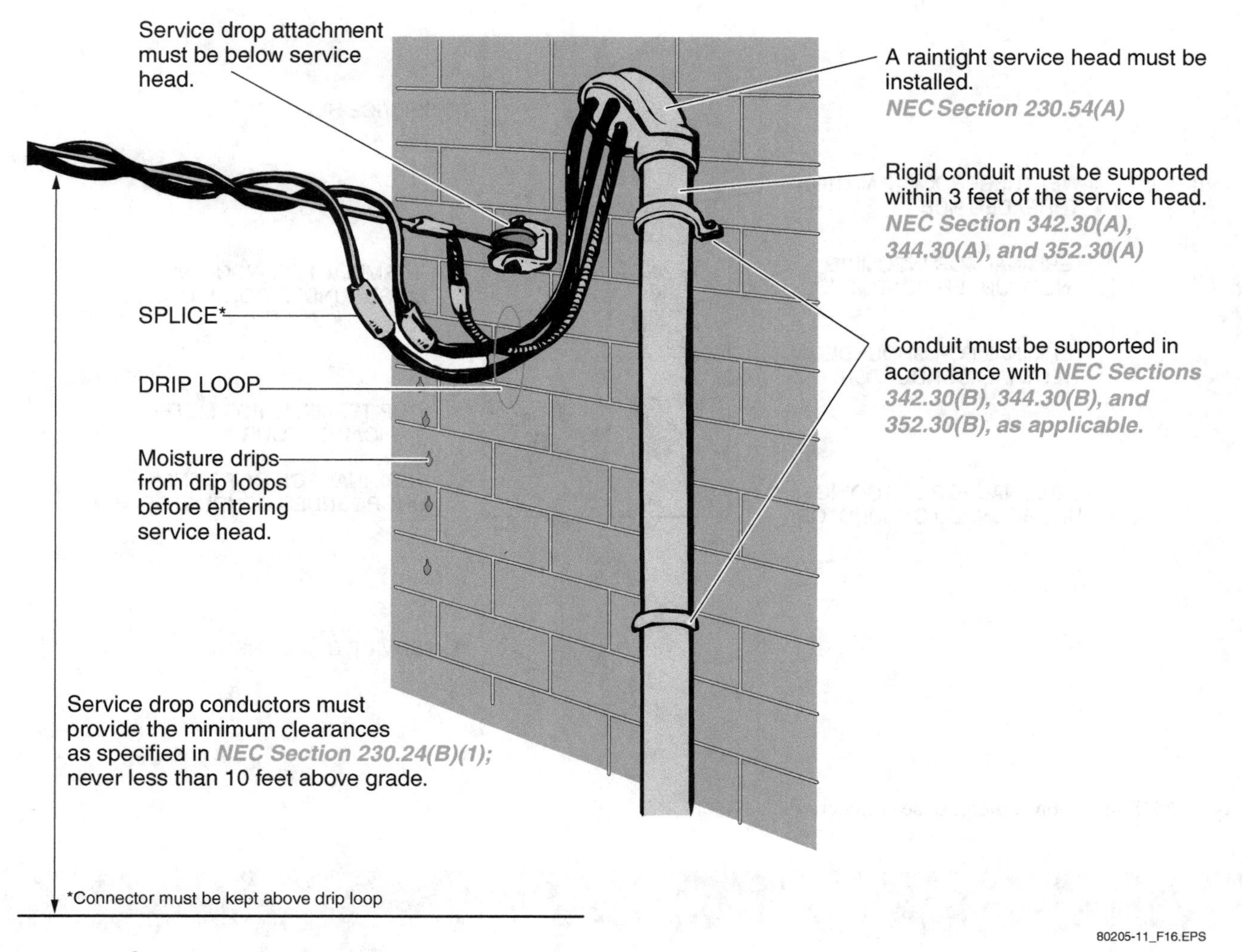

Figure 16 NEC® requirements for meter loop on building.

NOTE

NEC® requirements specify a minimum that must be met in order to satisfy the code. A utility or locality may have requirements that exceed *NEC®* requirements. Always comply with local codes.

WARNING!

It is a safety hazard and a code violation to install an overhead service drop over a swimming pool.

4.1.2 Meter Loop with Mast

A meter loop with mast (*Figure 18*) is a popular method for supplying power to a customer. Its major advantage is that it can be used on any building, regardless of its height. The mast can be made long enough to provide the required minimum ground clearances. The utility specifies a minimum mast height and a maximum height that can be used without a brace. The mast is **rigid metal conduit (RMC)** made of galvanized steel. The messenger (bare neutral) is attached to the mast using a clevis-type wire holder that is clamped around the mast. The messenger supports the triplex service drop.

A service head is placed on top of the conduit and the conduit is attached to the building with conduit straps. The conduit is attached to the meter base. Conductors within the conduit must protrude from the service head with enough length (18 to 24 inches) so that a **drip loop** (*Figure 19*) can be formed when the conductors are connected to the service drop cable. Drip loops must be formed at all service heads, regardless of the type of meter loop installation. Like all meter loop installations, the utility specifies various dimensions that must be maintained when installing the drip loop.

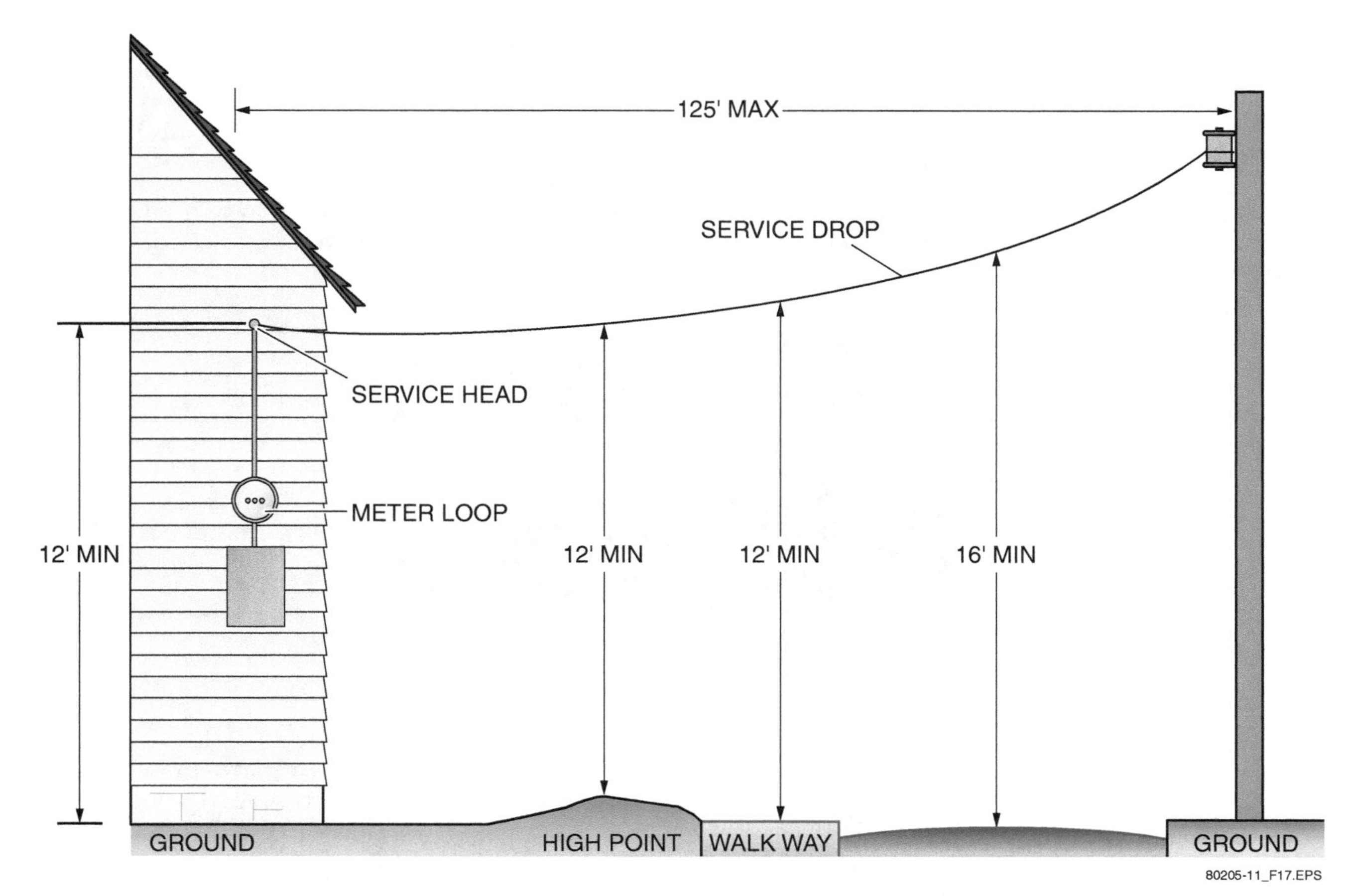

Figure 17 Ground clearance dimensions.

Figure 18 Meter loop with mast.

Figure 19 Drip loop.

4.1.3 Overhead Meter Pole

Overhead meter poles (*Figure 20*) are often used to supply power to mobile homes. Another use is on farms to supply power to remote irrigation pumps. Both of these installations are considered perma-

On Site

Service Mast Material

Most utilities do not allow electrical metallic tubing (EMT) or intermediate metal conduit (IMT) to be used for a service mast. Most utilities specify the stronger rigid metal conduit (RMC) made of galvanized steel. Of the three metal conduit types, RMC has a thicker wall, which gives it more strength.

RIGID METAL CONDUIT (RMC)

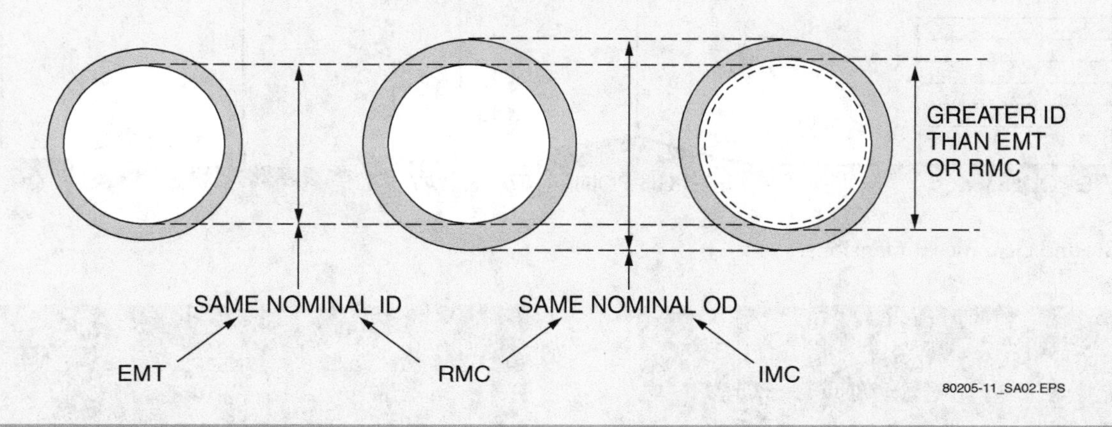

nent. A major difference between the other two types of overhead meter loop installations is the need for a pole on which to install the loop. The utility specifies the type of pole. The pole is made of treated wood, with a length that allows at least 15 to 20 feet of height after it is placed in the ground. That height provides enough ground clearance for the service drop so that it meets minimum distance requirements. Conduit with a service head is attached to the pole with conduit straps and connected to the meter base. The messenger (bare neutral) is attached to the pole using a clevis-type wire holder that is bolted through the pole. The messenger supports the triplex service drop.

NOTE

Temporary overhead meter poles are often installed at construction sites and for special or temporary events such as a lot that sells Christmas trees. Their installation requirements are different.

4.2.0 Installing an Overhead Service Drop

Before installing a service drop, determine the correct length of the cable. Lay the correct length of cable on the ground. One worker then dead-ends the cable at the building, service mast, or pole. Another worker raises the other end of the service drop up the utility pole and dead-ends the cable. This procedure is done by hand to prevent overstressing the connection at the building. However, a strap hoist (jack-strap) may be used if required by a long conductor run. Some utilities provide tension tables based on the length and weight of the cable and amount of sag in the service drop. The service drop must be tensioned per company policy. It is very important that connections to the building, mast, or pole be strong enough to support the service drop under normal and loaded conditions.

80205-11_F20.EPS

Figure 20 Overhead meter pole.

CAUTION

The clevis-type wire holder must be through-bolted to a structural member of the building. Do not use lag screws and do not attach the wire holder to the siding or sheathing. Bolting the wire holder to the building prevents the service drop and/or meter loop from being torn from the building under adverse conditions.

After the service drop is secured to the building, the conductors in the service head are connected to the service drop cable to form drip loops. Compression splices (*Figure 21*) are commonly used for this purpose. The hot conductor sleeves are insulated. The sleeve on the bare neutral does not require insulation. Compression sleeves are used to connect the service drop to the secondary mains at the pole.

WARNING!

Always check the polarity of the secondary mains before connecting the service drop cable. This prevents the accidental connection of the customer's ground to a hot conductor.

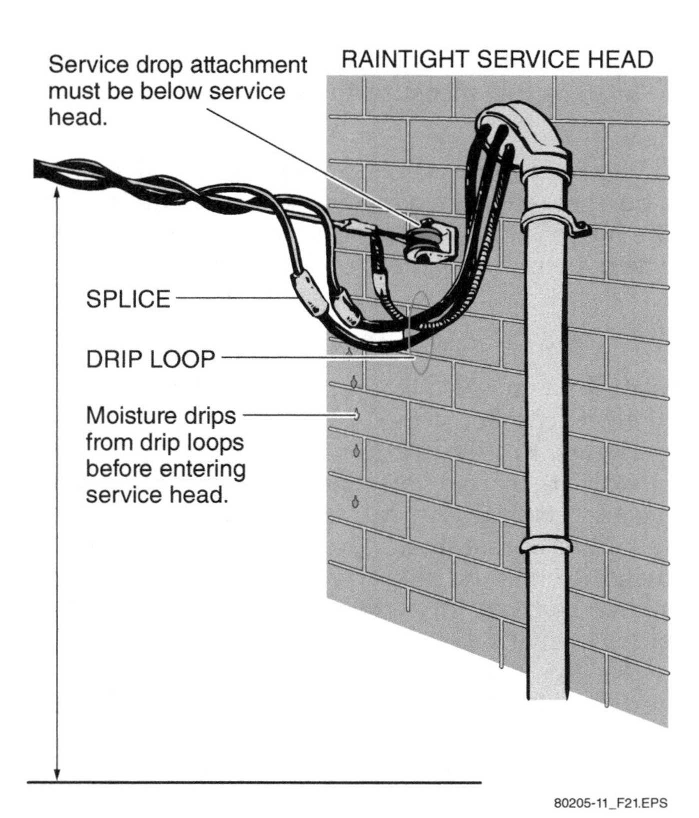

80205-11_F21.EPS

Figure 21 Service drop connections.

Before power can be turned on, the following must take place:

- The customer's service entrance and building wiring must be inspected and approved to ensure that it complies with all codes.
- The service drop may require inspection and approval by the utility.
- The meter must be installed in the meter base.

4.3.0 Commercial Overhead Service Installation

For commercial applications using single-phase power, installation is similar to residential installations. Commercial applications that require three-phase power are also similar but have differences including the following:

- The service drop conductors are larger to carry more current.
- Quadruplex (four-conductor) cable is used instead of triplex for the service drop.
- Delta-connected four-wire systems must have the high leg conductor permanently marked in orange.
- Conduit must be larger to handle four larger conductors.
- The service head may require greater clearance to the ground.

- The use of current transformers and/or a meter that can measure three-phase power is required.

Figure 22 shows three-phase power delivered to a business through a meter loop with a mast. Note the larger size of the conductors and the conduit. The mast is also braced for added support.

4.4.0 Watt-Hour Meters

Watt-hour meters are used to record the amount of electricity consumed by customers. The meter readings, in kilowatt-hours, are used to bill the customer for their electrical service. The meter base is generally installed by electricians when wiring the customer's premises. The watt-hour meter itself is the property of the utility and is installed by the utility when service is initiated. The older-style electro-mechanical dial meter shown in *Figure 23* has been around for many years and can still be found on many homes and businesses. Analog meters have largely been replaced by direct-readout digital meters such as the one shown. Analog meters are read from left to right as shown in *Figure 24*. The reading on the meter shown is 22,179 kilowatt-hours. These meters can automatically report power consumption to the utility, so it is no longer necessary to have meter readers go from building to building, or to have customers read their own meters. Special meters are used for three-phase service (*Figure 25*).

4.4.1 Meter Installation

Once an electrical service has been installed and inspected, the utility sends a technician to install the meter. *Figure 26* shows a typical electric meter base with the front cover removed, as well as a close-up of the meter base. There are basically two sides in the meter base: the line side and the load side. The line side is where the secondary cables enter the meter. The load side is where secondary power exits the meter and flows to the breaker panel. There are two hot secondary lines and one neutral line on both the line side and the load side of the meter.

The terminations that are used to connect the secondary cables on the line side and load side of a meter base can vary, depending on the meter. This meter uses mechanical connectors for all of the cable terminations. The insulating sheath on each secondary cable is removed far enough back to allow the bare conductor to be inserted into the mechanical connector. If the cable has an aluminum conductor, the conductor wiring should be wire brushed and an antioxidant should be applied. Then, the cable end should be inserted into the connector and the connector bolt should be tightened to secure the termination.

80205-11_F22.EPS

Figure 22 Three-phase meter loop with mast.

Regardless of the type of termination being used, it is important to select connectors that are appropriate for the conductor material, the conductor size, and the type of terminal in the meter.

> **WARNING!**
>
> When testing and installing meters, be sure to wear appropriate PPE, including eye and face protection, as well as gloves.

Before installing the meter, test the circuit to make sure it is safe. Use a voltmeter on the load side to verify that there is no backfeed from a generator or other source. Check line to line and line to neutral. Test the line side to make sure the meter is not installed into a fault. A fault current could injure the installer and seriously damage the meter.

It is common to install a meter into a live circuit. When installing the meter, always plug in the load side first, followed by the line side (*Figure 27*).

Figure 23 Single-phase watt-hour meters.

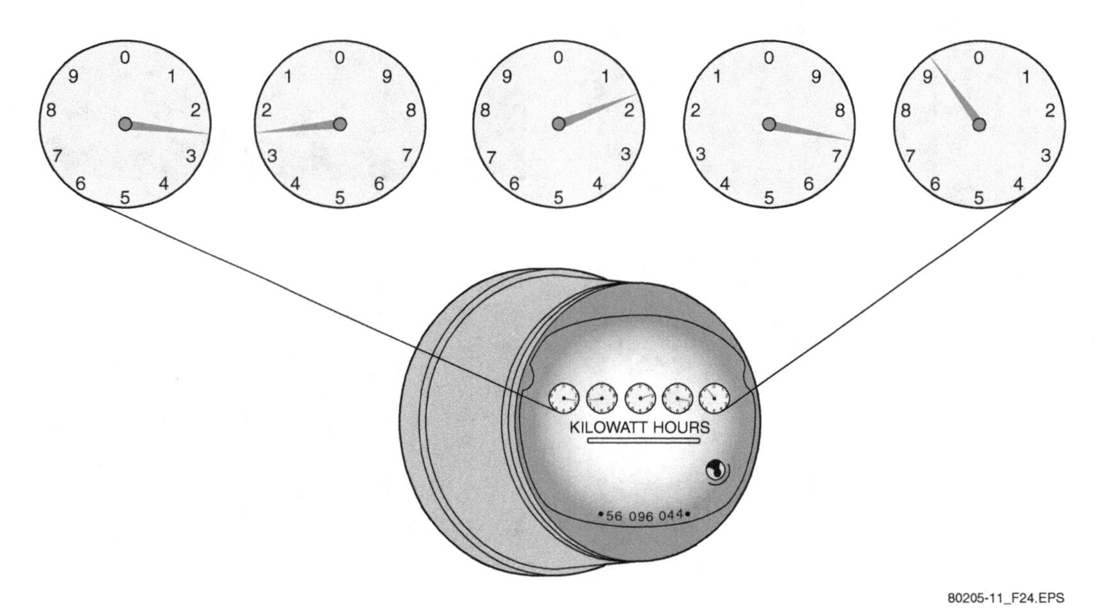

Figure 24 Reading an analog meter.

5.0.0 Installing and Terminating Underground Secondary Services

A typical underground distribution system consists of primary conductors that transition underground by way of a riser, pad-mounted transformers, and underground secondary conductors.

5.1.0 Underground Conductors

Triplex rated for direct burial is used as secondary conductors in residential systems (*Figure 28*). It differs from triplex used in overhead systems in two ways—the insulation is rated for direct burial, and the neutral conductor is insulated. A different insulation color or striping identifies the neutral conductor.

NOTE

Three-conductor underground cable is also made in a flat, parallel arrangement contained in a single jacket.

5.2.0 Pad-Mounted Transformers

Transformer types used in underground systems are pad-mounted, submersible, and direct-buried. All three types function like pole-mounted transformers. Pad-mounted transformers (*Figure 29*) are the most widely used in residential distribution systems. The internal wiring of a pad-mounted transformer is the same as that of an aerial transformer. Secondary terminals are identified as X1, X2, and X3. Spade terminals are commonly used on the secondary bushings. They allow more than one secondary cable to be connected to a single bushing. Ring terminals are crimped on the ends of cables so that they can be bolted to a spade terminal. The bayonet-style fuses that protect the transformer primary can be used to disconnect the secondary if required. They are removed using a hot stick after the oil pressure has been relieved.

80205-11_F25.EPS

Figure 25 Three-phase watt-hour meter.

NOTE

It is common to connect three single-phase aerial transformers to supply three-phase power. In an underground system, three-phase power is supplied to a customer by a single three-phase pad-mounted transformer.

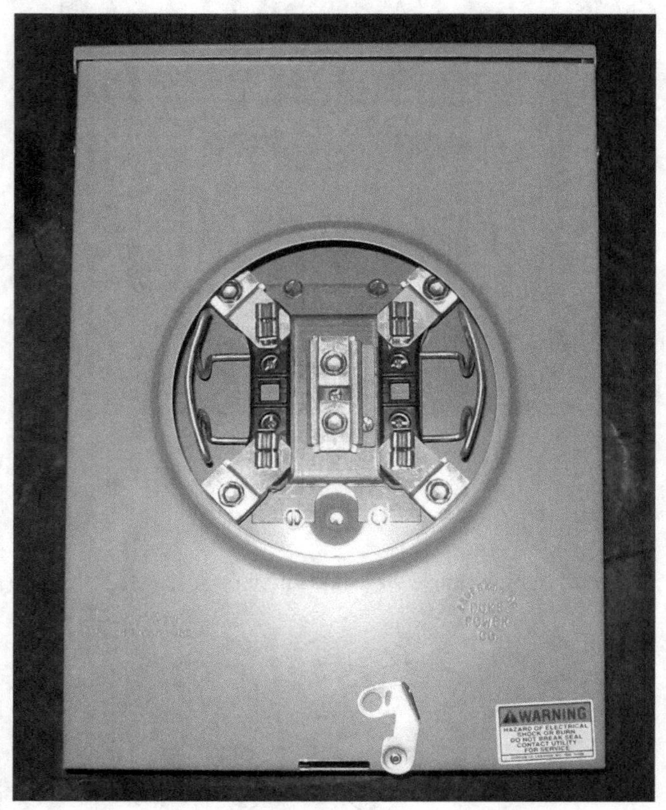

80205-11_F26A.EPS

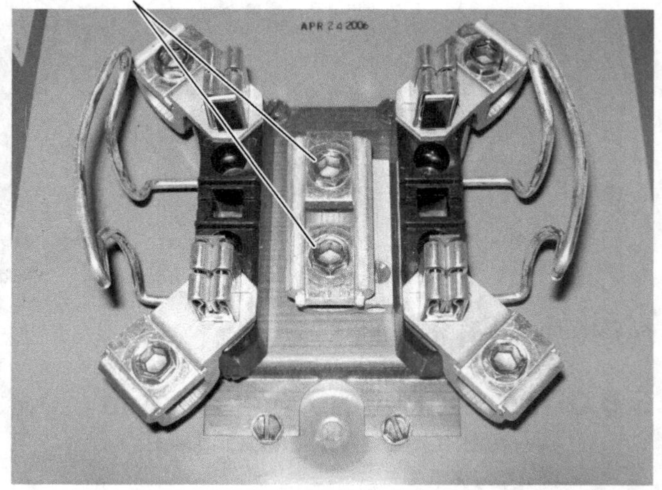

80205-11_F26B.EPS

Figure 26 Meter base.

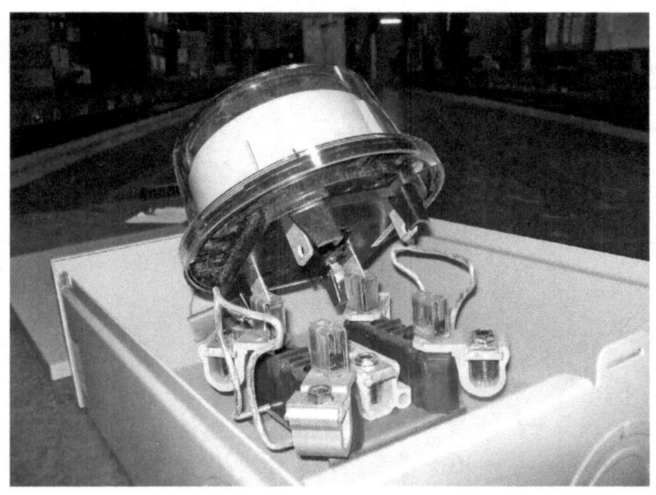

80205-11_F27.EPS

Figure 27 Meter installation.

> **WARNING!** For safety reasons, pad-mounted transformer cabinets are kept locked. An open cabinet creates the risk of exposure to dangerous potentials.

5.3.0 Underground Secondary Distribution Systems

In a housing tract, pad-mounted transformers are often placed along rear property lines (*Figure 30*). A single transformer typically serves four to six homes. A primary cable from a riser feeds all transformers in the tract. Secondary service can be delivered directly to homes from the transformer, or through a secondary pedestal (*Figure 31*). A secondary main from the transformer feeds each secondary pedestal. From there, a buried service drop delivers power to the home.

Cables must be buried deep enough to protect them from damage. *Table 1* lists burial depths based on the voltage the cable must carry. In areas where the ground freezes, cables must be buried below the frost line. The bottom of the trench in which the cable is to be buried must be fairly smooth. It may be necessary to add a layer of well-tamped soil to achieve a smooth surface. Soil used to backfill the trench must be free of rocks or other debris that might damage the cable. All backfill material must be tamped to prevent settling. Conduit with a long-radius elbow is used to protect the cable as it transitions into or out of the ground. Conduit is also required to protect the cable where it passes under any street or driveway.

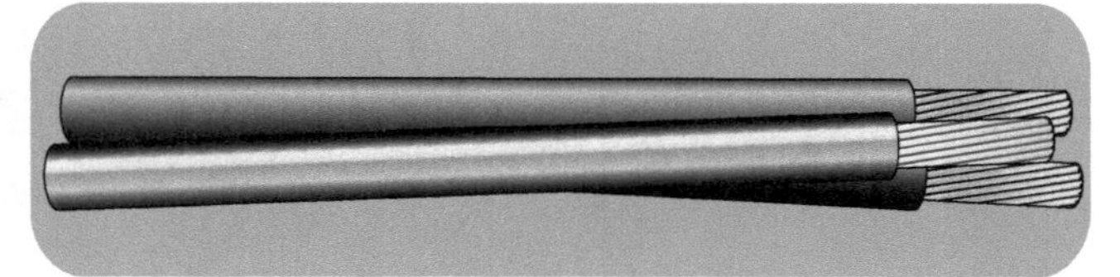

80205-11_F28.EPS

Figure 28 Underground triplex cable.

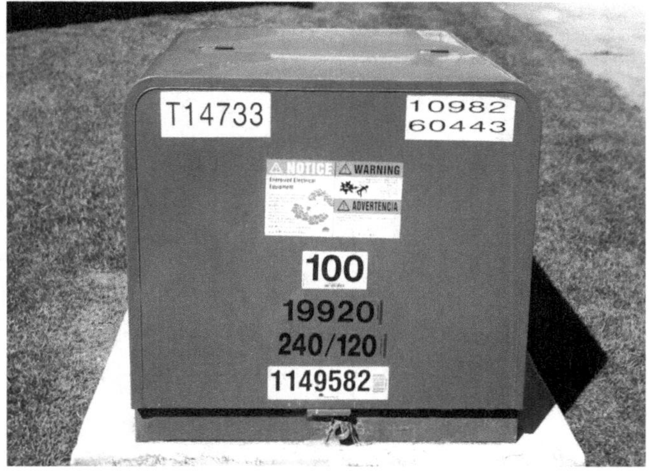

80205-11_F29.EPS

Figure 29 Pad-mounted transformer.

On Site

Meter Circuit Tester

The Super Beast is a test set designed to check the condition of secondary conductors and terminations. It plugs directly into the meter base and reads the supply voltage on each leg of the service.

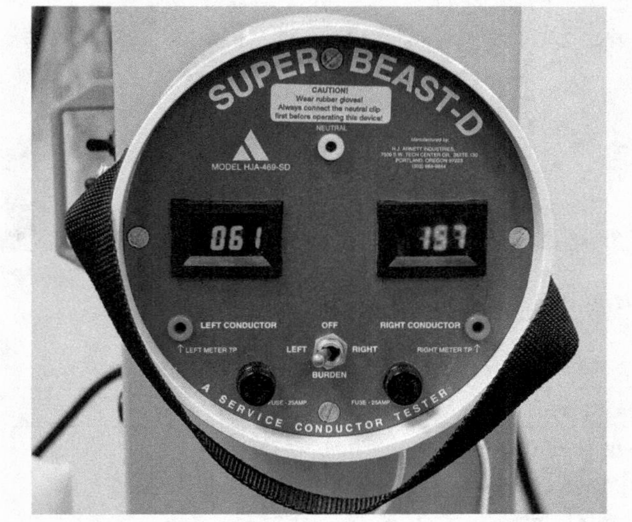

80205-11_SA04

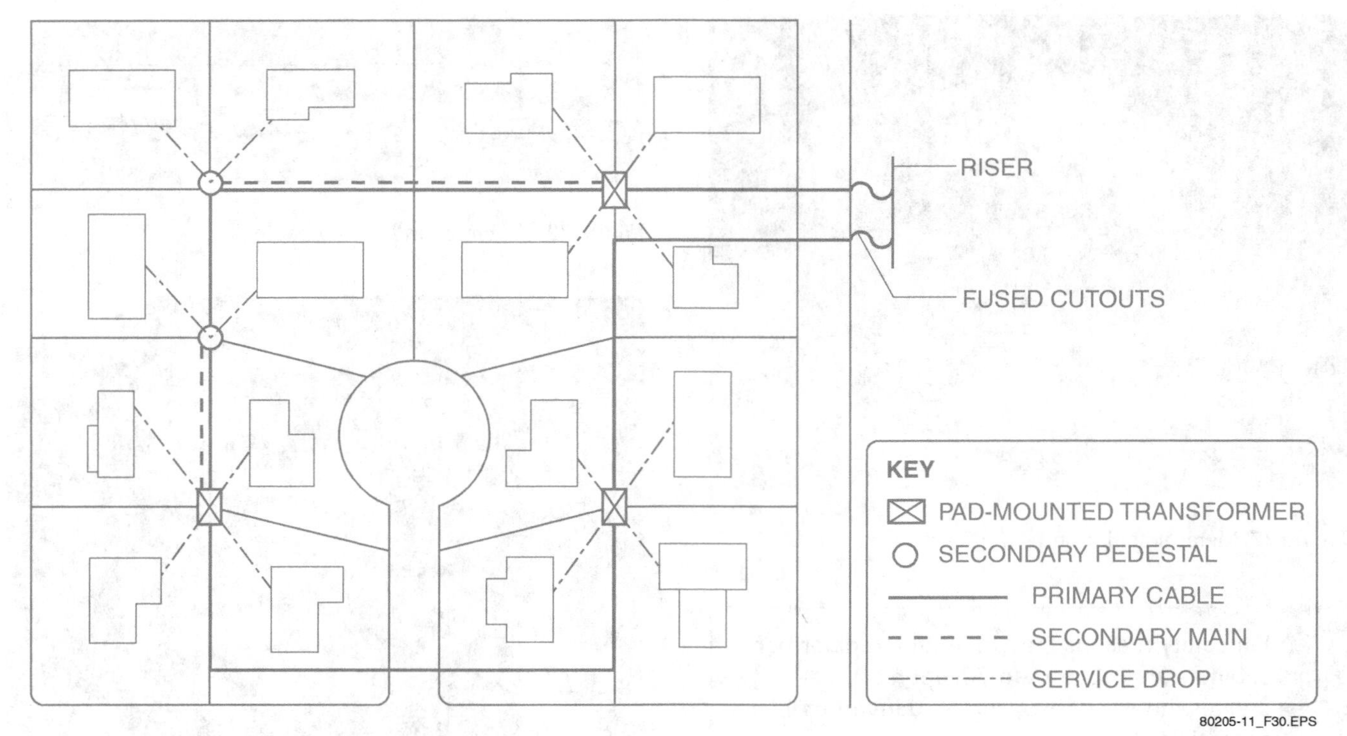

Figure 30 Residential underground distribution system.

Figure 31 Secondary pedestal.

Table 1 Underground Cable Burial Depths

Voltage Range	Burial Depth
Up to 600V	24"
600V to 22kV	30"
22kV to 40kV	36"
Over 40kV	42"

CAUTION

Only trained and authorized persons can operate trenching machines. Before digging in any area, have a locator service mark all buried utilities in the area where digging will occur. Failure to do this may result in legal action if an underground utility is damaged during digging. If other utilities, such as gas or water, are buried in the same area, maintain specified clearances to those other buried lines.

5.4.0 Installing Underground Meter Loops

The two basic types of underground meter loops used on homes are the underground meter loop on a building and the underground meter pole.

5.4.1 Underground Meter Loop On Building

An underground meter loop on a building (*Figure 32*) is attached to the side of the building. Triplex-rated cable for direct burial is used for the service drop. Conduit with a wide-radius elbow is used to protect the cable as it transitions out of the ground and into the meter loop. All other aspects of the installation, such as conduit size and distance dimensions, are similar to those of an overhead installation.

NOTE

All underground meter loop installations must comply with the *NEC*® or local code requirements.

5.4.2 Underground Meter Pole

Underground meter poles (*Figure 33*) are widely used to supply power in mobile home parks. Installation is similar to that of an underground meter loop on a building. The exception is that the meter base and service disconnect are mounted on a pole or other structure that is firmly embedded in the ground.

5.5.0 Street Lighting

Many residential neighborhoods that use underground residential distribution (URD) systems also have street lighting. The circuit that is used for these street lights is generally no different from the secondary circuit that supplies power to a customer's home. The street lighting circuit is likely to be a single-phase, three-wire 120/240V circuit that runs from the secondary terminals of a pad-mounted transformer to several street lights. At each street light, there is a junction box (*Figure 34*). Inside the box, separate wiring for that street light is connected to the secondary feed. Each street light circuit is fused at the transformer (the supply point) and at the junction box for each light. Often, conduit is laid between the junction box and the street light post.

The same cable terminations used for the secondary lines that run to a customer's home are used for the lighting circuit. In this case, they are two-hole compression terminal lugs. They are

80205-11_F32.EPS

Figure 32 Underground meter loop on a building.

80205-11_F33.EPS

Figure 33 Underground meter pole.

80205-11_F34.EPS

Figure 34 Junction box for a street lighting circuit.

connected to two of the four holes in the transformer's secondary spade lug. Fuses for the two hot legs of the secondary supply are included in the junction box.

Inside the junction box at one of the street lights in the circuit, the wiring for the light is connected to the secondary cables from the transformer. One common wiring size for street lights is AWG 10/2 UF (underground feeder) cable. Parallel groove tap connectors are used to connect the wiring together. These are mechanical connectors into which two cable ends are placed and a bolt is tightened to secure the connection. One hot leg from the secondary supply is connected to the hot leg of the street light wiring. A fuse is located at that connection. The secondary neutral is connected to the neutral wire from the street light. If necessary, the smaller AWG 10/2 UF wire can be folded several times to fill up the groove in the tap connector.

Once the connections are made, they need to be sealed using self-fusing tape. The tape should be applied in a half-lap manner so that there is enough overlap to adequately seal the splice.

5.6.0 Commercial Underground Service

Figure 35 shows examples of commercial underground services. The view on the bottom shows the bus bars that are used to feed the meter circuits. Services rated 400A and above normally use current transformers to feed the watt-hours meter. The current transformers may be housed in a separate enclosure (*Figure 36*).

80205-11_F35.EPS

Figure 35 Examples of commercial underground services.

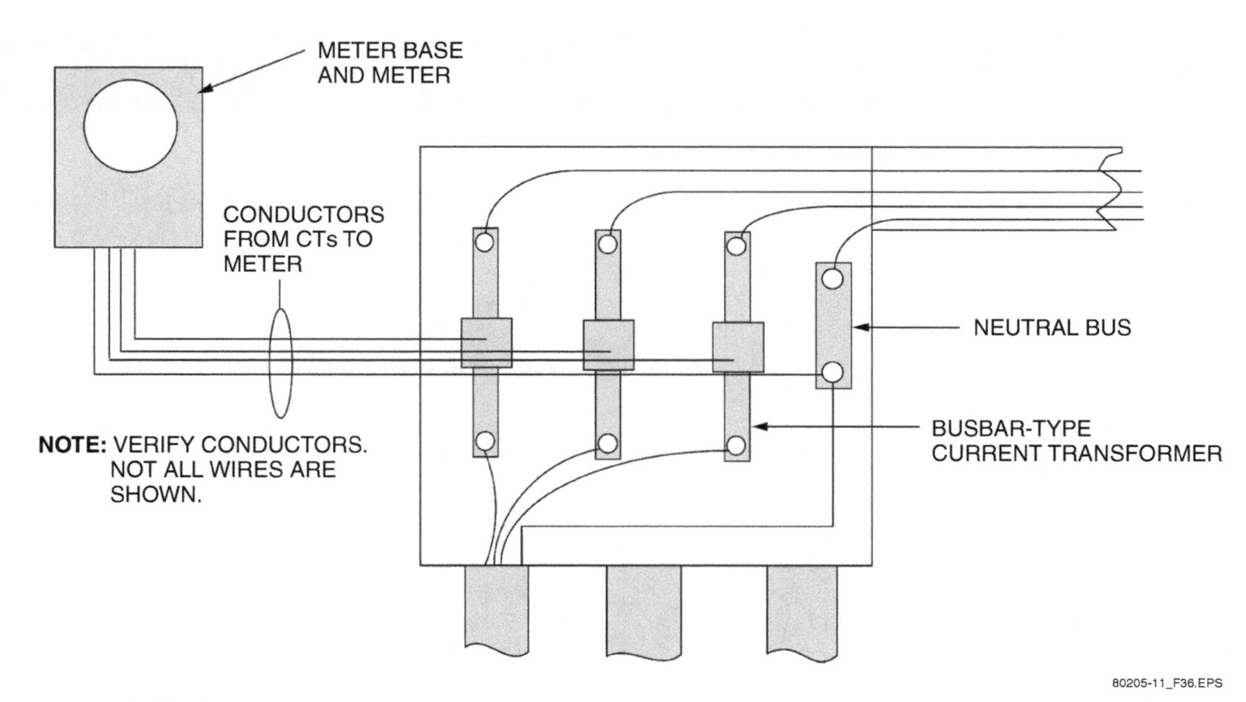

Figure 36 Typical CT cabinet arrangement.

SUMMARY

Systems that deliver power are designed so that the fewest number of customers are affected should an outage occur. Loop-fed and radial distribution systems are used. Loop-fed systems can supply power from different directions so that outages are minimized. Power supplied to customers is delivered by way of overhead or underground cables. Both delivery systems offer advantages and disadvantages. Cables are connected to a meter loop that is part of a customer's service entrance. In some areas, the utility installs the meter loop. In others, the meter loop may be installed by the customer. Electricity meters are typically installed by the utility. The older analog meters have largely been replaced with electronic meters.

Overhead systems deliver power to the meter loop using three- or four-conductor cable called a service drop. Overhead meter loops can be attached to a building, a service mast, or a pole. Primary power for an underground system transitions from overhead to underground at a riser pole. Secondary power is carried by three- or four-conductor cable rated for direct burial. Buried cables must be placed at the correct depth and protected from damage. Underground meter loops are placed on the sides of buildings or on underground meter poles.

Review Questions

1. A meter loop is part of a(n) _____.
   a. riser
   b. service entrance
   c. drip loop
   d. service drop

2. Which of these is an advantage of an overhead distribution system?
   a. Easy to repair
   b. Good aesthetics
   c. Low damage risk
   d. Safe to repair

3. The size of an aerial transformer (in kW) is based on _____.
   a. height of the pole
   b. its location in the loop
   c. its electrical load
   d. the primary voltage

4. In a residential subdivision, each home is powered by its own transformer.
   a. True
   b. False

5. The pole where overhead power transitions to underground distribution is called a(n) _____.
   a. transition
   b. downer
   c. down feeder
   d. riser

6. The maximum unsupported span length for a service drop is _____.
   a. 200 feet
   b. 150 feet
   c. 125 feet
   d. 100 feet

7. A mast that is part of a meter loop must be made of _____.
   a. galvanized steel
   b. aluminum
   c. PVC conduit
   d. metal tubing

8. The length of cable required to form a drip loop cannot be shorter than _____.
   a. 12 inches
   b. 18 inches
   c. 24 inches
   d. 36 inches

9. The cable used to connect a commercial overhead service is _____.
   a. quadruplex
   b. triplex
   c. orange
   d. duplex

10. When installing an electric meter into a meter base, the load side terminals should always be connected first.
    a. True
    b. False

11. How can the neutral conductor in triplex rated for direct burial be identified?
    a. It has no insulation.
    b. It has a much larger diameter.
    c. It has striped or colored insulation.
    d. It is marked N.

12. What action must be taken before pulling a bayonet fuse?
    a. The transformer must be drained.
    b. Primary power must be turned off.
    c. The transformer oil pressure must be relieved.
    d. The secondary cables must be disconnected from the transformer.

13. Secondary bushings on pad-mounted single-phase transformers are labeled _____.
    a. H1, H2, and H3
    b. SX, SY, and SZ
    c. T1, T2, and T3
    d. X1, X2, and X3

14. In a housing tract with an underground distribution system, power is delivered to all homes through secondary pedestals.
    a. True
    b. False

15. In a commercial service, separate current transformers are used to feed the meter circuits if the service is rated for more than _____.
    a. 100 amps
    b. 200 amps
    c. 300 amps
    d. 400 amps

Trade Terms Introduced in This Module

Drip loop: Cables formed in a loop as they enter the service head. Their purpose is to allow water to drip from the cables and not enter the service head and conduit in the meter loop.

Loop-fed distribution system: A power distribution system where power is available to a transformer through different routes or directions. If a system fault occurs, power can be easily rerouted to the transformer to restore power.

Meter loop: The part of a customer's electrical service consisting of the service head, service mast, conductors, conduit, meter base, and service equipment disconnect. It is the point at which the utility delivers power to the customer.

Quadruplex: A four-conductor cable used in three-phase service drops. For overhead service, the bare (neutral) conductor acts as the messenger to support the cable. For underground services, the neutral conductor is insulated and is identified with a different color or special markings.

Radial distribution system: A power distribution system with a central power source. Lines radiate from the power source like spokes of a wheel. If the central power source fails, all customers in the system lose power.

Rigid metal conduit (RMC): A thick-walled steel conduit that is the preferred material for conduit and service masts used in meter loops.

Riser: The pole on which overhead power distribution transitions to underground power distribution.

Secondary pedestal: A junction point in an underground secondary distribution system. Secondary power is fed to the secondary pedestal from the transformer. From there, power is fed to additional customers.

U-guard: A type of conduit used on risers. It is attached directly to the pole without the use of standoffs that are required with standard conduit.

Additional Resources

This module presents thorough resources for task training. The following resource material is suggested for further study.

Underground Power Cables. New York, NY: Longman Group.

Underground Transmission Systems Reference Book. Palo Alto, CA: Electric Power Research Institute (EPRI).

Figure Credits

Topaz Publications, Inc., Module opener, Figures 4, 7, 9–12, 14, 18–20, 22, 23, 25–27, 31–33, and SA02 (photo)

Used with permission of General Cable, Figures 5 and 28

Tony Vazquez, Figure 6

www.sandman.com, SA01

John Traister, Figures 15, 16, and 21

Itron Inc., SA03

H.J. Arnett Industries, SA04

GE Consumer & Industrial, Figure 29

Harry Wood, Figure 34

Al Hamilton, Figure 35

NCCER CURRICULA — USER UPDATE

NCCER makes every effort to keep its textbooks up-to-date and free of technical errors. We appreciate your help in this process. If you find an error, a typographical mistake, or an inaccuracy in NCCER's curricula, please fill out this form (or a photocopy), or complete the online form at **www.nccer.org/olf**. Be sure to include the exact module ID number, page number, a detailed description, and your recommended correction. Your input will be brought to the attention of the Authoring Team. Thank you for your assistance.

Instructors – If you have an idea for improving this textbook, or have found that additional materials were necessary to teach this module effectively, please let us know so that we may present your suggestions to the Authoring Team.

NCCER Product Development and Revision
13614 Progress Blvd., Alachua, FL 32615

Email: curriculum@nccer.org
Online: www.nccer.org/olf

❏ Trainee Guide ❏ AIG ❏ Exam ❏ PowerPoints Other ______________________________

Craft / Level: Copyright Date:

Module ID Number / Title:

Section Number(s):

Description:

Recommended Correction:

Your Name:

Address:

Email: Phone:

80206-11

Distribution Line Maintenance

Module Six

Trainees with successful module completions may be eligible for credentialing through NCCER's National Registry. To learn more, go to **www.nccer.org** or contact us at **1.888.622.3720**. Our website has information on the latest product releases and training, as well as online versions of our *Cornerstone* newsletter and Pearson's product catalog.

Your feedback is welcome. You may email your comments to **curriculum@nccer.org,** send general comments and inquiries to **info@nccer.org**, or fill in the User Update form at the back of this module.

V.1 12/11

80206-11
DISTRIBUTION LINE MAINTENANCE

Objectives

When you have completed this module, you will be able to do the following:

1. State the safety precautions associated with power line maintenance.
2. Describe the requirements for pole and distribution line inspections.
3. Describe the maintenance requirements for pole-mounted equipment and conductors.
4. Describe methods used to achieve load management and fuse coordination.
5. Re-conductor overhead lines.
6. Replace cross-arms, arrestors, switches, insulators, and associated hardware.
7. Replace an aerial transformer.
8. Describe the methods used to locate and correct faults in URD cabling systems.
9. Perform testing and inspection of aerial transformers.

Performance Tasks

Under the supervision of your instructor, you should be able to do the following:

1. Re-conductor overhead lines.
2. Replace cross-arms, arrestors, switches, insulators, and associated hardware.
3. Replace an aerial transformer.
4. Perform testing and inspection of aerial transformers.

Trade Terms

Aeolian vibration
Dynamometer
Thermovision

Industry Recognized Credentials

If you're training through an NCCER-accredited sponsor you may be eligible for credentials from NCCER's Registry. The ID number for this module is 80206-11. Note that this module may have been used in other NCCER curricula and may apply to other level completions. Contact NCCER's Registry at 888.622.3720 or go to nccer.org for more information.

Contents

Topics to be presented in this module include:

1.0.0 Introduction ... 1
2.0.0 Safety ... 1
2.1.0 Personal Protective Equipment (PPE) ... 1
2.2.0 Protective Grounds ... 1
2.3.0 Protective Insulating Covers ... 2
2.4.0 Effects of Fatigue ... 2
2.5.0 Working Near Traffic ... 3
3.0.0 Pole Inspection ... 3
3.1.0 Wood Pole Problems ... 4
3.1.1 Pole Decay ... 4
3.1.2 Inspecting for Pole Decay ... 5
3.1.3 Pole Testers ... 5
3.2.0 Inspecting Pole Guys ... 5
4.0.0 Inspection and Maintenance of Conductors ... 5
4.1.0 Conductor Inspection ... 5
4.2.0 Abrasion ... 6
4.3.0 Trees and Brush ... 7
4.4.0 Repairing and Replacing Conductors ... 7
4.4.1 Splices ... 7
4.5.0 Conductor Sag ... 7
4.6.0 Conductor Ties ... 8
4.7.0 Vibration Dampers ... 8
5.0.0 Inspection and Maintenance of Pole-Mounted Equipment ... 9
5.1.0 General Inspection Requirements ... 9
5.1.1 Thermal Testing ... 10
5.1.2 Oil-Filled Equipment ... 10
5.2.0 Field Maintenance ... 10
5.2.1 Reclosers ... 10
5.2.2 Sectionalizers ... 11
5.2.3 Capacitors ... 11
5.2.4 Pole-Mounted Switches ... 12
5.2.5 Distribution Transformers ... 12
5.2.6 Voltage Regulators ... 13
5.2.7 Insulators ... 13
5.3.0 Transformer Testing and Diagnostics ... 13
5.3.1 Visual Inspection ... 14
5.3.2 Operation of Switches or Breakers ... 14
5.3.3 Secondary Voltage Test ... 14
6.0.0 Locating and Correcting Faults in URD Systems ... 15
6.1.0 Transformer Faults ... 15
6.2.0 Cable Faults ... 16
6.3.0 Repairing Damaged Underground Systems ... 18

7.0.0 Load Management Overview ... 19
7.1.0 Voltage Regulators ... 20
7.2.0 Tap-Changing Transformers ... 20
7.2.1 Computing a Tap Change ... 21
7.3.0 Capacitor Banks ... 21
8.0.0 Protective Device Coordination ... 21

Figures and Tables

Figure 1 Safety tag ... 1
Figure 2 Fall arrest system ... 2
Figure 3 Protective grounds installed on de-energized power lines ... 3
Figure 4 Overhead lines with insulating blankets and line hoses ... 3
Figure 5 Examples of damaged poles ... 4
Figure 6 Guy wire connection to guy anchor ... 5
Figure 7 Tightening a pole guy ... 6
Figure 8 Wire wrap ... 6
Figure 9 Cross-arm extender ... 7
Figure 10 Butterfly takeup reel ... 8
Figure 11 Automatic splice ... 8
Figure 12 Automatic splices installed on distribution lines ... 8
Figure 13 Conductor tie ... 9
Figure 14 Handheld color thermovision camera ... 10
Figure 15 Congested pole ... 10
Figure 16 Electronic recloser ... 11
Figure 17 25KV electronic resettable sectionalizer ... 12
Figure 18 Pole transformer pressure-relief valve ... 13
Figure 19 Bucket truck boom hoist ... 13
Figure 20 Voltage regulator oil sight glass and drain valve ... 14
Figure 21 Damaged transformer ... 14
Figure 22 Transformer secondary circuit breaker handle ... 14
Figure 23 Measuring secondary voltage on a single-phase transformer ... 15
Figure 24 Loop-feed URD system with possible transformer fault ... 16
Figure 25 Capacitor discharge unit (thumper) ... 16
Figure 26 Fault indications in a loop-feed URD system ... 17
Figure 27 Time domain reflectometer (TDR) ... 18
Figure 28 Hydraulic cable-spiking tool ... 19
Figure 29 Encapsulated cable splice ... 19
Figure 30 Secondary tap-changing switch ... 20
Figure 31 Transformer decal showing tap percentages ... 20
Figure 32 Capacitor bank ... 21
Figure 33 Example of protective device coordination ... 22

Figures and Tables

1.0.0 Introduction

Maintenance of distribution lines primarily involves inspection of poles, conductors, and pole-mounted equipment for signs of damage or deterioration. It also requires replacement of damaged or failed parts. Line maintenance and repair should be performed on de-energized circuits whenever possible. Such situations typically occur when installing new lines and sometimes when repairing disabled lines.

Inspection of power lines to identify damage and other problems is important. While your company may not have an established inspection program, it is a good idea to check out nearby poles, conductors, and equipment any time you are working on power lines. Wind, lightning, and age can result in damage and deterioration. In the absence of a formal inspection program, the only way these problems are uncovered is through the alertness of the crews performing installation and repair work.

2.0.0 Safety

Utility companies will generally not shut off power to customers for maintenance or repair work. For that reason, power line workers often work around energized lines. They may work on one phase of a three-phase system while the other two phases are live, for example. If the power is off due to weather, accident, or equipment failure, it is generally left off until repairs are completed. In such cases, switching devices are opened in order to prevent accidental re-energizing. Switches are locked out and safety tags (*Figure 1*) are installed.

Work crews must follow all company procedures to prevent the accidental energizing of a circuit while work is in progress. Companies typically require line workers to use all protective gear, and install safety grounds and rubber blankets, even when working on a de-energized circuit. Strict adherence to established procedures by everyone can prevent injury or death from an accidental re-energizing of a circuit.

2.1.0 Personal Protective Equipment (PPE)

All power line workers are required to wear PPE to prevent on-the-job injuries. Companies typically require line workers to wear all PPE at all times, install insulating blankets, line covers, and safety grounds, and use hot sticks to operate equipment. These are wise precautions, even

80206-11_F01.EPS

Figure 1 Safety tag.

when working on de-energized lines. There is always a danger of induced voltages from adjacent lines, from static electricity, or from accidental re-energizing of the line. Strict adherence to established procedures by everyone involved helps to prevent injury or even death from the accidental re-energizing of a circuit.

If climbing or using a bucket lift, always use a fall arrest system anchored to the provided attachment point, as shown in *Figure 2*.

2.2.0 Protective Grounds

Protective grounds are installed when working on de-energized lines. Even though a line may be de-energized, there is still the danger from accidental energizing of the line, as well as induced voltages, and the buildup of static electricity. Protection from these hazards is achieved by installing protective grounding cables at the work site. One major concern is the possibility that emergency generators put into use during a power outage will cause a backfeed onto a de-energized line.

A transformer works in either direction. It will step voltage down or step-up the voltage depending on which side of the transformer windings the voltage is applied to. For example, if 240 volts from a home generator is applied to a de-energized transformer on the secondary bushings, the transformer will step-up the voltage to primary

80206-11_F02.EPS

Figure 2 Fall arrest system.

voltage on the high-side bushings and on the previously de-energized power line.

Induced voltages occur from mutual induction when a de-energized line lies parallel to an energized line. Expanding and collapsing magnetic fields from the energized line pass through the conductor of the de-energized line. This induces a voltage in the de-energized line in the same manner in which a transformer works. If the induced voltage is given a path to ground, current flows. Induced voltage can be deadly. The amount of induced voltage depends on factors such as the length of the lines, the amount of current on the energized line, and the angles and proximity of the two lines. Long conductors that are parallel and close together generally have more induced voltage

Static electricity is similar to induced voltage, but the amount of the voltage can vary considerably. The amount depends upon factors such as the length of the parallel line, the proximity of the two lines, and temperature and humidity. Static electricity can be caused by wind blowing across a line. Temperature and humidity have a large influence on the amount of static electricity produced. Static electricity can be discharged by the application of a ground.

Before installing any grounds, always test for the presence of voltage using a voltmeter rated for the voltage being worked on. First test the voltmeter, then test for voltage. Finally, re-check the meter to make sure it is still working.

Figure 3A shows protective grounds attached to de-energized power lines. The lower line is the neutral and those above it are phase conductors. The protective grounds are connected to a cluster bar (*Figure 3B*), which is in turn connected to a ground rod driven into the earth (*Figure 3C*). The following sequence should be followed when installing protective grounds:

- Install the cluster bar.
- Connect the cluster bar to earth ground or pole bond.
- Connect the cluster bar to the neutral conductor neutral.
- Connect the neutral to the nearest phase conductor (phase A).
- Connect protective grounds between the phases (A-B, B-C).

The ground or neutral end of the protective ground is first connected to the cluster bar, which is bonded to the pole ground. Upon completion of the work, the current-carrying conductor line end of the short-circuiting connection is disconnected first. The leads at the end of the conductor are always installed where the work is being performed. Placing the leads at the other end, or any distance from the work location, can still leave sufficient length of parallel lines for static to accumulate or induced voltages to occur. Grounding is the first thing done on a job.

> **NOTE**
>
> Protective grounding procedures vary from one company to another. For example, after attaching a ground cable from the cluster bracket to the system neutral, some companies prefer to attach the ground cable from the cluster bracket to the center phase conductor on the basis that this method reduces the chance of circulating currents should the lines become re-energized.

2.3.0 Protective Insulating Covers

After connecting the protective grounds, the line worker installs the protective covers, line hoses, and insulating blankets over all areas where there is a risk of coming into contact with energized lines or terminals (*Figure 4*). If a crew is working on a de-energized phase, protective covers are installed on the de-energized line, as well as on the other phases, which may still be live.

2.4.0 Effects of Fatigue

After working long hours or repairing storm damage in the early morning hours, a tired and exhausted line worker will not be as alert as normal. Conditions may not appear as they seem,

80206-11_F03.EPS

Figure 3 Protective grounds installed on de-energized power lines.

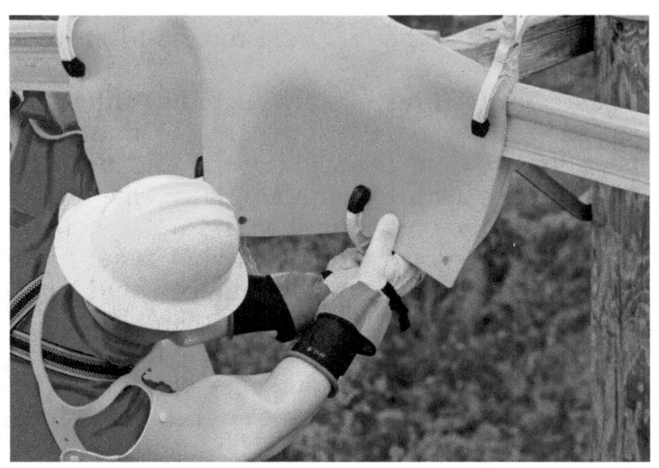

80206-11_F04.EPS

Figure 4 Overhead lines with insulating blankets and line hoses.

especially when working around damaged lines. Ice and storm damage can cause lines to sag and pull in unexpected ways. Excess moisture can create unexpected paths of current flow. Procedures that are routine on a warm sunny day can be very hazardous on a cold, wet, windy night. Think each step through. Talk over procedures with co-workers and work just a little slower. There is no emergency so great that line workers cannot take some extra time to be cautious.

2.5.0 Working Near Traffic

Most of the tasks line workers perform occur on or near a highway. Moving traffic is a dangerous hazard. Cones, signs, or barricades must be placed in accordance with company policy and federal, state, and local regulations to alert drivers that utility work is in progress. Workers must wear protective traffic vests and be on the alert for distracted drivers.

3.0.0 Pole Inspection

Every line worker is involved in the inspection and maintenance of utility poles. It may be a scheduled inspection or a task as simple as looking up to check for unusual conditions while on a job.

Bad weather, traffic accidents, and vandalism are the cause of most damage to poles and related equipment. To repair the damage, an experienced line worker requires a total commitment to safety, along with the knowledge needed to accomplish repairs to the system quickly and economically. Strict adherence to safety is critical, especially when making repairs. If the equipment is not in perfect working order, line workers must be especially careful when making repairs to prevent injury or further damage.

Distribution lines are inspected at intervals that should not exceed ten years. Normally, inspections are done by walking the line. Binoculars are very useful in enabling an inspector to closely examine pole equipment from the ground. The inspector submits a report of his or her findings in enough detail so that engineers and work planners can create a useful work plan.

Inspectors look for damaged poles, as well as leaning poles caused by soft soil. Construction in the area may have changed the environment since the original pole installation. The inspector must be observant to notice if changes have created a hazardous location for the pole.

Observant line workers always keep an eye out for broken insulators, leaking oil, broken cross-arms, large sags in a line, wood rot, corrosion, or anything unusual.

3.1.0 Wood Pole Problems

Maintenance required on the poles, timbers, and cross-arms in a power distribution system is minimal. A wood pole lasts for many years; twenty years or more is not unusual. A common problem occurs when a pole settles unevenly into the ground, requiring straightening. Also, wood shrinkage can cause hardware to come loose. Cross-arm brackets and mounts should be tight and secure. Bolts should not wobble in the holes from pole shrinkage. The best way to prevent these problems is through good installation work. *Figure 5* shows example of damaged poles. The pole in *Figure 5C* has already been replaced.

3.1.1 Pole Decay

Wood poles are treated with preservatives to prevent decay. However, small organisms, insects, and fungi can cause wood preservatives to break down. A pole must be bored to determine the presence of internal voids. Poles with internal decay can be treated with insecticides. External decay is removed, and the area is treated with preservatives and wrapped with a moisture-proof barrier. Utility companies reinforce or replace poles weakened excessively by internal or external decay. If caught soon enough, the life of a pole can be extended by prompt treatment of damage and decay.

A

B

C

80206-11_F05.EPS

Figure 5 Examples of damaged poles.

> **NOTE**
> At one time, wood utility poles were coated with creosote to prevent decay and discourage insects. Today, the poles are treated with wood preservatives like those used in pressure-treated lumber.

3.1.2 Inspecting for Pole Decay

Rot normally occurs where the pole enters the ground. Check for ground level pole rot by probing the pole with a screwdriver. Dig around the base 6 to 12 inches deep. Probe the pole with a large screwdriver. Use a hammer to tap the screwdriver. If the pole is rotten, there will be very little resistance. With a rotted pole, the screwdriver may pierce anywhere from several inches to all the way through the pole.

A pole inspection normally includes sounding the pole by hitting it with a hammer from just below ground level to approximately 6 feet above ground. If the pole is compromised, a hollow sound is emitted. This method can detect defects from scaling, rot, and bird or insect damage. Insect colonies can eat away at a pole until it will no longer support its own weight.

3.1.3 Pole Testers

Non-destructive testing of wood poles is possible using test equipment developed for that purpose. The line worker enters the wood type and pole diameter, then strikes the pole with a steel ball. A microprocessor analyzes the vibrations. The test equipment manufacturer's instructions must be followed to ensure an accurate test.

3.2.0 Inspecting Pole Guys

Over time, pole guys can stretch and require retensioning. When inspecting pole guys, check to see if guy wires are slack or too close to conductors, that guy insulators are in place, and that the guy is properly connected to the system ground and neutral.

Guy wires should not be buried because moisture and chemicals in the ground can cause them to deteriorate. The guy anchor is designed to be underground and the guy wire should be above the ground (*Figure 6*).

If slack is found in a guy wire with a turnbuckle, tighten the turnbuckle. Most newer installations do not use turnbuckles. If a guy wire without a turnbuckle is slack, it can be tightened by removing the wire wrap, then using a guy grip, come-along, and pulling eye (*Figure 7*) or other method to take up the slack. A new wire wrap is then used to secure the guy wire (*Figure 8*).

80206-11_F06.EPS

Figure 6 Guy wire connection to guy anchor.

> **CAUTION**
> Never reuse a preform guy tie. Be sure to use the correct size preform guy wrap for the guy wire. The preform guy wraps are color-coded per wire size to prevent mistakes.

> **NOTE**
> Be sure to inspect guys at ground level. Distracted drivers sometimes drive into utility equipment. It is also common to find guy wires and pedestals damaged by lawn equipment. Vibration from a vehicle collision can damage equipment higher up on a pole or loosen mountings. Guy wires can be pulled loose, pole grounds broken, or even the pole itself may be damaged. Guy wires are rarely protected at the point where they enter the ground and are particularly vulnerable there.

4.0.0 Inspection and Maintenance of Conductors

Conductors normally last a long time with minimal maintenance. However, they are subject to damage from wind-induced vibration, oscillation, unbalanced loads, short circuits, and lightning.

4.1.0 Conductor Inspection

During inspections, and any time when working on a line, look for excessive slack, loose conductors, foreign objects hanging on the line, and conductors burning in trees. Both vertical and horizontal clearances are critical. Check the conductors to ensure there is sufficient clearance from other wires, buildings, signs, waterways,

80206-11_F07.EPS

Figure 7 Tightening a pole guy.

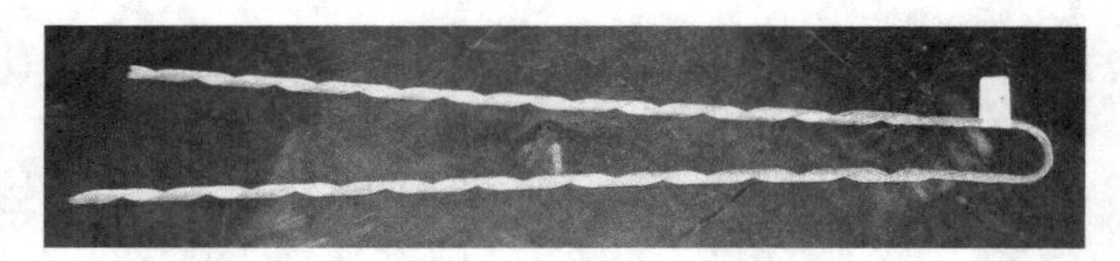

80206-11_F08.EPS

Figure 8 Wire wrap.

railroads, and swimming pools. Pay particular attention to areas under conductors that are easily accessible to people, such as roads, loading docks, parks, playgrounds, and pedestrian areas. Check splices to verify they are still secure.

4.2.0 Abrasion

Abrasion is surface damage. It is recognized by black deposits on either the tie or the conductor. Abrasion is often the result of chafing caused by unwanted movement of the conductors. The moving conductor rubs against conductor hardware or loose ties, resulting in abrasion of the conductor. Abrasion can usually be repaired with an armor rod if the damage is slight or a splice if the damage is significant. A conductor splice is used when the steel core of the conductor is in good condition, but there is damage to the outer layers of aluminum.

4.3.0 Trees and Brush

Look for tree limbs that are too close to the lines. In high winds, they can sway excessively or break off, taking the line to the ground with them. All danger trees must be removed or pruned. A danger tree is one with cavities or rotted parts in the trunk or large branches; cracks or splits in the trunk; or large dead branches near the pole or conductors. Unwanted tree branches are removed with pruning saws or the tree pruner attachment of a hot stick.

Be alert for tall brush under distribution lines. When lines sag lower in hot weather, they can contact tall bushes and create a fire hazard.

4.4.0 Repairing and Replacing Conductors

Normally, conductors are repaired rather than replaced. Replacing conductors, known as reconductoring, only occurs on a system upgrade or when there is severe storm damage. Repair typically consists of cutting out sections of conductor and allowing them to be lowered to the ground after establishing safe conditions to do so. A line worker in a bucket truck may tie a rope to the end of the conductor and slowly lower it to the ground. The bucket truck and conductor must always be grounded in such cases.

When it is necessary to replace conductors for reasons such as increased load demand, it is generally done without disrupting power. Cross-arm extenders (also called layout arms) with insulators are attached to the inboard side of the cross-arm (*Figure 9*). Before the conductors can be removed, the ties that secure the conductor to insulators, as well as conductor spacers, must be removed. They are replaced when the new conductors are installed. The existing conductors are lifted from the cross-arm insulators using a hot stick, or lifted by hand by a line worker wearing protective gloves and sleeves, and relocated to the cross-arm extender. The new conductors can then be installed while the old ones continue to carry the load.

Once the load is switched over to the new conductors, the old ones are removed. A short length is lowered to the ground and then rolled onto a takeup reel using the capstan winch on the utility service vehicle. Longer conductors are cut into smaller sections that fit onto the takeup reel. Reels commonly used to take up conductors are called butterfly reels, or simply butterflies (*Figure 10*). One side of the reel can be collapsed to allow the spooled conductor to be dropped into a recycling container.

4.4.1 Splices

When a section of conductor is damaged, it is usually repaired by cutting out the section and splicing in a new section. An automatic splice (*Figure 11*) is often used to make a quick repair. Other types of conductor splices were covered in the *Cable and Conductor Installation and Removal* module. Instructions for proper application of splices are provided by the splice manufacturer. *Figure 12* shows automatic splices installed.

Always clean the splice area to remove accumulated oxides and apply a corrosion inhibitor before installing the splice around the conductor.

Crimp-on splices restore conductivity and full strength to conductors. If corrosion is present under the splice connector, or if the crimp is not tight against the conductor, a hot spot can register when performing infrared scanning on the line. Many times, a splice shunt is applied over the crimp-on splice to restore conductivity.

A repair sleeve is similar to a crimp-on splice, but it has two halves that fit around the conductor. The sleeve is placed over the damaged area of the conductor and crimped in place.

A full-tension splice is installed when it is necessary to restore strength as well as conductivity. Cut the conductors on both sides of the repair to expose the core. Tape the cut ends to prevent unraveling. Wrap the steel core splice around the core. Affix filler rods around the splice to return the conductor to its original diameter.

4.5.0 Conductor Sag

Be sure to consider the temperature when replacing conductors. As temperature increases, conductors sag closer to the ground. When installing

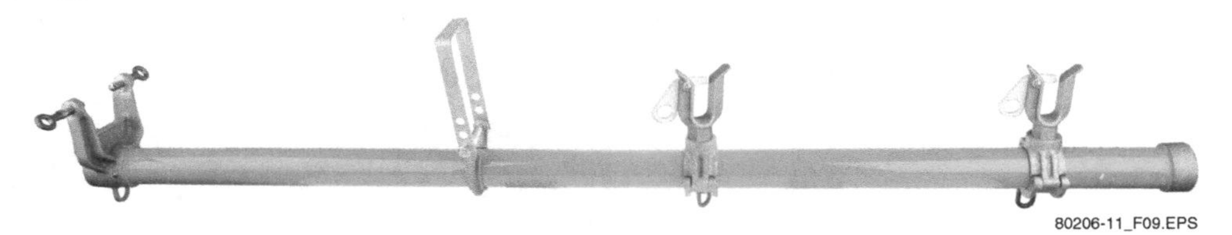

80206-11_F09.EPS

Figure 9 Cross-arm extender.

Figure 10 Butterfly takeup reel.

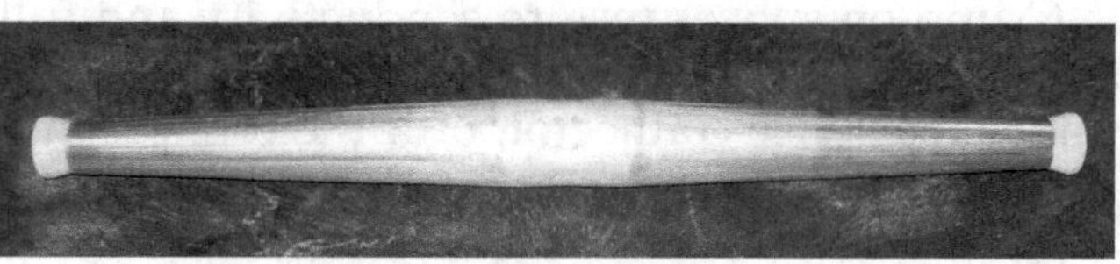

Figure 11 Automatic splice.

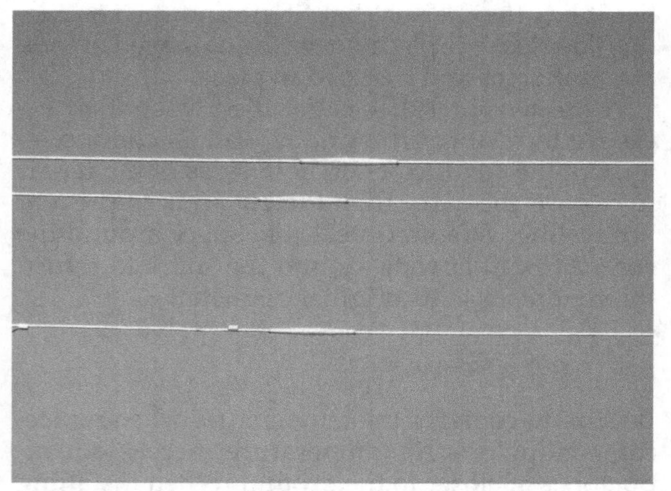

Figure 12 Automatic splices installed on distribution lines.

a conductor in warm weather with insufficient sag, the line will contract and tighten when the weather turns cold.

A **dynamometer** is used to apply the correct tension and sag to the conductor, and the sag is checked using a sag scope. Sagging and tensioning of replacement conductors were covered in a previous module.

4.6.0 Conductor Ties

Always use new conductor ties when replacing conductors. Use a tie of the correct length and strength (*Figure 13*). Apply the tie by hand. Tools can damage and weaken the tie.

4.7.0 Vibration Dampers

Vibration dampers are installed when a line is subject to prolonged periods of **Aeolian vibration**. Aeolian vibration occurs when air flowing across a conductor creates alternating pressures above and below a conductor. The alternating pressures cause the conductor to move at right angles to the airflow. Aeolian vibration is greatest at wind speeds below 15 mph (24.1 kph).

80206-11_F13.EPS

Figure 13 Conductor tie.

If permitted to continue, excessive vibration may cause a line to fatigue and break. Vibration can so weaken a conductor that it will have to be derated or replaced—both of which can result in tremendous expense.

5.0.0 Inspection and Maintenance of Pole-Mounted Equipment

The general condition of the pole-mounted equipment should be checked whenever the opportunity arises. In addition, companies have a preventive maintenance plan and schedule to make sure that all equipment is regularly inspected. Pole-mounted equipment requiring maintenance includes hardware, conductors, reclosers, sectionalizers, switches, circuit breakers, voltage regulators, and transformers. Insulators may crack or get dirty and require cleaning, especially near coastal areas where there is salt in the air. As connections age, they can become loose and require retorquing to prevent hot spots.

NOTE

Manufacturers of pole-mounted equipment, such as transformers, voltage regulators, reclosers, and capacitor banks, provide installation, operation, and maintenance instructions for their equipment. These instructions should be followed when maintenance of the equipment is required. Utility companies establish inspection and maintenance schedules for equipment consistent with the manufacturers' requirements.

GOING GREEN

Hazardous Materials

Line workers must always be careful when working with distribution equipment containing dielectric fluids, and other liquids or gases. Some of these substances are hazardous chemicals that can have significant impact on health and the environment. For example, the dielectric fluids and other chemicals routinely used in the power industry can cause serious environmental and health problems. Leaks, spills, and fires can spread toxic substances into the atmosphere or into the ground water system. Careless handling of these substances can harm people and can place utility companies at a great financial risk from cleanup, lawsuits, and fines. Observe all handling instructions and comply with all regulatory instructions when handling these dangerous chemicals. Many companies have hazmat teams to deal with cleanup and disposal of hazardous materials.

5.1.0 General Inspection Requirements

Look closely at pole top assemblies for broken wires, burn marks, damaged pins and crossarms, and broken skirts on insulators. Also check that pole grounds are secure. Make sure cutout gates and fuses fit securely. Check that conductor connections are clean, tight, and free of corrosion.

Inspect the equipment on the pole for leaking oil, blown fuses, switch contacts not properly closed, blown arrestors, and isolation devices. Look for oil leaks around inspection covers and valves. If available, check the oil level indicators for the correct amount of oil. Check the sight glass to ensure that the level is satisfactory. Take operation counter readings on reclosers and circuit breakers. Replacement is necessary if they have exceeded the manufacturer's recommended maximum of switch closures.

Inspect all areas of the pole and equipment for burn residue that could have come from a lightning strike. Blackened areas can also mean that conductor connectors are loose, frayed conductors are arcing to other conductors, or there is a failed component.

Many older poles do not have animal guards installed. When an inspection reveals their absence, install them to prevent damage from small animals and birds. Inspect riser poles connected to underground circuits the same way as aerial distribution poles.

5.1.1 Thermal Testing

Inspectors and maintenance crews use special cameras designed to react to heat. These **thermovision** cameras (*Figure 14*) are used to detect hot spots in equipment, splices, conductors, and fittings that require maintenance. The inspection can be done by helicopter, specially equipped trucks, or by a line worker on the ground. Problem areas show up as a bright white image on a black and white thermovision camera. On color cameras, hot spots vary in intensity from black at low temperatures through red, orange, and yellow, to white at the highest temperature.

5.1.2 Oil-Filled Equipment

Check oil-filled equipment such as transformers, voltage regulators, reclosers, and circuit breakers for evidence of leakage. If a leak cannot be field repaired by tightening hardware, contain the leak and replace the equipment.

> **WARNING!**
>
> Never apply power to any oil-filled equipment unless it is filled with oil. Also, be careful not to spill the oil, as it is an environmental hazard. Years ago, in the 1970s and before, transformers and other equipment used lubricants that may contain PCB-contaminated oil. A quick field test can be done with a simple test kit. If the oil contains PCBs, follow all HAZMAT procedures for safe removal and disposal of the oil-filled equipment. An EPA-licensed disposal company is needed to dispose of PCB oil and PCB oil-filled equipment.

5.2.0 Field Maintenance

Changing equipment on poles is very similar to installing new equipment. Drilling mounting holes and other preparations for new equipment is not usually necessary, since the original mounting holes are generally reusable. However, it is important to inspect the pole and check the mounting holes to make sure they are still usable. Different size bolts and fittings may be needed to compensate for variations caused by expansion of the hole or wood shrinkage. Updated equipment may require new brackets and drilling as if it were a new installation.

Take special care on poles with a great deal of mounted equipment, such as the one shown in *Figure 15*. With a congested a pole, the task is more complex. If line workers are not careful, the risk of personnel injury, damage to expensive equipment, or a power outage increases.

> **WARNING!**
>
> Before any maintenance is performed on de-energized equipment, temporary safety grounds must be used to isolate the lines being worked on. This must be done by qualified individuals wearing the required PPE and using insulated tools.

5.2.1 Reclosers

A recloser is a circuit breaker that automatically interrupts a circuit when a high current condition exists. It is called a recloser because the circuit recloses automatically, restoring power to the distribution line. Reclosers are generally set to trip at twice the normal line current, thus preventing spurious tripping for a temporary overload. A recloser opens when it senses a fault. After a

80206-11_F14.EPS

Figure 14 Handheld color thermovision camera.

80206-11_F15.EPS

Figure 15 Congested pole.

predetermined time, it recloses. If the fault is still present, it repeats this process for up to three more times before it locks open. This gives the fault time to clear.

There are several types of reclosers, including hydraulically controlled oil reclosers, electronically controlled oil reclosers, and vacuum reclosers. Each has its own maintenance schedule specified by the manufacturer.

Line reclosers normally have bypass switches that permit the recloser to be removed from the circuit for testing and maintenance without disrupting power.

Electronic reclosers are microprocessor-controlled and come equipped with a USB port that permits programming and monitoring with a laptop computer. If desired, power companies can specify connecting a radio to the USB port. The radio mounted on the recloser receives its power from the USB port. Each recloser has its own unique address and workers can monitor and make changes to the recloser while it remains in service. The radio permits remote monitoring from the truck at a range of 500 feet. *Figure 16* shows an electronic recloser with the USB port, lockout beacon, and battery compartment.

Some electronic reclosers have a lockout beacon, which is a high intensity, daylight-visible indicator to quickly enable line workers to identify a locked-out recloser. The beacon uses a lithium battery that requires replacement every eight years.

The service literature provided by manufacturers of reclosers includes a recommended maintenance schedule and maintenance instructions. The schedule could be based on time in service, number of closures, or both. For example, a manufacturer may recommend that a recloser be serviced every two years or 50 operations. Others may recommend more frequent or less frequent servicing. Depending on company policy, maintenance of a recloser generally requires removing it from service and replacing it with another recloser. Vacuum reclosers typically require less frequent maintenance; intervals of five to six years are common.

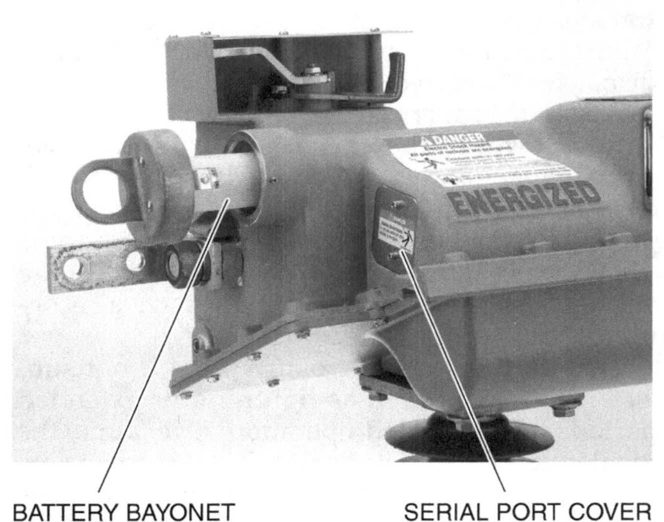

Figure 16 Electronic recloser.

5.2.2 Sectionalizers

Sectionalizers are slave devices. They cannot interrupt when fault current exists, but work in coordination with an upstream recloser to isolate faults to the smallest possible section of a distribution line.

Sectionalizers operate during the time a recloser or circuit breaker is open. If the fault is permanent, the sectionalizer will open the circuit after the upstream device operates two, three, or four times, depending on how it is set. When the recloser restores, power is available to the circuit except for that portion protected by the sectionalizer.

A hydraulic sectionalizer contains an oil switch that automatically opens after fault currents flow through the coil a specified number of times. Line workers can manually operate a sectionalizer by pulling the operating lever down with a hot stick. The operating lever is under the sleet hood.

Electronic sectionalizers like those shown in *Figure 17* fit into a cutout much the same way that a fuse fits into a holder. Inside is a copper tube with bronze castings on both ends. When the electronic sectionalizer is closed, current flows through the copper tube. A current transformer monitors current flow through the copper tube. The current transformer secondary controls an actuator. When the current flows, activating the actuator, the copper tube drops out of the holder. Because an electronic sectionalizer is like a solid blade knife switch, the cutout opens between the top and bottom contacts much the same way a fuse holder is open with no fuse installed. Line workers use a portable load break tool to manually interrupt sectionalizer load current.

5.2.3 Capacitors

Capacitor banks are installed on distribution lines to help balance the effect of inductive loads. Capacitors retain a charge, even when disconnected from the power source. They can retain

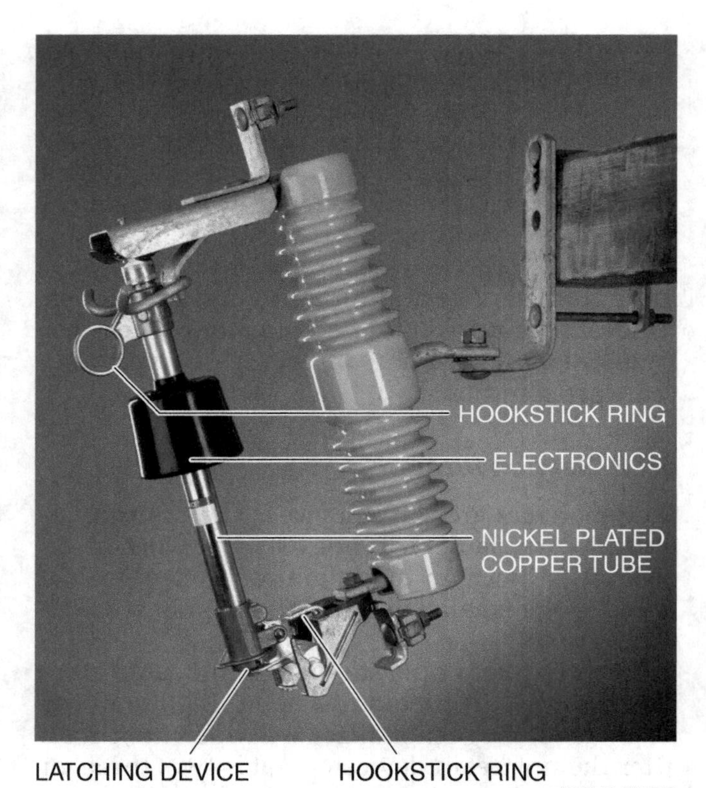

Figure 17 25KV electronic resettable sectionalizer.

the charge for a considerable amount of time. Capacitors used on distribution lines have an internal discharge (bleeder) resistor that bleeds off the charge when the capacitor is removed from the circuit. When the capacitor is disconnected, the resistor provides a low-resistance path for current flow across the terminals. Current flows from the negative plate of the capacitor through the bleeder resistor to the positive plate. After about five minutes, the charge on the capacitor plates should have been bled off. However, always test the capacitor with a high voltage tester to make sure it is discharged. If any voltage exists, wait another five minutes. Once the voltage is 0 (some manufacturers recommend less than 50 volts), short across the terminals with a hot stick, short across the terminals with copper wire, and ground them to the capacitor case.

Pole-mounted capacitor banks are removed for shop maintenance on a schedule that is based on either the elapsed time in operation or the number of switching operations performed. The maintenance schedule might be anywhere from three to eight years, depending on the type of On/Off control installed. In most cases, the maximum number of switch operations should not exceed 2,500. If a capacitor tank or casing shows signs of swelling, bulging, or oil leakage, the capacitor has failed and must be replaced.

> GOING GREEN
>
> **Recycling Used Oil**
>
> Part of the repair shop maintenance procedures is restoring the dielectric strength of the oil. Maintainers do this by filtering the oil to remove any carbon and moisture. It may be necessary to run the oil through a filter several times may before reusing it. The filtering process is repeated until the oil tests satisfactorily. The equipment manufacturer's specifications for the desired oil purity must be followed.

5.2.4 Pole-Mounted Switches

Each time a line worker operates a switch, it should be inspected for loose hardware, proper alignment, worn parts, and burned contacts. Also look for damaged interrupters, arcing horns, and insulators. Verify that there are proper ground connections and sufficient lightning arrester protection.

Normally, switches are inspected every five years. Temporary line modifications to route current flow around the switch is usually necessary in order to perform maintenance.

5.2.5 Distribution Transformers

The inspection of transformers is limited to visual observation to find oil leaks or obvious damage to the transformer bushings or tank. The transformer load management system monitors the load on distribution transformers monthly. The load is compared with the KVA rating of the transformer in order to identify problems. If the report indicates a problem, line workers will be dispatched to correct it.

Pole transformers do not generally have a drain valve. However, they do have a pressure-relief valve (*Figure 18*) that can be used to check the oil condition after an event. Pulling the pin from the pressure-relief valve provides the opportunity to smell the oil. If it has been affected by a lightning strike or fault current, the oil will have a recognizable burnt odor.

Transformers last for many years if not subjected to surges such as lightning or extended periods of overloaded operation. Over time, the insulating oil can deteriorate due to exposure to heat, oxygen , or moisture, which can occur if the unit is not tightly sealed. If the deterioration is not caught in time, the transformer will short internally. When a transformer needs to be replaced, it must be isolated from the circuit, then discon-

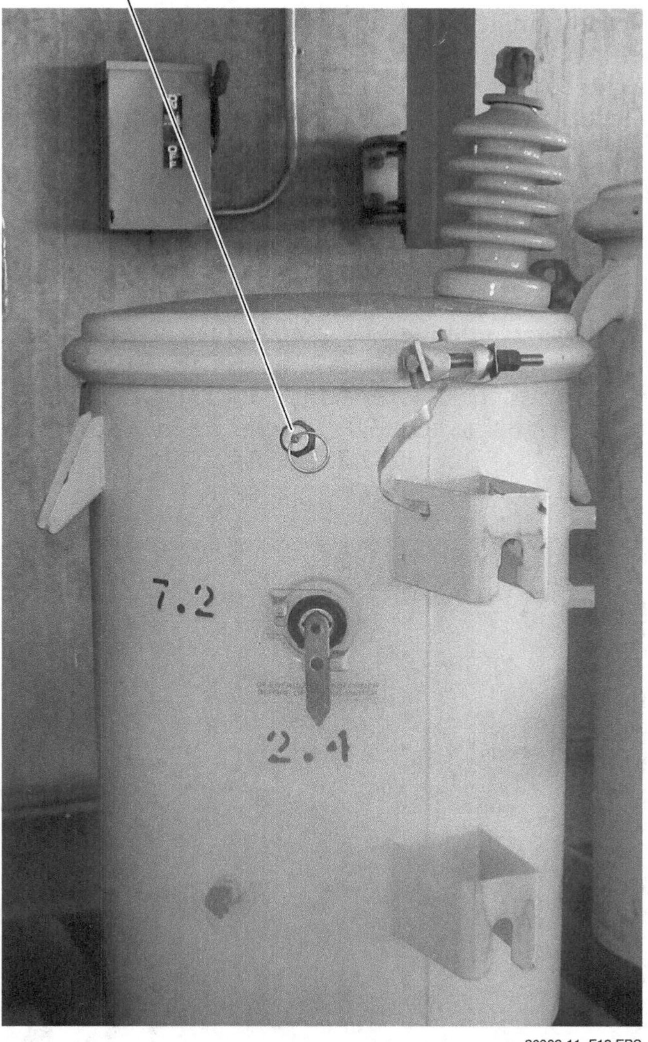

80206-11_F18.EPS

Figure 18 Pole transformer pressure-relief valve.

nected from the pole and lowered using a block and tackle or pole-mounted capstan winch. Some bucket trucks have a winch attached to the boom that is designed for the purpose of hoisting equipment such as transformers (*Figure 19*). Be sure the hoisting mechanism is secured to the transformer and the line is pulled taut before the transformer is released from the pole. Once the transformer is returned to the maintenance shop, it will be opened and examined. If it is not seriously damaged, it will be refurbished at the shop and returned to inventory.

5.2.6 Voltage Regulators

Regular voltage regulator inspections are part of a distribution line maintenance patrol. Listen for unusual operating noise and look for broken insulators, leaking oil, and visible damage. If there has been a fault current, the terminals are usually discolored. Record the value of the operation counter. If the count exceeds the manufacturer's recommendation, the regulator is replaced and sent to the maintenance shop.

80206-11_F19.EPS

Figure 19 Bucket truck boom hoist.

Like transformers, the components inside voltage regulators are immersed in oil, which can deteriorate over time. Voltage regulators are usually equipped with an oil sight glass (*Figure 20*) so line workers can periodically check the oil level. The unit shown also has an oil drain valve, which allows the oil to be periodically sampled and tested.

5.2.7 Insulators

Most damage to insulators is caused by wind, lightning, and vandalism. Always replace damaged insulators when noticed. Damaged insulators can leak voltage and cause corona discharge. This is a condition in which high potential from the conductor ionizes the surrounding air. If allowed to continue, corona further weakens the insulator. Corona usually has a purple glow. It can cause power loss and radio frequency interference.

Corona is a problem in heavily populated areas because it produces ozone, which contributes to air pollution. Ozone can interfere with lung function and can cause irritating respiratory problems.

5.3.0 Transformer Testing and Diagnostics

A line worker is not expected to perform an in-depth analysis of a transformer to determine how or why it failed. The line workers, main task is to remove and replace transformers that have failed.

Figure 20 Voltage regulator oil sight glass and drain valve.

Troubleshooting and diagnostic tests are limited to the following:

- Visual inspection
- Operation of switches or breakers (if provided)
- Secondary voltage test

Further testing may be done at a central location once a transformer is removed from service. There, it can be determined if the transformer can be rebuilt or if it should be scrapped.

5.3.1 Visual Inspection

The appearance of a failed transformer can often give a clue about its condition (*Figure 21*). Things to check for in a visual inspection include the following:

- Overload signal light On
- Cracked bushings
- Bullet holes or bullet damage
- Oil leaks
- Discolored areas caused by overheating
- Burnt smell
- Rusted areas

5.3.2 Operation of Switches or Breakers

Some transformers may contain a secondary circuit breaker (*Figure 22*) and/or a tap-changing switch. If the transformer contains a secondary circuit breaker or a tap-changing switch, check that they function freely.

5.3.3 Secondary Voltage Test

A line worker may have to check the secondary voltage of a transformer to verify if it is providing the correct voltage to the customer. This is done with a voltmeter placed across the secondary terminals X1, X2, and X3 (*Figure 23*). On a single-phase transformer, 120V should be measured between X1 or X3 and X2. Across X1 and X3, 240V should be measured.

Figure 21 Damaged transformer.

Figure 22 Transformer secondary circuit breaker handle.

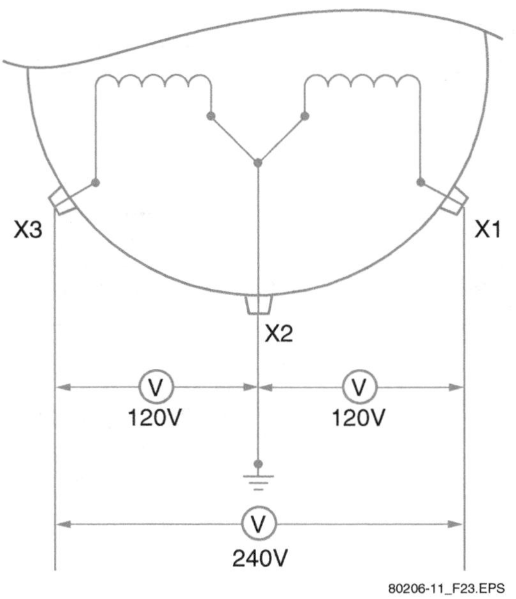

Figure 23 Measuring secondary voltage on a single-phase transformer.

WARNING!

When checking transformer secondary voltage, always wear the correct PPE, use a voltmeter rated for line work, and follow all safety procedures.

6.0.0 Locating and Correcting Faults in URD Systems

Because the cables are buried, locating and correcting problems in a URD system can be a challenge. Usually, customers who lose power call in to report an outage. As calls are compiled by operators or computer systems, the area being affected by the outage begins to materialize. But even when the outage area is identified, it can be time consuming to locate and determine the cause of the problem. Because URD cables are buried, it can take a long time to dig up the cable, make the repairs, and bury the cable again. As more and more URD systems are put into use, it is increasingly important for power line workers to know how to quickly locate faults and make the needed repairs so that power can be restored to customers. This section describes some common methods used to locate and correct transformer faults and cable faults in a loop-feed URD system.

6.1.0 Transformer Faults

As a general rule, transformer faults are easier to locate and identify than cable faults in a URD system. If all of the customers who report an outage are supplied by the same transformer, then the logical place to begin troubleshooting is at that transformer. Only the customers who are supplied from that transformer will be out of power. The remaining customers on the URD loop should still have power (*Figure 24*).

GOING GREEN

Transformer Recycling

When a transformer is taken from service, it can be rebuilt or recycled. When recycled, the oil is first drained. PCB oil must be disposed of in accordance with federal regulations and company policy. Used oil that does not contain PCBs can be refined to like-new purity for reuse. The steel, copper, and other valuable metals that are recovered can provide the utility with a cash stream that adds to their bottom line.

First, a visual inspection of the transformer's exterior should be done to look for any obvious signs of damage. These could include bulges in the transformer casing, burn marks, or leaking oil. Damage like this could be caused by high voltages and arcing related to a transformer fault. If no obvious damage is noticed, the transformer housing should be opened and the interior should be checked for damaged load break elbows, deformed bushings, and any other signs of damage.

If the transformer's fuse is blown, it might make sense to try simply replacing the fuse. Keep in mind, though, that company policies can dictate when and how a transformer is re-fused. A better approach is to first disconnect the secondary loads from the transformer before re-fusing. This is because a problem with one of the secondary loads could possibly cause the fuse to blow. If the transformer is re-fused and the fuse blows again, it is almost certain that the transformer is faulted.

If the transformer is re-fused and the fuse does not blow, then the problem could have been caused by a fault or overload on one of the secondary loads. If that is the case, it will be necessary to troubleshoot the secondaries to determine the cause of the problem before reconnecting them to the transformer.

Another situation that sometimes occurs in URD systems is when a riser pole fuse blows. If this occurs, troubleshooters do not know whether the fuse blew because of a transformer fault or a cable fault. If a riser pole fuse blows because of a transformer fault, there is typically some visual indication of a problem. This is because a transformer fault usually causes an enormous amount of heat to be generated before a riser pole fuse

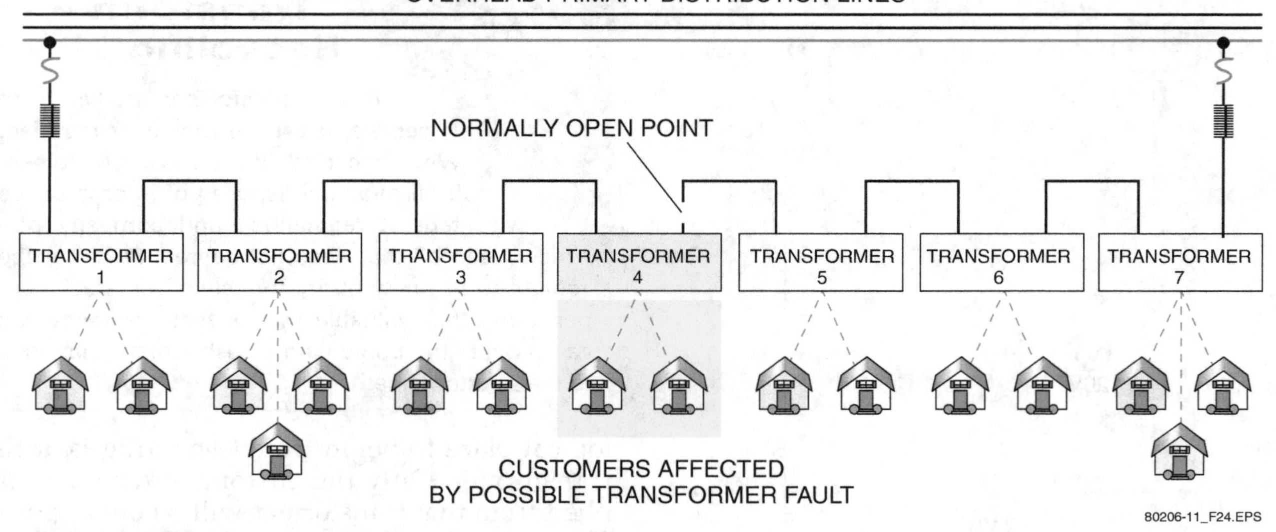

Figure 24 Loop-feed URD system with possible transformer fault.

blows. So, a logical approach is to check each transformer in the URD system first.

Do a visual inspection of each transformer. Check for bulged housings, burn marks caused by arcing, leaking oil, blown fuses, damaged load break elbows, deformed bushings, and any other obvious signs of damage. If an inspection of each transformer in the URD loop does not turn up any problems, the next logical step is to troubleshoot for a cable fault.

6.2.0 Cable Faults

Underground systems can be damaged in different ways. Damaged insulation that is exposed to moisture can break down under the stress of high voltage. Ground movement caused by frost heaving, vehicle traffic, and seismic events can cause breaks. One of the greatest causes of damage is digging in an area that has not had buried utilities clearly marked.

A cable failure interrupts the normal flow of electric current. URD cable faults can include electrical shorts, grounds, or open circuits. Ground faults are the most common form of cable fault. This section will focus on locating and correcting ground faults in a loop-feed URD system.

The most common cause of ground faults in a URD system is a break in the cable's insulation. Damaged cable insulation can occur for a number of reasons. Sometimes it simply deteriorates with age. In other cases, it is damaged by overvoltage conditions caused by lightning, or may have been compromised during installation. The insulation can also be damaged by machinery or digging tools.

If a ground fault occurs in a primary URD cable, it will usually cause a fuse, disconnect switch, or circuit breaker to de-energize part of the circuit. The response of the protective device prevents any further damage to the system. Customers who are serviced by that part of the circuit lose power. Before power can be restored, the fault has to be located and repaired.

Figure 25 Capacitor discharge unit (thumper).

Different methods can be used to locate a fault in a primary URD cable. One method involves isolating the suspected cable and then using a capacitor discharge unit (*Figure 25*) to apply a high test voltage to the cable. A capacitor discharge unit is sometimes called a thumper because the test voltage it applies to the cable causes an audible thump to occur underground at the fault. A pickup device and detector can be used to listen for the fault along the length of the cable.

A common URD cable fault situation is shown in *Figure 26*. This is a loop-feed URD system with a normally open point about midway through the loop. The two halves of the system are fed from the same overhead primary line. Riser pole fuses are used on both primary feeders. One riser pole fuse has blown. Each pad-mounted transformer in the system has a fault indicator. Three of the fault indicators just down from the blown riser pole fuse are indicating a fault. The fourth fault indicator in that section is not indicating a fault.

Based on the fault indicators, a logical starting point for troubleshooting would be the section of primary cable between the last tripped fault indicator and the first non-tripped indicator. That section of cable should be de-energized and isolated so that it can be tested. Then, a capacitor discharge unit is attached to one end of the cable so that two tests can be performed.

To verify that the cable is actually faulted, the capacitor discharge unit is set to do a high potential (hi-pot) test. This test involves applying a high test voltage to see if the cable can withstand the voltage. The voltage level should be set for slightly less than the cable's rated voltage to avoid doing any further damage. If the cable is faulted, it fails to conduct the applied voltage before the voltage reaches the set level.

If the cable is faulted, a capacitance discharge test is done to locate the fault. This test applies pulses of high voltage to the cable. The test voltage is discharged to ground at the cable fault. Sometimes, the thumps caused by the energy releases can be felt and heard at the site of the fault. If not, a pickup device and detector can be used to listen along the length of the cable until the exact location of the thump can be heard.

Once the cable fault is located, line workers can replace the section of cable or cut out the fault and splice together the two ends of the cable using an appropriate splicing method.

Another method that can be used to locate cable faults involves the use of a radar cable fault locator. This type of device is known as a time domain reflectometer, or TDR (*Figure 27*). This method basically involves sending low voltage signals through the faulted cable and interpreting a visual display that is based on reflections back to the fault locator. The graph, or trace, that is displayed provides a visual representation of things like splices, transformers, taps, and opens in the cable section. By carefully analyzing the graph, it is possible to identify and locate faults in the cable.

Often, a cable fault locator is used in conjunction with a capacitor discharge unit. The cable fault locator is used to narrow down the area of cable that is faulted, and the capacitor discharge unit is used to pinpoint the fault location.

Faults also occur in secondary URD cable. The procedures for locating a fault in a secondary URD cable are similar to those used for primary cable fault location, but with much lower voltages involved. A transmitter is used to pulse a test voltage along an isolated section of secondary cable that is suspected of having a fault. Then, a

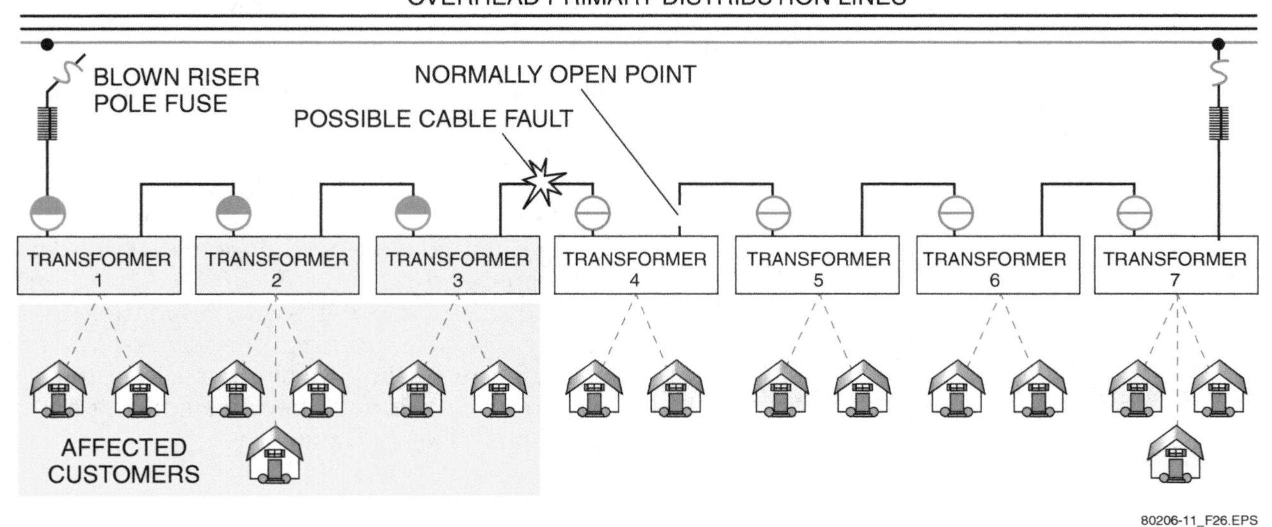

Figure 26 Fault indications in a loop-feed URD system.

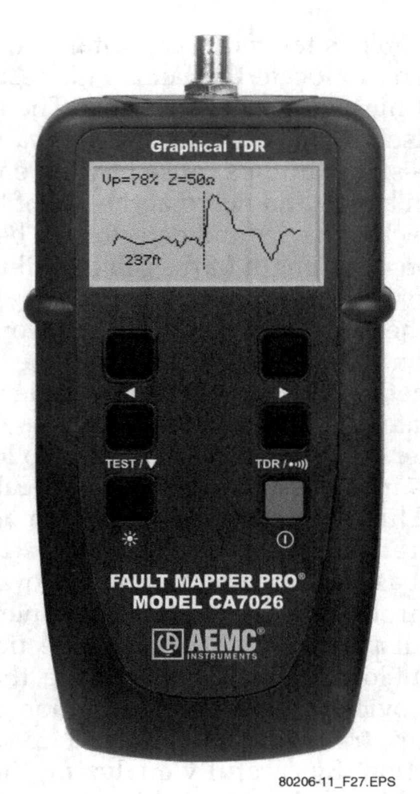

80206-11_F27.EPS

Figure 27 Time domain reflectometer (TDR).

pickup and detector are used to pinpoint the location of the fault. With this type of system, the detector measures changes in the voltage along the cable. A significant change indicates the location of the fault. For instance, the detector might measure little or no voltage along the cable until it nears the fault. Then, the voltage level increases significantly until the detector passes directly over the fault. At that point, the measured voltage drops back to zero. Then, on the other side of the fault, the voltage level increases significantly again before tapering off as the detector moves farther away from the fault. The area between the two high voltage readings, where the detector measured no voltage, is where the fault is located. The secondary cable can be dug up at that location and repaired as needed.

6.3.0 Repairing Damaged Underground Systems

Once the affected cable is located, the next step is to de-energize the affected circuit. Care must be taken when digging up the damaged cable. Other utilities in the area should be located and marked before digging starts. Once the cable is exposed, make sure it is the right cable and that it is truly de-energized. When working with primary cable, this is generally done with a cable-spiking tool.

On Site

Before You Dig

Always have underground utilities identified and marked before you dig. Not only is it the smart thing to do, it is required by law in most states. If you dig and damage underground utilities, you can be fined and may be subject to legal action for any damage you cause. Never dig until you are 100 percent sure that any and all underground utilities have been marked.

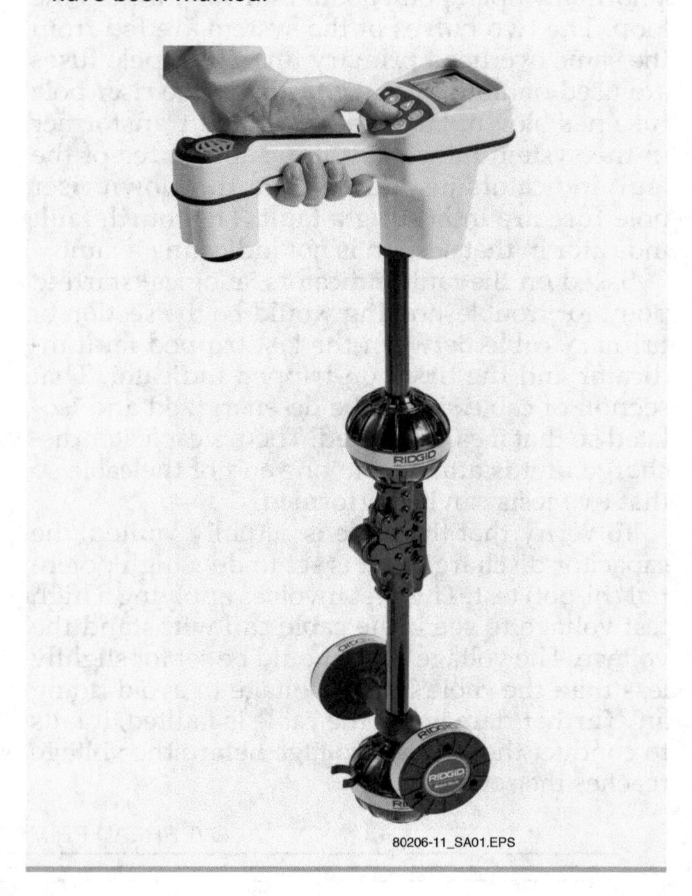

80206-11_SA01.EPS

This tool contains a spike that penetrates the cable and comes into contact with a conductor. If the circuit is still live, there will be a sudden release of energy in the form of a fault current.

Manual spiking tools are still available, but are being replaced with powder-actuated or hydraulically operated versions (*Figure 28*). These versions are safer because the worker operates the tool from a distance, pulling on a rope or lanyard to activate the spiking mechanism. Keeping a safe distance is important because spiking an energized cable is very dangerous.

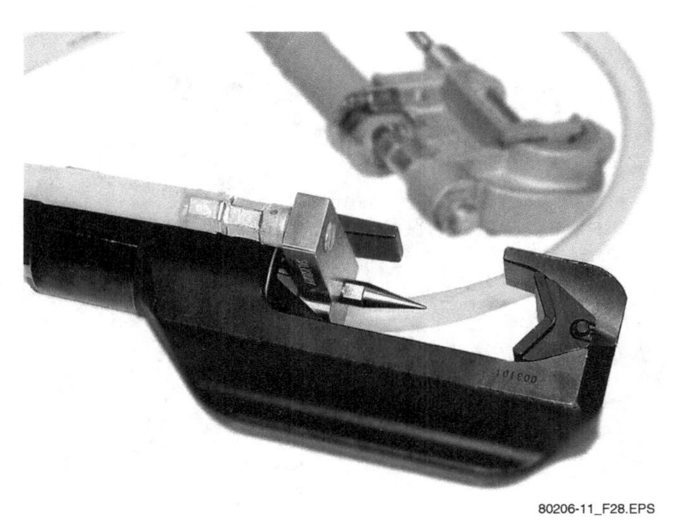

80206-11_F28.EPS

Figure 28 Hydraulic cable-spiking tool.

Once the cable has been verified, the defective section can be cut out. This should be done with a grounded remote cutting tool.

NOTE

Powder-actuated and hydraulically operated cable-cutting tools can be fitted with a spike point so they can be used to spike cables.

WARNING!

Only trained and authorized persons can operate powder-actuated tools. Cable-spiking tools should only be used by individuals who have carefully read the tool manufacturer's safety procedures and operating instructions and have received training in the use of the tool.

Severe damage to an underground cable may require that the cable be replaced. If damage is not too severe, the cable can be repaired using a splice kit. Splice kits are available to repair two-conductor concentric cable and triplex cable. Crimp-on and setscrew type connectors are used to join the conductors. Various methods are used to seal the splice to keep out water. Shrink-wrap tubing is widely used. Some splice kits encapsulate the splice in epoxy for a watertight seal. The epoxy is mixed and poured into a small capsule around the splice where it cures (*Figure 29*). When applying a splice kit, use the components that come with the kit and follow the installation instructions. Failure to do so can result in a failed splice. After the repair is made, prepare the trench to receive the cable for burial as if you were burying a new cable. Firmly tamp the soil over the buried cable.

80206-11_F29.EPS

Figure 29 Encapsulated cable splice.

NOTE

It is very important that underground splice repairs be made watertight. Failure to do this will cause the splice to corrode and fail over time.

7.0.0 Load Management Overview

The two primary goals in power distribution are to make sure there is sufficient, affordable, uninterrupted power to meet current demand, and to make sure the voltage levels are stable. Demand management is done by system operators who monitor demand and adjust the supply to satisfy the demand. This can involve bringing additional generating capacity on line or transferring surplus power back to the power grid. Voltage regulation equipment is used to fine-tune the supply voltage at substations and distribution points in order to ensure that the voltage delivered to customers remains within the specified tolerance.

All electrical equipment is designed to operate within defined voltage limits. Operating outside those limits for an extended period can result in inefficient operation and, ultimately, damage to the equipment. Low voltage can cause motors to overheat; high voltage can cause premature failure of appliances and lighting.

Engineering studies have shown that the voltage supplied to electronic equipment such as

On Site

Home Improvements

Homeowners may inadvertently damage the insulation on buried cables while doing yard improvement projects such as installing fencing or setting piers for a deck. It may take time for a fault to appear, as it will take a while for moisture to penetrate to the conductors. The fault could affect only one leg of the supply voltage, so the homeowner will experience a partial loss of power. If a trouble call involves only one customer, it is a good idea to ask about any recent improvement projects.

On Site

Peaker Plants

Some power plants, called peaker plants, operate only during peak power demand periods. Such periods typically occur in the afternoon during hot summer months when air conditioners are cranked up. This type of plant typically is a gas turbine plant that burns natural gas. There are also some that burn diesel oil. A peaker plant generally operates for only a few hours a day and may be offline for extended periods.

computers must not fluctuate more than 5 percent because the equipment is very sensitive to voltage variations. For these reasons, power distribution systems are designed to maintain the voltage supplied to customers at a constant level that remains within that 5 percent tolerance, or 114 to 126 volts.

Voltage regulation is generally accomplished at the substation level using tap-changing transformers and voltage regulators. However, on long runs, and in areas where the load is likely to vary significantly, voltage regulators, pole-mounted tap-changing transformers, and capacitor banks may be used to compensate for voltage fluctuations and minor disruptions.

7.1.0 Voltage Regulators

Power line voltage regulators can be used to extend the length of a distribution line by compensating for the losses created by long feeder lines. A voltage regulator responds to load changes by sensing voltage variations and raising or lowering line voltage to compensate. An increased load means greater impedance, which results in a higher voltage drop on that part of the circuit. This reduces the voltage available on the feeder line. Sensors detect the increased load and a motor operates mechanical contacts to change taps to different transformer windings. This changes the turns ratio of the transformer, thereby increasing the output voltage. When the load decreases, the motor repositions the taps to reduce the output voltage. Step-voltage regulators use a tap-changing rotary switch to adjust the output voltage. They are designed to adjust the voltage in increments of ⅝ of a percent over a ±10 percent range. They use bridging resistors to maintain circuit continuity during the switch.

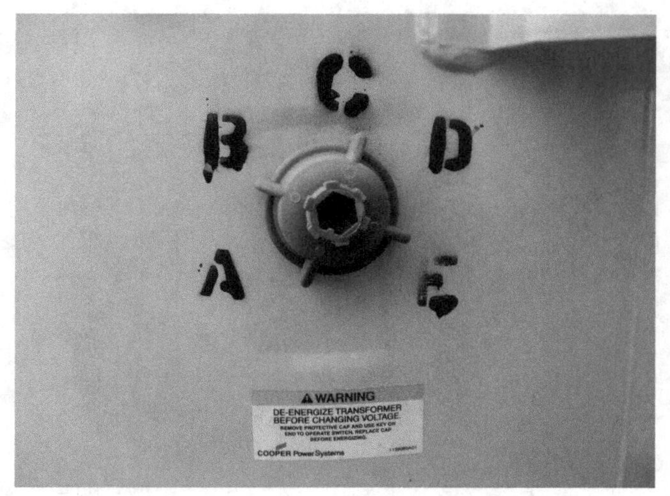

80206-11_F30.EPS

Figure 30 Secondary tap-changing switch.

7.2.0 Tap-Changing Transformers

Substation transformers operate automatically to adjust the load using switchable taps controlled by voltage-sensing equipment. Some pole-mounted transformers have secondary tap-changing switches (*Figure 30*) installed that allow raising or lowering the output voltage manually. Normally, the taps are set at installation before energizing the transformer. If load requirements change, the transformer must be de-energized in order to change the taps. A primary tap changer is used to change the voltage in the substation transformer to meet system requirements. A secondary tap changer matches the transformer output to the customer need.

The standard for the secondary taps in transformers used in industrial systems is two 2½-percent taps above and below rated voltage, for a total of five taps. A decal attached to the transformer shows the output for each tap position.

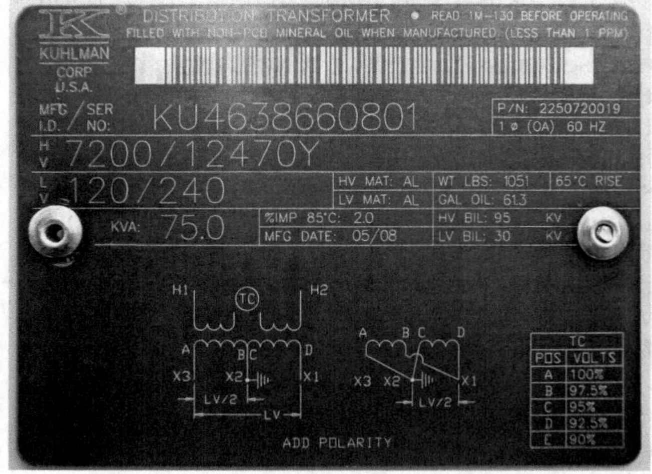

80206-11_F31.EPS

Figure 31 Transformer decal showing tap percentages.

Some decals show the output amps and voltage for each tap position, as well as the contacts used. Other decals are little more than a simple chart like that in *Figure 31*, which shows the percentage of change from each tap position.

Tap changing must be carried out with the transformer disconnected from the circuit.

While all tap adjusters are similar, there are some variations among manufacturers. Always follow the manufacturer's instructions to adjust a tap. Some have locking screws that must be loosened; others simply require turning a handle. Tap changers on the side of pole-mounted transformers require the use of a hot stick to change the tap position. Some handles have a fitting that permits the line worker to lock the handle in position.

Line voltages change during the course of the day depending on loads. Changing a tap may not always be the best option. If a tap is changed when the feeder voltage is at a low value, the voltage to the customer's premises may be too high when the line voltage returns to normal. In that situation, it is better to install a voltage regulator on the pole or at the substation.

If the change is due to increased electrical load variation at the customer's premises, it may be necessary to change out the transformer.

7.2.1 Computing a Tap Change

A percentage of change charts require the line worker to perform math calculations to determine the correct setting. To determine the amount to raise the voltage, divide the present output by the desired output. The answer is the percentage the voltage has to be raised. If a customer requires 240VAC, but is only getting 224VAC, divide 224 by 240. Multiply the answer by 100 to get a percentage of 93.3 percent. Adjust the tap switch to the position closest to that percentage.

To lower the output, divide the desired output by the present output. If the customer is receiving 249VAC, but needs 240VAC, divide 240 by 249. Multiply the answer by 100 to get a percentage of 96.4 percent. Move the tap switch to the closest position to that percentage.

7.3.0 Capacitor Banks

Distribution line capacitor banks are used to compensate for the current lag created by inductive loads such as motors. They are often installed close to inductive loads to prevent the loads from affecting other services on the distribution line. Capacitor banks are installed to reduce line losses, provide additional capacity when needed, and improve operating efficiency.

80206-11_F32.EPS

Figure 32 Capacitor bank.

Capacitor banks (*Figure 32*) can be automatically or manually switched on and off. They may be switched on during hot periods to compensate for increased use of air conditioning equipment, which uses AC induction motors.

One of the main functions of the capacitor bank is to provide power factor (VAR) correction. The concept of power factor was introduced in the *Alternating Current and Three-Phase Systems* module. Power factor is the ratio of true power to apparent power. Inductive loads, such as AC motors, reduce the power factor because the current is out of phase with the voltage. A circuit with a low power factor uses more current than a circuit with a higher power factor. Capacitors are used to obtain power factor correction. The capacitors supply reactive power that is opposite that of the inductive loads, thus canceling the effect of the inductive load. The capacitor bank contains a current transformer that senses the current and switches capacitors into the circuit in steps to maintain the power factor at the desired level.

8.0.0 Protective Device Coordination

Protective devices such as circuit breakers, fuses, cutouts, and reclosers are installed at key points in a distribution line. Protective devices are designed to carry normal line current, and to interrupt fault currents that exceed their interrupting ratings. *Figure 33* is a highly simplified example of a distribution system with protective devices installed. Note that the interrupting ratings of the devices increase as the distribution line moves toward the substation, and that the fuses protecting the load circuits are time-delay (T) devices. This reflects the basic idea of fuse coordination, which is to have the protective device closest to the fault

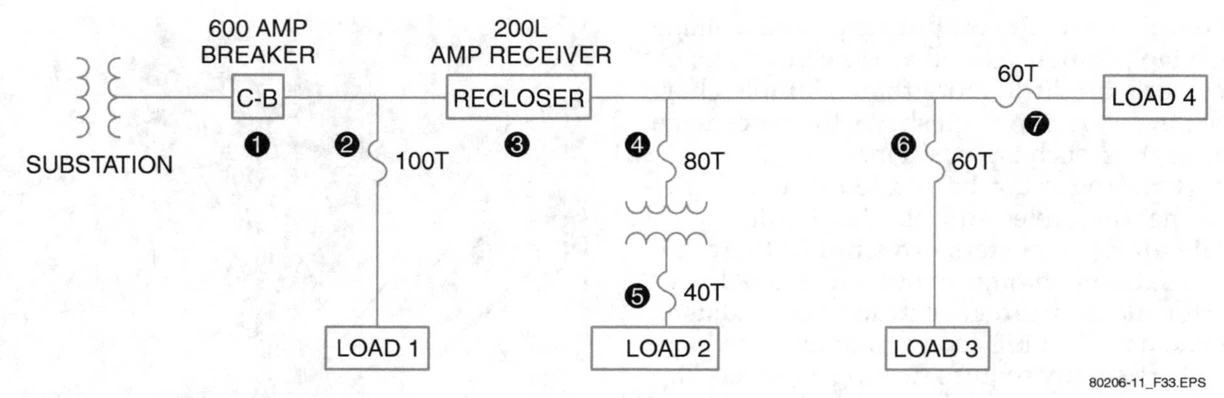

Figure 33 Example of protective device coordination.

trip early so that power to customers upstream of a fault is not disrupted.

Distribution systems are designed to keep power disruptions to the smallest number of customers. If for example, Load 4 were to draw excessive fault current, Fuse 5 would melt and open after a short time delay to allow the fault to clear itself. Thus, the only customer affected would be Load 4. If a major fault, such as a phase-to-phase or phase-to-ground short, were to occur, it would be reflected back up the line and cause the recloser to open. After a delay of milliseconds to allow the fault to clear itself, the recloser would close again and remain closed for a short period. If the fault did not clear in that interval, the recloser would open again and remain open. The substation circuit breaker (CB1) is the last line of defense. Its purpose is to protect the substation in case a major fault event overwhelms the downstream protective devices.

This has been only one simplified example of protective device coordination. Coordination can be accomplished in a variety of ways.

A strategy used in protective device coordination is to select fuse ratings so that a device such as a fused cutout actually trips when a fault occurs. Such a trip would be visible from the ground. If the fuses have too high a trip rating, a recloser might lock open before a fuse blows, taking many customers off line. In addition, the troubleshooter would spend a lot of time trying to track down the problem due to a lack of visual clues.

NOTE

Fuses subject to fault currents can be damaged, even though they do not open. Each occurrence wears down the fuse and will reduce its current-handling capacity. Fuses may need to be replaced if they have been subjected to fault currents.

SUMMARY

Power poles, conductors, and pole-mounted equipment are subject to damage from a variety of sources, including bad weather, vehicle accidents, and vandalism. Wood poles may decay over time, or may be damaged by birds or insects. Lightning strikes and power surges can damage conductors, transformers, and other components. For all these reasons, it is important to check the equipment whenever the opportunity presents itself. Problems such as leaking transformers, damaged insulators, loose or broken terminations, and blown fuses can be identified by a visual inspection from the ground using binoculars. Overhead and underground distribution systems require repair when damaged. Damage to overhead lines is often visible and easily repaired. Underground systems can be damaged in different ways. One of the greatest causes of damage is digging in an area that has not had buried utilities clearly marked. Hidden underground cable faults can be located with special equipment designed for that purpose. Once located, the cable must be isolated from the power source. Generally, the damaged section of cable is removed and a new section is added using approved splices.

Utility companies use pole-mounted voltage regulators, tap-changing transformers, and capacitor banks to regulate voltage on feeder lines in order to ensure that customers receive the correct supply voltage. Pole-mounted voltage regulators and capacitor banks operate automatically, but line workers may occasionally be required to change the settings on tap-changing transformers. Coordination of protective devices is an important part of distribution line design. The interrupting ratings of protective devices are selected in order to ensure that power interruptions are limited to the fewest customers.

Review Questions

1. Which step should follow installation of a cluster bar when installing protective grounds?
   a. Connect the cluster bar to earth ground or pole bond.
   b. Connect the neutral to the cluster bar.
   c. Connect the cluster bar to the nearest phase conductor.
   d. Connect the neutral to the nearest phase conductor.

2. Which of the following is true about static electricity and induced voltage?
   a. Static electricity is always a constant voltage.
   b. Static electricity is weaker when the parallel conductors are closer.
   c. Induced voltage is caused by nearby power lines.
   d. Induced voltage will vary with environmental conditions.

3. Which of these is a useful accessory when inspecting poles and pole-mounted equipment?
   a. Multimeter
   b. Ladder
   c. Helicopter
   d. Binoculars

4. When stabbing a wooden pole with a screwdriver, what determines if the pole is good?
   a. The screwdriver encounters resistance and pierces the pole.
   b. The screwdriver encounters resistance and does not pierce the pole.
   c. The screwdriver pierces the pole and a cloud of insects appear.
   d. The screwdriver encounters resistance and a cloud of insects appear.

5. The preferred method of repairing a slack guy wire without a turnbuckle is to _____.
   a. remove and replace the entire guy wire
   b. install a turnbuckle and tighten the guy wire
   c. use a come-along to tighten the guy wire
   d. install new anchors in the ground as in a new guy wire installation

6. The black deposits sometimes found on a conductor tie or on the conductor are caused by _____.
   a. corrosion
   b. abrasion damage
   c. vibration damage
   d. uncleaned splice damage

7. The purpose of using a dynamometer when installing conductors is to determine the _____.
   a. extent of abrasion damage
   b. strength of a full tension splice
   c. correct amount of sag
   d. amount of vibration on a conductor

8. Aeolian vibration is caused by _____.
   a. excessive sag
   b. wind
   c. corrosion
   d. loose insulators

9. The color in a thermovision camera that indicates the highest temperature is _____.
   a. white
   b. red
   c. yellow
   d. orange

10. The purpose of the beacon on electronic reclosers is to _____.
    a. indicate excessive operations
    b. identify a locked-out recloser
    c. provide a warning to aircraft
    d. indicate a low oil condition

11. The health hazard from corona discharge is _____.
    a. skin cancer
    b. eye problems
    c. heart problems
    d. respiratory problems

12. When checking a single-phase transformer, approximately 240 volts should be measured across _____ .
    a. X1 and X2
    b. X2 and X3
    c. X1 and X3
    d. X3 and X4

13. How does increased impedance affect load regulation?
    a. The current flow to the customer increases.
    b. The current flow in the distribution line increases.
    c. The voltage drop at the customer increases.
    d. The voltage drop in the distribution line increases.

14. If a customer is receiving 255 volts, by what amount should the voltage be lowered in order to drop it to 240 volts?
    a. 94.1 percent
    b. 91.4 percent
    c. 96.4 percent
    d. 94.6 percent

15. The purpose of adding capacitors across a load is to _____.
    a. adjust the inductive reactance
    b. have reactive power one-half to the capacitor value
    c. improve the power factor
    d. increase the length of the distribution line

Trade Terms Introduced in This Module

Aeolian vibration: When air flows across a conductor, it creates alternating high and low pressures above and below the conductor. The alternating pressures cause the conductor to vibrate at right angles to the airflow.

Dynamometer: A device for measuring force, torque, or power. In the power industry, it measures the force or tension on a conductor during installation so that line workers can determine the correct amount of sag.

Thermovision: Thermovision is a camera used in thermography. It is the measurement of a temperature by recording the emission of infrared radiation from the area or object. It is a non-contact and non-destructive testing method that permits equipment evaluation without interrupting the process being tested.

Additional Resources

This module presents thorough resources for task training. The following resource material is suggested for further study.

Electric Power System Basics for the Non-Electrical Professional. Steven W. Blume. Hoboken, NJ: Wiley IEEE Press.

Electrical Machines, Drives and Power Systems, 6th Edition. Theodore Wildi. Upper Saddle River, NJ: Prentice Hall.

Figure Credits

Topaz Publications, Inc., Module opener, Figures 1–3, 5–8, 11, 12, 15, 18–22, and 30–32

Salisbury Electrical Safety, Figures 4 and 28

Hubbell Power Systems, Figures 9, 16, and 17

Reelstrong International, Figure 10

Photo, Courtesy of Preformed Lines Products, Figure 13

FLIR Systems, Inc., Figure 14

RIDGID®, SA01

High Voltage, Inc., Figure 25

Photograph provided by AEMC® Instruments, Figure 27

Uraseal, Inc., Figure 29

NCCER CURRICULA — USER UPDATE

NCCER makes every effort to keep its textbooks up-to-date and free of technical errors. We appreciate your help in this process. If you find an error, a typographical mistake, or an inaccuracy in NCCER's curricula, please fill out this form (or a photocopy), or complete the online form at **www.nccer.org/olf**. Be sure to include the exact module ID number, page number, a detailed description, and your recommended correction. Your input will be brought to the attention of the Authoring Team. Thank you for your assistance.

Instructors – If you have an idea for improving this textbook, or have found that additional materials were necessary to teach this module effectively, please let us know so that we may present your suggestions to the Authoring Team.

NCCER Product Development and Revision

13614 Progress Blvd., Alachua, FL 32615

Email: curriculum@nccer.org
Online: www.nccer.org/olf

❑ Trainee Guide ❑ AIG ❑ Exam ❑ PowerPoints Other ________________________________

Craft / Level: Copyright Date:

Module ID Number / Title:

Section Number(s):

Description:

Recommended Correction:

Your Name:

Address:

Email: Phone:

Glossary

Additive polarity: Internal construction of a transformer that allows current in two adjacent primary and secondary terminals to flow in opposite directions. It can be detected if the voltage measured across unconnected high- and low-voltage transformer terminals is greater than the voltage applied across the high-voltage terminals.

Aeolian vibration: When air flows across a conductor, it creates alternating high and low pressures above and below the conductor. The alternating pressures cause the conductor to vibrate at right angles to the airflow.

Air break switch: A switch used in power distribution systems that can be opened while the line is energized. Arc horns on the switch help break the arc that forms when the energized switch is opened.

Aluminum conductor steel reinforced (ACSR): A widely used aluminum conductor that has an alumoweld steel core for added strength.

Alumoweld: Steel wire that is coated with aluminum to increase conductivity and prevent rust.

Amorphous metal: A metal with a noncrystalline structure that gives it unique magnetic properties. When this metal is used in the core of a transformer, losses are reduced up to 60 percent over laminated steel cores.

Ampacity: The maximum amount of current that a cable can carry before it begins to deteriorate. Sometimes called current rating or current-carrying capacity.

Armor rods: Spiraled metal rods that are placed around a conductor on both sides of the insulator. They are used to stiffen the conductor to prevent bending and flexing. They are often called line guards.

Basic impulse level (BIL): A measure of a transformer's ability to withstand a current surge such as what might occur when lightning strikes.

Bushing: A terminal on a transformer where high- and low-voltage conductors are connected. Bushings are often made of porcelain.

Capacitance: The storage of electricity in a capacitor; capacitance produces an opposition to voltage change. The unit of measurement for capacitance is the farad (F) or microfarad (μF).

Common trench (joint trench): A trench in which multiple utilities, such as electrical cable, gas, communications, and cable TV, lines are installed.

Completely self-protected (CSP) transformer: A transformer with built-in secondary current overload protection. The type of protective equipment that is standard on a CSP transformer has to be added on to other transformers. This type of transformer is being replaced with conventional transformers.

Conductor splice: A type of conductor repair used to restore conductivity. It does not add much strength to the conductor.

Copperweld: Steel wire that is coated with copper to increase conductivity and prevent rust.

Crimp-on splice: A type of conductor repair that is squeezed (crimped) around the conductor with a tool. Crimp-on splices can restore conductivity and/or strengthen the conductor.

Dead front: A switchgear or transformer configuration in which energized components are shielded in some manner so that they are not exposed and accessible when users open access doors or panels.

Disconnect switch: A type of air break switch that is used to isolate a component. A disconnect must be opened with the line de-energized. For that reason, it does not contain arc horns.

Double feed: A type of URD system design in which two separate URD systems are connected via switchgear to a single customer who cannot tolerate being without electric power. If a problem occurs in one URD system, the switchgear automatically transfers the source of power from the failed system to the other system.

Drip loop: Cables formed in a loop as they enter the service head. Their purpose is to allow water to drip from the cables and not enter the service head and conduit in the meter loop.

Dual-rated: A type of electrical terminal or connector that is designed to be used on both aluminum and copper conductors.

Dynamometer: A device for measuring force, torque, or power. In the power industry, it measures the force or tension on a conductor during installation so that line workers can determine the correct amount of sag.

Faulted circuit indicator: A device that provides line workers with a visual indication that a fault has occurred in a line.

Feeder voltage regulator: A regulator used to maintain a constant voltage in distribution feeder circuits even as loads change. A motor-driven tap-changing switch in the secondary windings is used to make changes as small as ⅝-percent increments. Sometimes called a step-voltage regulator.

Frequency: The number of cycles an alternating electric current, sound wave, or vibrating object undergoes per second.

Full-tension splice: A type of splice meant to restore strength and conductivity to a conductor. Crimp-on and non-crimp styles are available.

Fused cutout: A type of switch that contains a fuse cartridge.

Fusing ratio: A ratio that reflects the relationship between the current-carrying capacity of a fuse and the transformer full-load current.

Hertz (Hz): A unit of frequency; one hertz equals one cycle per second.

Impedance: The opposition to current flow in an AC circuit; impedance includes resistance (R), capacitive reactance (X_C), and inductive reactance (X_L). Impedance is measured in ohms.

Inductance: The creation of a voltage due to a time-varying current; also, the opposition to current change, causing current changes to lag behind voltage changes. The unit of measure for inductance is the henry (H).

Lightning arrester: A type of surge arrester that is typically installed on overhead primary feeder lines and at pad-mounted switchgear and transformers in a URD system to protect system components from damage caused by lightning-induced current surges.

Live front: A switchgear or transformer configuration in which energized components are exposed and accessible when users open access doors or panels.

Loading districts: Areas of the United States established by the Bureau of Standards for the purpose of calculating snow and ice loads on power line conductors. Loading areas are designated light, medium, and heavy.

Loop feed: A type of URD system design in which both ends of the primary feeder line are connected to the same phase of the overhead primary, thereby forming a loop.

Loop-fed distribution system: A power distribution system where power is available to a transformer through different routes or directions. If a system fault occurs, power can be easily rerouted to the transformer to restore power.

Luminaire: A lighting unit consisting of a lamp, together with parts designed to distribute the light, protect lamps, and connect them to the power source.

Mast arm: The structure that extends the luminaire out from pole and serves as a conduit for the wiring that powers the lamp.

Messenger: A line that serves as a support for conductors. In a service drop, the neutral conductor serves as the messenger.

Metal-oxide varistor (MOV): An electronic component in which the resistance to current flow drops as the voltage applied to it increases. Also, a resistive device commonly used in lightning arrestors. The varistor offers a high resistance to normal system current and voltage. When a surge occurs, the resistance of the varistor breaks down at a preset voltage and conducts the surge to ground. Once the surge passes, the varistor returns to a high value.

Meter loop: The part of a customer's electrical service consisting of the service head, service mast, conductors, conduit, meter base, and service equipment disconnect. It is the point at which the utility delivers power to the customer.

Micro: Prefix designating one-millionth of a unit. For example, one microfarad is one-millionth of a farad.

Oil switch: A type of disconnect switch used to open energized lines. Switch contacts are immersed in oil to suppress the arc that forms when the switch opens. Oil switches can be opened manually or automatically.

Peak voltage: The peak value of a sinusoidally varying (cyclical) voltage or current is equal to the root-mean-square (rms) value multiplied by the square root of two (1.414). AC voltages are usually expressed as rms values; that is, 120 volts, 208 volts, 240 volts, 277 volts, 480 volts, etc., are all rms values. The peak voltage, however, differs. For example, the peak value of 120 volts (rms) is actually $120 \times 1.414 = 169.68$ volts.

Puller/tensioner: A dual-purpose pulling machine that can be used as either a puller or a tensioner.

Quadruplex: A four-conductor cable used in three-phase service drops. For overhead service, the bare (neutral) conductor acts as the messenger to support the cable. For underground services, the neutral conductor is insulated and is identified with a different color or special markings.

Radial distribution system: A power distribution system with a central power source. Lines radiate from the power source like spokes of a wheel. If the central power source fails, all customers in the system lose power.

Radial feed: A type of URD system design in which the primary feeder line is connected to an overhead primary distribution line at only one end. The other end of the system is terminated.

Reactance: The imaginary part of impedance. Also, the opposition to alternating current (AC) due to capacitance (X_C) and/or inductance (X_L).

Recloser: An overload device that operates like a circuit breaker. It resets itself after a preset time when a fault occurs. A magnetic solenoid in the recloser opens contacts to stop current flow if fault current occurs. Once current flow stops, the solenoid de-energizes to restore current flow. If the fault is still present, the recloser opens again. If the fault is not temporary, the recloser locks the circuit open after a predetermined number of reset attempts.

Repair sleeve: A type of crimp-on conductor repair that can be applied over a conductor without having to slip it on over the end of a conductor. They are available as two halves, or with a slit and are placed over the damaged area.

Resonance: A condition reached in an electrical circuit when the inductive reactance neutralizes the capacitance reactance, leaving ohmic resistance as the only opposition to the flow of current.

Rigid metal conduit (RMC): A thick-walled steel conduit that is the preferred material for conduit and service masts used in meter loops.

Riser: The pole on which overhead power distribution transitions to underground power distribution.

Root-mean-square (rms): The square root of the average of the square of the function taken throughout the period. The rms value of a sinusoidally varying voltage or current is the effective value of the voltage or current.

Secondary pedestal: A junction point in an underground secondary distribution system. Secondary power is fed to the secondary pedestal from the transformer. From there, power is fed to additional customers.

Sectionalizer: A type of switch that is used with reclosers to isolate faults in distribution circuits. If a fault continues to occur that results in many openings and resets of a recloser, the sectionalizer activates to isolate (sectionalize) the faulty section of line. Sectionalizers must be reset manually.

Self-inductance: A magnetic field induced in the conductor carrying the current.

Self-protected (SP) transformer: A transformer that has limited protective features such as a lightning arrestor.

Service drop: The cables going between a utility pole and the customer's electrical service. For a residential service, a service drop consists of two insulated conductors and one bare conductor that serves as the neutral and the messenger.

Shell: The metal housing containing the transformer. Sometimes called the tank.

Shell-type construction: A type of transformer construction in which the core surrounds the transformer windings instead of the coils surrounding the core.

Splice shunt: A repair meant to restore conductivity to an existing splice in a conductor. It is applied over the existing splice to form a parallel path (shunt) for current flow.

Subtractive polarity: Internal construction of a transformer that allows current in two adjacent primary and secondary terminals to flow in the same direction. It can be detected if the voltage measured across unconnected high- and low-voltage transformer terminals is less than the voltage applied across the high-voltage terminals.

Tensile strength: The resistance of material to being pulled apart. Conductors must have enough tensile strength to withstand the various forces placed on them.

Tension stringing method: A method used to string conductors between poles. The conductor is pulled under tension over stringing blocks on each pole. The method keeps conductors off the ground to avoid damage. It is the preferred method for installing new conductors.

Thermovision: The measurement of a temperature by recording the emission of infrared radiation from the area or object. It is a non-contact and non-destructive testing method that permits equipment evaluation without interrupting the process being tested.

Transformer taps: Points on a transformer's windings that can be connected to vary voltage. The taps increase or decrease the number of windings available in the transformer, thus varying voltage. Taps can be in the primary or secondary windings.

U-guard: A type of conduit used on risers. It is attached directly to the pole without the use of standoffs that are required with standard conduit.

Western Union tie: A popular method used to tie conductors to insulators using a standard tie.

Index

A

Abrasion, (Module 6): 6
AC. *See* Alternating current
Accidents, (Module 1): 31, (Module 2): 21, (Module 3): 20. *See also* Fatalities; Shock, electrical
ACSR. *See* Aluminum conductor steel reinforced
Aeolian vibration, (Module 6): 8, 26
Aerial distribution. *See* Power distribution, overhead
Air, (Module 1): 14, 22, (Module 4): 19, (Module 5): 6
Air conditioners, (Module 1): 22, (Module 4): 1, (Module 6): 20
Alternating current (AC)
- capacitance in circuits, (Module 1): 12–17
- inductance in circuits, (Module 1): 9–12, 13
- overview, (Module 1): 1
- phase relationships, (Module 1): 6–7, 8, 13, 18, 25, 26
- power distribution, (Module 1): 1, 2
- power in circuits, (Module 1): 19–22, 23
- resistance in circuits, (Module 1): 7–9
- RL, RC, LC, and RLC circuits, (Module 1): 17–19, 21
- sine wave of an, (Module 1): 1–6
- symbol, (Module 1): 8
- *vs.* direct current, (Module 1): 1, 2, 5

Aluminum
- in cable, (Module 4): 8, 25, 26, (Module 5): 14. *See also* Aluminum conductor steel reinforced
- cable connector, (Module 4): 10
- characteristics, (Module 3): 2, 3
- conductor tie, (Module 3): 12
- purity, (Module 3): 1
- transformer terminal, (Module 2): 8, 9, (Module 4): 26

Aluminum conductor steel reinforced (ACSR), (Module 3): 2, 3, 5–6, 17, 24
Aluminum oxide, (Module 2): 9, (Module 3): 17–18, (Module 4): 25, 26, (Module 5): 14
Alumoweld, (Module 3): 2, 24
American National Standards Institute (ANSI), (Module 2): 1, 17, (Module 3): 1, (Module 5): 1
American Wire Gauge (AWG), (Module 3): 3, 4, 12, 15, (Module 4): 8
Ammeter, (Module 1): 28
Amorphous metal, (Module 2): 5, 8, 30
Ampacity (current rating; current-carrying capacity), (Module 2): 14, (Module 4): 8, 30
Ampere, (Module 1): 9, 28, (Module 5): 7
Ampere-turn ratio, (Module 1): 28
Amplitude, (Module 1): 4, 5
Anchor, guy, (Module 6): 5, 6
ANSI. *See* American National Standards Institute
ANSI Standard Z87.1, (Module 2): 1, (Module 3): 1, (Module 5): 1
ANSI Standard Z89.1, (Module 2): 1, (Module 3): 1, (Module 5): 1
Antioxidant, (Module 4): 25, 26, (Module 5): 14. *See also* Oxide inhibitor
Appliances, (Module 2): 15, (Module 4): 1, (Module 6): 19–20
Arcing, electrical, (Module 2): 16, 21, (Module 3): 17, (Module 4): 19, (Module 6): 15, 16
Arresters, (Module 2): 6. *See also* Lightning/surge arresters
ASTM B230, (Module 3): 5–6
ASTM B232, (Module 3): 5–6
ASTM B498, (Module 3): 5–6
ASTM F711, (Module 2): 21
ASTM International (ASTM), (Module 2): 21, (Module 3): 5–6
Auto-boost regulator, (Module 2): 18. *See also* Voltage regulator
Average value, of a sine wave, (Module 1): 5–6

B

Backfeed, (Module 5): 14, (Module 6): 1
Backfill. *See* Trenching and backfilling methods
Backhoe, (Module 4): 6
Backstrap, for disconnect switch, (Module 2): 18, 19
Ballast, electronic, (Module 1): 28
Banding tool, (Module 3): 11
Bands, on Kellems grip, (Module 3): 9, 11
Banks
- capacitor, (Module 2): 17, (Module 6): 11, 12, 20, 21
- three-phase transformer, (Module 5): 4–5

Basic impulse level (BIL), (Module 2): 2, 30
Battery, (Module 4): 9, (Module 6): 11
Beacon, lockout, (Module 6): 11
Bedding, sand, (Module 4): 7
BIL. *See* Basic impulse level
Binoculars, (Module 6): 4
Blanket, rubber or insulating, (Module 2): 1, (Module 3): 1, (Module 5): 1, (Module 6): 1, 3
Bleeder (internal discharge resistor), (Module 6): 12
Block, (Module 2): 24, (Module 3): 9, 10, 13, 16
Boom, (Module 2): 2, (Module 3): 1, (Module 6): 13
Brace, (Module 2): 17, 24, (Module 3): 9, (Module 5): 14
Bracket, (Module 2): 5, 19, (Module 4): 15, 20, (Module 6): 2, 4, 10
Brass, (Module 4): 14
Breaker panel, (Module 5): 14. *See also* Circuit breaker
Bronze, (Module 6): 11
Brush, wire, (Module 2): 17, (Module 4): 25, 26, (Module 5): 14
Burndy® LLC, (Module 4): 10, 12–13
Bus, (Module 5): 4, 21
Bus bar, (Module 4): 25, (Module 5): 21
Bushings, on a transformer
- damaged, (Module 6): 14, 16
- definition, (Module 2): 30
- and elbows, (Module 4): 24
- lightning/surge arrester which fits in, (Module 4): 15, 16, 17
- overview, (Module 2): 5–6, (Module 4): 2
- single-phase transformer connections, (Module 2): 8
- spark gap for surge protection, (Module 2): 11

C

Cabinet, transformer, (Module 5): 6, 17, 21
Cables
bends in, (Module 4): 7
duplex, (Module 4): 9
installation, (Module 3): 9–15, 16, (Module 4): 22–23
jacketed, (Module 4): 1, 8, 11, (Module 5): 16
neutral, (Module 4): 8, 9, 25, (Module 5): 3, 10, 12
overhead, (Module 3): 14–15, 16
overview, (Module 3): 2, (Module 4): 11, (Module 5): 3
primary
burial depth, (Module 4): 7
components, (Module 4): 8, 11
in a housing tract, (Module 5): 17
installation, (Module 4): 22–23
transition from overhead to underground, (Module 5): 5–6
troubleshooting, (Module 6): 15, 16–17
quadruplex, (Module 5): 4, 7, 13, 24
repair and replacement, (Module 3): 15–19, 20, (Module 4): 11–12
secondary
burial depth, (Module 4): 7
components, (Module 4): 8, 9, 11
connections, (Module 4): 25–26
to electric meter, (Module 5): 14
installation, (Module 4): 22
overview, (Module 4): 3
in radial-feed system, (Module 4): 3
troubleshooting, (Module 6): 17–18
television service, (Module 4): 6
triplex. *See* Triplex
types, (Module 4): 7–9
underground, (Module 4): 7–9, 22, 25–26, (Module 5): 16, 17–18
vs. conductor, (Module 3): 2
Cable-pulling equipment
bullwheel tensioner, (Module 3): 9, 11, 20
pulling machine (cable puller), (Module 3): 9, 10, 20, 21, 24
pulling rope, (Module 3): 19, 20
take-up reel, (Module 3): 16, 20, (Module 6): 7, 8
Cable-spiking tool, (Module 6): 18–19
Camera, (Module 3): 18, (Module 6): 10, 26
Capacitance, (Module 1): 12–17, 37
Capacitors
connecting multiple, (Module 1): 14. *See also* Banks, capacitor
current and voltage characteristics, (Module 2): 16, (Module 6): 12
overview, (Module 1): 12–13, (Module 2): 16, (Module 6): 11–12, 21
safety, (Module 2): 16
specifications, (Module 1): 14–15
zero power consumed by, (Module 1): 20
Capacitor discharge unit (thumper), (Module 6): 16, 17
CEMF. *See* Counter electromotive force
Ceramic, (Module 1): 15. *See also* Porcelain
Certification, (Module 3): 20
Charts. *See* Sag chart; Tap percentages chart
Circuits
AC, (Module 1): 7–12, 19–22, 23
feeder, (Module 2): 15–16
parallel, (Module 1): 14
power distribution
coordination of protective devices, (Module 6): 21–22
de-energizing and safety, (Module 6): 1
isolating device for safety, (Module 2): 16–20, 21, (Module 4): 18
power factor in, (Module 1): 23, (Module 6): 21
prevention of electrical disturbances, (Module 1): 30
RL, RC, LC, and RLC, (Module 1): 17–19, 21
series, (Module 1): 14, 19, 28
short, (Module 2): 2, 11, 12, (Module 3): 4, 7, (Module 6): 2
street lighting, (Module 5): 19
testing, (Module 5): 14, 17
Circuit breaker, (Module 4): 18, (Module 6): 14, 16, 22. *See also* Breaker panel
Circular mil, (Module 3): 3–4, (Module 4): 8
Clamps
clamp-type fault indicator, (Module 4): 16
dead-end, (Module 3): 10, 12
on insulator, (Module 3): 12, 14
luminaire mast arm, (Module 2): 24, 25
Clamp stick (shotgun stick), (Module 2): 22
Clearance
conductor, (Module 3): 10, (Module 5): 8, 10, 11, 12, (Module 6): 5–6
service head, (Module 5): 13
Clothing, (Module 2): 1, 6, (Module 3): 1, (Module 5): 1
Cluster bar, (Module 6): 2, 3
Codes, local, (Module 4): 7, 21, (Module 5): 10, 19
Coil, (Module 1): 9, 10, 22, 25, 30, (Module 2): 6, (Module 5): 4
Color coding, (Module 3): 14, 15, (Module 4): 11, 24, 26, (Module 5): 13
Come-along, (Module 6): 5
Commercial operations
farms, (Module 5): 3, 11
overhead service, (Module 5): 12, 13–14
service drop, (Module 5): 13–14
single-phase light and power system, (Module 2): 8
three-phase power system, (Module 1): 31, 33, (Module 5): 3, 4, 14
typical voltage, (Module 2): 1
underground service, (Module 4): 1, (Module 5): 20–21
Common-trench approach, (Module 4): 5–6, 30
Compensator, line-drop, (Module 2): 18. *See also* Voltage regulator
Compression tool and die, (Module 4): 11, 12, 13, 14
Compressor, air, (Module 1): 22, (Module 2): 1, (Module 3): 1, (Module 5): 1, 4
Computer, (Module 6): 11, 20
Concrete, (Module 4): 1, 21
Conductivity, (Module 3): 2, 8, 17, 18
Conductors
clearance, (Module 3): 10, (Module 5): 8, 10, 11, 12
conductivity, (Module 3): 2
copperweld and alumoweld, (Module 3): 2, 24
dead-ending process, (Module 3): 10, 12, 13
drip loop in. *See* Drip loop
energizing process, (Module 3): 14, 16
inspection, (Module 6): 5–7, 9
installation into transformer terminal, (Module 2): 9, 10
installation onto poles, (Module 3): 9–15, 16
for luminaire, (Module 2): 24, 26
messenger, (Module 3): 15, 24, (Module 5): 3, 7, 10, 12
movement relative to magnetic field, (Module 1): 1–3, 9, 11, (Module 6): 2
overview, (Module 3): 2
repair and replacement, (Module 3): 15–19, 20, (Module 6): 7, 8
selection, (Module 3): 4–9, (Module 4): 23
self-inductance by, (Module 1): 9–10

for service drop, (Module 3): 14–15, 24, (Module 4): 2, (Module 5): 3
sizes, (Module 3): 3–4, 5–6
stranded, (Module 3): 3, 5–6, (Module 4): 7, 8, 12, 25. *See also* Cables, underground
strength, (Module 3): 2, 3–4
tensioning and sagging process, (Module 3): 10–11, 13, (Module 5): 12, (Module 6): 7–8
thermal expansion/contraction, (Module 3): 3, 8, 10
underground, (Module 5): 15
vs. cables, (Module 3): 2. *See also* Cables
Conduit
for commercial service, (Module 5): 13
encased in concrete, (Module 4): 1
intermediate metal (IMT), (Module 5): 12
on overhead meter pole, (Module 5): 12
rigid metal (RMC), (Module 5): 10, 12, 24
on riser pole, (Module 5): 5–6
on service head, (Module 5): 7, 10
service mast, (Module 5): 10, 11, 12
for street lighting, (Module 5): 19
for underground meter loop on building, (Module 5): 18
for underground secondary service, (Module 5): 17
for underground service drop, (Module 5): 6
Conduit strap, (Module 5): 10, 12
Configurations, winding
delta-delta arrangement (closed delta), (Module 1): 31, 32, (Module 5): 4, 5
delta-wye arrangement, (Module 1): 31, 33, (Module 5): 4
wye-wye arrangement, (Module 1): 31, 32, (Module 2): 8, 10, (Module 5): 4
Connections, cable-to-transformer, (Module 4): 23–26
Connectors
compression, (Module 4): 11, 12–13, 24
dual-rated, (Module 4): 10, 11, 30
mechanical, (Module 5): 14
parallel groove tap, (Module 5): 20
split-bolt mechanical, (Module 4): 11–12
Construction sites, (Module 5): 12, (Module 6): 4. *See also* Excavations
Copper
characteristics, (Module 3): 2, 3
conductor tie, (Module 3): 12
cost, (Module 3): 9
purity, (Module 3): 1
recycled, (Module 3): 1, (Module 6): 15
transformer terminal, (Module 2): 8, 9
tube in sectionalizer, (Module 6): 11, 12
wire, (Module 2): 24, (Module 3): 24, (Module 4): 8
Copperguard, (Module 2): 6
Copperweld, (Module 3): 2, 24
Core
aluminum conductor steel reinforced, (Module 3): 3, 17, 18
transformer, (Module 1): 22, 23, 24, (Module 2): 5, 8
Corona, (Module 6): 13
Corrosion
aluminum oxide, (Module 2): 9, (Module 3): 17–18, (Module 4): 26
of conductor/cable, (Module 4): 8, 25, (Module 6): 7. *See also* Corrosion, aluminum oxide
of electrical contact, (Module 1): 34
rust, (Module 2): 5, (Module 3): 2, (Module 6): 14
underground systems prone to, (Module 5): 2, (Module 6): 7, 19
Cost considerations
common-trench approach, (Module 4): 5–6
of conductor materials, (Module 3): 9
electricity, billing protocols, (Module 5): 9, 14
underground *vs.* overhead power distribution, (Module 4): 1, 3, (Module 5): 2
Coulomb, (Module 1): 13
Counter electromotive force (CEMF), (Module 1): 11
Creep, of a newly-installed conductor, (Module 3): 10, 11
Creosote, (Module 6): 5. *See also* Wood preservatives
Crimping tool (crimper), (Module 3): 17, 18, (Module 4): 12–13, 26
Cross-arm, (Module 2): 17, (Module 3): 9, (Module 6): 4
Cross-linked polyethylene (XLPE), (Module 4): 8, 11, 12, 25
CSP. *See* Transformers, completely self-protected
Current
in an inductive AC circuit, (Module 1): 9, 10–11, 16–17
in an inductive DC circuit, (Module 1): 9
backfeed, (Module 5): 14, (Module 6): 1
in a capacitive AC circuit, (Module 1): 12, 15–16
in a capacitor, (Module 2): 16
and conductor selection, (Module 3): 4, 5–6
eddy, (Module 1): 22
exciting, (Module 1): 25
fault, (Module 5): 14, (Module 6): 18, 22. *See also* Faulted circuit indicator
imbalance in three-phase systems, (Module 1): 33–34, (Module 5): 4
lag, (Module 6): 21
leakage, (Module 1): 15, (Module 2): 2, (Module 3): 1, (Module 5): 1, (Module 6): 13
in Ohm's law formulas, (Module 1): 7–8, 12
rating (ampacity), (Module 2): 14, (Module 4): 8
relationship with apparent power, (Module 1): 20
surge, (Module 2): 2, 9, 11–15
Current-carrying capacity (ampacity), (Module 2): 14, (Module 4): 8, 30
Current-limiter, (Module 2): 22
Cutouts, fused, (Module 2): 3, 12–14, 30, (Module 5): 6, (Module 6): 22
Cutter, cable, (Module 4): 9
Cycle, of a sine wave, (Module 1): 3–4

D

DC. *See* Direct current
Dead-ending process, (Module 3): 10, 12, 13
Dead front, (Module 4): 15, 19, 24, 30
Delta-delta arrangement (closed delta), (Module 1): 31, 32, (Module 5): 4, 5
Delta-wye arrangement, (Module 1): 31, 33, (Module 5): 4
Diagrams, schematic, (Module 1): 29, 30
Dielectric material, (Module 1): 12, 13, 14, 15, (Module 6): 9, 12
Digital technology, (Module 1): 5, (Module 2): 12, (Module 3): 11, (Module 5): 14, 15
Direct current (DC), (Module 1): 1, 2, 5, 9, 19
Disconnect attachments, (Module 2): 22, 23
DMM. *See* Multimeter, digital
Double-feed design, (Module 4): 4–5, 18, 30
Drilling, horizontal directional (HDD), (Module 4): 6, 7
Drip loop, (Module 2): 24, (Module 5): 4, 10, 11, 13, 24
Duplex, (Module 4): 9
Dynamometer, (Module 3): 10, 11, 13, (Module 6): 8, 26

E

Edison, Thomas, (Module 1): 1
Effective value, of a sine wave, (Module 1): 4–5
Elbows
deadbreak, (Module 4): 18
lightning/surge arrester, (Module 4): 16, 17

Elbows *(continued)*
load break, (Module 4): 19, 24, (Module 6): 15, 16
long-radius, (Module 4): 21, 22, (Module 5): 17, 18
ninety-degree, (Module 4): 24
test point fault indicator, (Module 4): 17, 18
Electricity. *See also* Power distribution
cost structure, (Module 5): 9
power generation, (Module 1): 1–3, 4
speed, (Module 1): 4
static, (Module 3): 9, 10, (Module 6): 1, 2
Electromotive force (EMF), (Module 1): 10–11, 25, 34
Emergency response, (Module 6): 9
EMF. *See* Electromotive force
EMT. *See* Tubing, electrical metallic
Engineer, responsibilities, (Module 2): 14, (Module 3): 8, 21
Environmental considerations. *See* Going Green
Epoxy, (Module 4): 14, (Module 6): 19
EPR. *See* Ethylene propylene rubber
Equipment
boring, (Module 4): 6, 7
cable-pulling. *See* Cable-pulling equipment
capacitor discharge unit (thumper), (Module 6): 16, 17
electrical shock from chassis, (Module 1): 27, 29
excavation, (Module 4): 5, 7
fall protection, (Module 2): 2, (Module 3): 1, (Module 5): 1, (Module 6): 1
horizontal directional drilling (HDD), (Module 4): 6, 7
oil-filled, (Module 6): 9, 10
protective. *See* Personal protective equipment
Erosion control, (Module 4): 7
Ethylene propylene diene monomer rubber (EPDM), (Module 4): 14
Ethylene propylene rubber (EPR), (Module 4): 8, 12
Excavations
erosion control, (Module 4): 7
and existing utility lines, (Module 5): 1, 18, (Module 6): 16, 18, 19
planning, (Module 4): 5
trenching and backfilling methods, (Module 4): 5–7, (Module 5): 17, 18, (Module 6): 19
Excavator, (Module 4): 6
Extender, cross-arm (layout arm), (Module 6): 7
Extendo stick, (Module 2): 21, 22, 23
Eye, pulling, (Module 4): 14, 16, 20, 24, (Module 6): 5
Eyebolt (clamp-type) terminal, (Module 2): 8, 9, 10, (Module 4): 24
Eye protection, (Module 2): 1, (Module 3): 1, (Module 5): 1, 14

F

Face protection, (Module 5): 14
Fall protection equipment, (Module 2): 2, (Module 3): 1, (Module 5): 1, (Module 6): 1
Farad, (Module 1): 13
Farms, (Module 5): 3, 11
Fatalities, (Module 1): 28, 29, 31, (Module 3): 9, (Module 6): 1
Fatigue (metal), (Module 3): 17, (Module 6): 9
Fatigue (tiredness), (Module 6): 2–3
Faulted circuit indicator (FCI), (Module 2): 19–20, 21, 30, (Module 4): 15–18, (Module 5): 6, (Module 6): 17
Faults, in URD systems, (Module 4): 3–4, 5, (Module 5): 2, (Module 6): 15–19
FCI. *See* Faulted circuit indicator
Feeder lines/cables, primary
layout. *See* Double-feed design; Loop-feed design; Radial-feed design
location of lightning/surge arrester on, (Module 4): 14
overview, (Module 4): 1, 2, 3, 22
on riser with fused cutouts, (Module 5): 6
splicing procedure, (Module 4): 12
Feeder voltage regulator, (Module 2): 15–16, 30, (Module 6): 20
Fence, silt, (Module 4): 7
Fiberglass, (Module 2): 21, (Module 4): 1
Fiberglass-reinforced plastic (FRP), (Module 2): 21
Filter (LC), (Module 1): 19
Filter (oil), (Module 6): 12
Finger (pilot) rope, (Module 3): 10
Fire, (Module 2): 5, (Module 6): 5, 7, 9
First aid, (Module 2): 2, (Module 3): 2
Flux, magnetic, (Module 1): 2, 3, 9, 10, 11, 22, 25
Footwear, (Module 2): 1, (Module 3): 1, (Module 5): 1
Frequency, (Module 1): 3–4, 17, 19, 37
Friction, (Module 1): 11
Frost heaving, (Module 6): 16
Frost line, (Module 5): 17
FRP. *See* Fiberglass-reinforced plastic
Fuses
bayonet, (Module 4): 20, 21, 22, (Module 5): 16
function, (Module 2): 12, (Module 4): 2, (Module 6): 21–22
link type and response time, (Module 2): 12
location on power distribution system, (Module 6): 21
mounted in cutouts, (Module 2): 3, 12–14, 30, (Module 5): 6
ratings, (Module 2): 12, (Module 6): 22
replacement, (Module 6): 15, 22
riser pole, (Module 4): 19, (Module 6): 15, 17
selection, (Module 2): 14–15
for street lighting circuit, (Module 5): 19, 20
in switchgear, (Module 4): 18
transformer, (Module 4): 20, 21, 22, (Module 5): 16, (Module 6): 15, 16
troubleshooting, (Module 6): 15–16
vs. recloser, (Module 2): 18
Fusing ratio, (Module 2): 14, 30

G

Gap, spark, (Module 2): 11
Gas (SF6 switchgear medium), (Module 4): 19
Gas (service), (Module 4): 6, (Module 6): 20
Generator, for emergency power, (Module 5): 14, (Module 6): 1
Geometry, (Module 1): 22
Gin pole, (Module 2): 24
Glasses, safety, (Module 2): 1, (Module 3): 1, (Module 5): 1
Gloves, (Module 2): 1, 2, (Module 3): 1, 12, (Module 5): 1, 14, (Module 6): 7
Going Green
energy-efficiency, (Module 1): 23, (Module 2): 8
erosion control, (Module 4): 7
hazardous materials safety, (Module 6): 9, 15
LED street lights, (Module 2): 26
oil recycling, (Module 6): 12, 15
recycling, (Module 6): 12, 15
Smart Grid power distribution technology, (Module 2): 12
transformer, (Module 2): 8, (Module 6): 15
underground *vs.* overhead power distribution, (Module 5): 4
wildlife considerations, (Module 2): 2, 5, 6, (Module 5): 4
Gold, (Module 3): 2
Grid (grounding), (Module 3): 20
Grid (power distribution), (Module 2): 12, (Module 6): 19
Grips, (Module 3): 9, 10, 11, (Module 6): 5

Ground (electrical)
for cable/conductor work, (Module 3): 10, 15, 16, (Module 5): 13
conductor, for meter loop, (Module 5): 8
grounding grid, (Module 3): 20
and lightning arrester function, (Module 2): 11, (Module 4): 13
protective, for power line maintenance, (Module 6): 1–2, 3, 10
rod, (Module 2): 24, (Module 4): 20, (Module 5): 8, (Module 6): 2
role of isolation transformer, (Module 1): 27, 29
running, (Module 3): 10, 16, 20
street light mast arm, (Module 2): 24
transformer, (Module 2): 8, (Module 4): 20
Ground (soil), (Module 4): 7, (Module 5): 17, (Module 6): 4, 16
Gunfire, (Module 3): 17, (Module 6): 14

H

Hard hat or hard cap, (Module 2): 1, (Module 3): 1, (Module 5): 1
Hazardous materials, (Module 2): 6, (Module 5): 4, (Module 6): 4–5, 9, 10, 13, 15
HAZMAT procedures, (Module 6): 9, 10
HDD. *See* Drilling, horizontal directional
Health considerations, (Module 6): 2–3, 13. *See also* Hazardous materials
Hearing protection, (Module 2): 1, (Module 3): 1, (Module 5): 1
Heat exchanger, (Module 2): 2, 3
Heat production, (Module 1): 5, 20, 25, 34, (Module 2): 2, 3, (Module 6): 19
Helicopter, (Module 6): 10
Helmet, safety, (Module 2): 1, (Module 3): 1, (Module 5): 1
Henry, (Module 1): 9
Hertz, (Module 1): 4, 37
High-leg delta system, (Module 5): 4, 13
Historical background, (Module 1): 1, (Module 4): 1–2, (Module 5): 8
Hoisting operations. *See* Lift operations
Hook, loadbreak, (Module 2): 19
Hook stick, (Module 2): 17, 18, 19, 20
Hose, line, (Module 2): 1, (Module 3): 1, (Module 6): 2, 3
Hospital, (Module 4): 4
Hot stick
capacitor maintenance, (Module 6): 12
to change tap, (Module 6): 21
elbow removal from bushing, (Module 4): 24
fault indicator installation, (Module 4): 17, 18
fuse and fuse holder removal, (Module 4): 20, (Module 5): 16
lightning/surge arrester installation, (Module 4): 16
overview, (Module 2): 21
power line maintenance, (Module 6): 1, 7
sectionalizer operation, (Module 6): 11
for tree pruning, (Module 6): 7
HYLUG™ terminals, (Module 4): 10

I

Identification marks on system components, (Module 3): 15, (Module 4): 8, 10, 11, 13, 24, (Module 6): 20–21
IEEE 386 Standard, (Module 4): 16
Impedance, (Module 1): 17, 19, 29, 31, 37, (Module 2): 8, (Module 6): 20
IMT. *See* Conduit, intermediate metal
Indicators
faulted circuit (FCI), (Module 2): 19–20, 21, 30, (Module 4): 15–18, (Module 5): 6
oil level, (Module 6): 9, 13, 14
on a transformer, (Module 2): 5
Inductance, (Module 1): 9–12, 13, 28, 37
Inductor, (Module 1): 9, 11, 12, 20, 22
Industrial operations, (Module 1): 31, 33, (Module 4): 1, 4, (Module 5): 4, (Module 6): 20
Infrared scan, (Module 6): 7
Insecticides, (Module 6): 4
Inspections
building wiring, (Module 5): 13
conductor (power distribution line), (Module 6): 5–7, 9
guy wire, (Module 6): 5
insulated tool, (Module 2): 21
pole-mounted equipment, (Module 6): 1, 9–10
schedule for, (Module 6): 9
service drop, (Module 5): 13
service entrance, (Module 5): 13
switch, (Module 6): 12
transformer, (Module 6): 14, 15, 16
utility pole, (Module 6): 1, 3–5, 6
voltage regulator, (Module 6): 13
Inspection report, (Module 6): 4
Installation
air break switch, (Module 2): 17–18
cables and conductors, (Module 2): 9, 10, (Module 3): 9–15, 16, 20–21, (Module 4): 23
conductor tie, (Module 3): 12
electric meter, (Module 5): 14–15, 17
fault indicator, (Module 2): 19, 20, (Module 4): 16–18, (Module 5): 6
lightning/surge arrester, (Module 4): 16–18
luminaire, (Module 2): 24–25, 26
meter loop, underground, (Module 5): 18–19
overhead cables and services, (Module 3): 14–15, 16, (Module 5): 7–15, 16
service drop, (Module 3): 14–15, 16, (Module 5): 3, 12–14
transformer, pad-mounted, (Module 4): 21–23, (Module 5): 16
trenching and backfilling methods, (Module 4): 5–7, (Module 5): 17, 18, (Module 6): 19
underground secondary services, (Module 5): 15–21
Western Union tie, (Module 3): 14, 16, 24
wire holder, (Module 5): 13
Instruments
ammeter, (Module 1): 28
camera, thermovision, (Module 6): 10, 26
dynamometer, (Module 3): 10, 11, 13, (Module 6): 8, 26
multimeter, digital, (Module 1): 5
reflectometer, time domain, (Module 6): 17, 18
sag scope, (Module 3): 10, 14
voltmeter, (Module 1): 30, (Module 2): 7, (Module 5): 14, (Module 6): 2, 14–15
Insulation
in cable, (Module 4): 8, 11, 12, (Module 5): 14, (Module 6): 16
in conductor, for direct burial, (Module 5): 15
dielectric permittivity, (Module 1): 14
in lightning/surge arrester, (Module 4): 14
oil , in a transformer, (Module 2): 2, 5, 7
rubber, in a transformer, (Module 1): 24
Insulators
on cross-arm extender, (Module 6): 7
dead-end, (Module 3): 10, 12
identification marks on, (Module 3): 15

Insulators *(continued)*
maintenance, (Module 6): 9, 13
tie process to install conductor on, (Module 3): 11–14, 15, 16
types, (Module 3): 12, 14
Iron, (Module 1): 9, 22, 23
Isolating devices, (Module 2): 16–20, 21

J

Jack-strap, (Module 5): 12
Joint-trench approach, (Module 4): 5–6, 30
Jumper, (Module 2): 6, 7, 8, (Module 3): 14, 16
Junction box, (Module 5): 19, 20

K

Kilowatt-hour, (Module 5): 14
Kits. *See* Splice/termination kit; Tie kit

L

Layout arm, (Module 6): 7
Lead (connector). *See* Connectors
Lead (metal), (Module 4): 11
Leaks
current, (Module 1): 15, (Module 2): 2, (Module 3): 1, (Module 5): 1, (Module 6): 13
oil, (Module 6): 9, 10, 12, 13, 14
PCBs from old transformer, (Module 2): 6
Leak resistance, capacitor, (Module 1): 15
LED, street lights, (Module 2): 26
Legal action, (Module 5): 18, (Module 6): 18
Lift bucket, (Module 2): 2, (Module 3): 1, 15, (Module 5): 1, (Module 6): 1, 2
Lift operations, (Module 2): 17, 24, 25, (Module 6): 13
Lighting, (Module 1): 28, (Module 2): 23–26, (Module 4): 25, (Module 5): 3, 19–20, (Module 6): 19
Lightning, (Module 2): 2, (Module 3): 17, (Module 6): 9, 12, 13, 16
Lightning/surge arresters
definition, (Module 2): 30, (Module 4): 13, 30
installation, (Module 4): 16–18
overview, (Module 2): 11, (Module 4): 2, 13–14
primary feeder connection, (Module 4): 3
on transformer, (Module 2): 3, (Module 4): 20
types, (Module 4): 15
Linear low-density polyethylene (LLDPE), (Module 4): 8, 12
Line guard. *See* Rods, armor
Lines (rope). *See* Finger (pilot) rope; Pulling rope
Lines (utility service), (Module 4): 6. *See also* Excavations, and existing utility lines
Lines of flux, (Module 1): 2, 3, 10, 22, 25
Live front, (Module 4): 19, 20, 24–25, 30
LLDPE. *See* Linear low-density polyethylene
Load (electrical)
inductive, (Module 6): 21
load break tool, (Module 2): 17, 22–23, 24
management
overview, (Module 6): 19–20
and protective devices, (Module 2): 9, 11–20, 21, (Module 6): 21–22. *See also* Load, surge current
no-load condition, (Module 1): 24–25
surge current, (Module 2): 2, 9, 11–15
tap-changing considerations, (Module 6): 21
Load (physical), on conductor, (Module 3): 3, 4, 7, 10, 13. *See also* Tensioning and sagging process
Load break tool, (Module 2): 17, 22–23, 24, (Module 6): 11
Loading districts, U.S., (Module 3): 4, 7, 24
Lock out, (Module 6): 1, 11
Loop. *See* Drip loop; Meter loop; Service loop
Loop-feed design
definition, (Module 4): 30, (Module 5): 24
locating and correcting faults, (Module 4): 17, (Module 6): 15–19
location of fault indicating device, (Module 4): 16, 17
location of lightning/surge arrester, (Module 4): 14
overview, (Module 4): 4, 5, (Module 5): 2
Lubrication, (Module 4): 16, 17
Luminaire, (Module 2): 24–25, 26, 30

M

Magnetic field, (Module 1): 1–3, 9, 10, 22, 25, 29, (Module 6): 2
Maintenance
cables and conductors, (Module 3): 15–19, 20, (Module 6): 5–9, 18–19
insulator, (Module 6): 9, 13
isolating devices to protect worker, (Module 2): 16–20, 21, (Module 4): 18
load management. *See* Load (electrical), management
locating faults, (Module 6): 15–18
overview, (Module 6): 1
pole, (Module 6): 3–5, 6, 10
pole-mounted equipment, (Module 6): 9–15
recloser, (Module 6): 11
repair of underground systems, (Module 6): 18–19. *See also* Splices
safety, (Module 6): 1–3
schedule, (Module 6): 9
street light, (Module 2): 23
three-phase power system, monitoring, (Module 1): 33
tool, (Module 2): 21
transformer, (Module 1): 30–31, (Module 6): 12–13
underground *vs.* overhead power distribution, (Module 4): 3, (Module 5): 2
Markings. *See* Identification marks
Mask, dust, (Module 2): 1, (Module 3): 1, (Module 5): 1
Mast, meter loop (service), (Module 3): 15, (Module 5): 10–11, 12, 14
Mast arm, (Module 2): 24, 30
Mast arm base, (Module 2): 24, 25
Mat, grounding, (Module 3): 9
Math, (Module 1): 2–3, 22
Messenger, (Module 3): 15, 24, (Module 5): 3, 7, 10, 12
Metal, (Module 3): 2, 17. *See also* Amorphous metal; *specific metals*
Meters, electric (watt-hour)
installation, (Module 5): 7, 13, 14, 17
location in service drop, (Module 4): 2, (Module 5): 5
reading, (Module 5): 14, 15
smart, (Module 5): 9
special for three-phase service, (Module 5): 16
Super Beast meter circuit tester, (Module 5): 17
types, (Module 5): 14, 15, 16
Meter base, (Module 5): 9, 14, 16, 19
Meter loop, (Module 5): 1, 6, 7–12, 13, 14, 18–19, 24
Mica, (Module 1): 14, 15
Micro, (Module 1): 10, 37
Microprocessor, (Module 6): 5, 11
Mnemonic device, ELI the ICE man, (Module 1): 11, 13, 16, 17, 22
Motor, (Module 1): 31, (Module 5): 4, (Module 6): 19, 21
Motor starter, (Module 1): 28, 30
MOV. *See* Varistor, metal-oxide
Multimeter, digital (DMM), (Module 1): 5

N

Nameplate information, (Module 1): 5
National Electrical Code® (NEC®), (Module 5): 8, 10, 19
National Institute for Occupational Safety and Health (NIOSH), (Module 1): 31
NEC®. See National Electrical Code®
NEC Section 230.24(B)(1), (Module 5): 10
NEC Section 230.54(A), (Module 5): 10
NEC Section 342.30(A), (Module 5): 10
NEC Section 342.30(B), (Module 5): 10
NEC Section 344.30(A), (Module 5): 10
NEC Section 344.30(B), (Module 5): 10
NEC Section 352.30(A), (Module 5): 10
NEC Section 352.30(B), (Module 5): 10
Neoprene, (Module 3): 13, 15
Nickel, (Module 6): 12
NIOSH. *See* National Institute for Occupational Safety and Health

O

Occupational Safety and Health Administration (OSHA), (Module 2): 20–21
Ohm's law, (Module 1): 7–8, 12
Oil
- diesel, (Module 6): 20
- insulating, in transformer, (Module 2): 2, 5, 7, (Module 4): 20, 22, (Module 6): 12, 13
- leaks, (Module 6): 9, 10, 12, 13, 14
- in oil switch, (Module 2): 18
- PCB-contaminated, (Module 2): 6, (Module 6): 10, 15
- to prevent or extinguish arcing, (Module 4): 19
- recycling, (Module 6): 12, 15
- in voltage regulator, (Module 6): 13, 14

OSHA Standard 29 CFR 1910.269(j), (Module 2): 21
Overhead distribution. *See* Power distribution, overhead
Overload, (Module 1): 33, (Module 2): 6, (Module 6): 14
Oxide inhibitor, (Module 2): 9, (Module 3): 17–18. *See also* Antioxidant
Ozone, (Module 6): 13

P

Pad, transformer, (Module 4): 1, 21, 22
Paper, (Module 1): 9, 14, 15, 24
Parallel. *See* Circuits, parallel
PCBs. *See* Polychlorinated biphenyls
Peak value, of a sine wave, (Module 1): 4, 5, 6
Peaker plant, (Module 6): 20
Pedestal, secondary, (Module 4): 21, (Module 5): 17, 18, 24, (Module 6): 5
Permittivity, dielectric, (Module 1): 14
Personal protective equipment, (Module 1): 31, (Module 2): 6, 7, (Module 4): 7, (Module 5): 14, (Module 6): 1
Personnel
- cable-spiking tool operator, (Module 6): 19
- lifting to work area. *See* Lift bucket
- meter reader, (Module 5): 9, 14
- puller/tensioner machine operator, (Module 3): 20
- qualified observer, (Module 2): 2, (Module 3): 2, (Module 5): 1
- trenching machine operator, (Module 5): 18

Phase
- in alternating current, (Module 1): 6–7, 8, 13, 18, 25, 26
- three-phase power, (Module 1): 31–34, (Module 2): 8, (Module 5): 3–5, 14, 16. *See also* Transformers, three-phase

Phase angle, (Module 1): 7, 8, 19, 20
Phase shift, (Module 1): 20
Photocell, (Module 2): 25, 26
Pin, blade stop, (Module 2): 19
Plastic
- clamp-style insulator, (Module 3): 14
- conductor ties, (Module 3): 12
- conduit, (Module 5): 5
- fiberglass, (Module 2): 21, (Module 4): 1
- hard hat, (Module 2): 1, (Module 3): 1, (Module 5): 1
- polyvinyl chloride, (Module 5): 5, 6
- transformer pad, (Module 4): 1, 21
- U-guard, (Module 5): 6

Plates, in a capacitor, (Module 1): 12, 13, 14, 17, (Module 6): 12
Plow, tractor-mounted, (Module 4): 6
Polarity
- additive, (Module 2): 6, 7, 30
- in a capacitor, (Module 1): 12, 16
- factors which affect, (Module 1): 10
- positive/negative within one cycle, (Module 1): 1, 3
- secondary mains, (Module 5): 13
- subtractive, (Module 2): 6, 7, 8, 31
- in a transformer, (Module 1): 26, (Module 2): 6–7

Poles (tools). *See* Clamp stick; Extendo stick; Gin pole; Hook stick; Hot stick
Poles (utility)
- cable to residence. *See* Service drop
- congested, (Module 6): 10
- cutting or drilling, (Module 2): 1, (Module 3): 1, (Module 5): 1
- distance between, (Module 3): 3, (Module 5): 8
- inspection and maintenance, (Module 6): 3–5, 6, 10
- installation of conductors onto, (Module 3): 9–15, 16, (Module 5): 3
- mounting holes, (Module 6): 10
- overhead meter, (Module 5): 11–12, 13
- riser
  - definition, (Module 5): 5, 24
  - inspection, (Module 6): 9
  - overview, (Module 5): 5–6
  - in radial-feed system, (Module 4): 3
  - surge arrester on, (Module 4): 14
  - troubleshooting, (Module 6): 15–16, 17
  - underground secondary distribution system, (Module 5): 17
- street light, (Module 2): 24
- transformer mounting strategies, (Module 2): 3, 5
- underground meter, (Module 5): 19
- wood preservatives used on, (Module 5): 4, (Module 6): 4–5

Polycarbonate, (Module 2): 1, (Module 3): 1, (Module 5): 1
Polychlorinated biphenyls (PCBs), (Module 2): 6, (Module 6): 10, 15
Polyethylene. *See* Cross-linked polyethylene; Linear low-density polyethylene
Polypropylene, (Module 2): 1, (Module 3): 1, (Module 5): 1
Polyvinyl chloride (PVC), (Module 5): 5, 6
Porcelain, (Module 3): 12, (Module 4): 24
Power
- in alternating circuits, (Module 1): 19–22, 23
- apparent, (Module 1): 19–20, 21, 22, 23, (Module 6): 21
- demand, (Module 5): 9, (Module 6): 19, 20
- for farm irrigation pump, (Module 5): 11
- inrush, (Module 1): 30–31
- metered. *See* Meters, electric
- reactive, (Module 1): 20, 22
- surplus, (Module 6): 19
- three-phase, (Module 1): 31–34, (Module 2): 8, (Module 5): 3–5, 14, 16
- true, (Module 1): 8, 20, 21, 22, 23, (Module 6): 21

Power distribution. *See also* Power lines
design. *See* Double-feed design; Loop-feed design; Radial-feed design
direct-buried. *See* Power distribution, underground residential
efficient transfer of power, (Module 1): 23
high-leg delta system, (Module 5): 4, 13
light and power systems, (Module 2): 8, 9
location of protective devices, (Module 6): 21–22
overhead
in combination with underground system, (Module 4): 1
current and voltage, and conductor selection, (Module 3): 4, 5–6
historical, (Module 4): 1, (Module 5): 8
installation of services, (Module 5): 7–15, 16
overview, (Module 4): 2, (Module 5): 2–5. *See also* Service drop
transformers, (Module 2): 2–4, 5
vibration, (Module 3): 4, (Module 6): 8–9
vs. underground distribution, (Module 4): 1, 3, (Module 5): 2, 4
prevention of electrical disturbances, (Module 1): 30
rural areas, (Module 5): 2, 3, 4, 11
Smart Grid technology, (Module 2): 12
three-wire secondary service, (Module 4): 1, 25
underground
cable burial depth, (Module 5): 17, 18
faults, locating, (Module 6): 15–18
installation of secondary services, (Module 5): 15–21
overview, (Module 5): 5–6
pad-mount transformers for, (Module 1): 33, (Module 2): 4, 5, 23, (Module 4): 1
repair, (Module 6): 18–19. *See also* Splices
total, (Module 5): 7
vault transformers, (Module 2): 4, (Module 4): 1
vs. overhead distribution, (Module 4): 1, 3, (Module 5): 2, 4
underground residential (URD)
advantages, (Module 4): 3, (Module 5): 2, 5
cable types, (Module 4): 7–9
design types, (Module 4): 3–5, 14, 16, 17
history and significance, (Module 4): 1–3
installation, (Module 5): 17–18
lightning protection and fault-indicating devices, (Module 4): 13–18
overview, (Module 4): 1, (Module 5): 5, 17–18
pad-mounted switchgear and transformers, (Module 4): 1, 18–23
termination methods, (Module 4): 9–13
transformer connections, (Module 4): 20, 23–26
trenching and backfilling methods, (Module 4): 5–7, (Module 5): 17, 18
troubleshooting and maintenance, (Module 4): 3, 13, (Module 6): 15–19
urban areas, (Module 5): 5
voltages. *See* Voltages, typical, along power distribution system
Power factor, (Module 1): 20–22, 23, 30, (Module 6): 21
Power generation, (Module 1): 1–3, 4, 31
Power lines. *See also* Cables; Conductors
aesthetics, (Module 4): 1
burning in trees, (Module 6): 5
cost considerations, (Module 1): 2
current and voltage, and conductor selection, (Module 3): 4, 5–6
foreign objects hanging on, (Module 6): 5
historical background, (Module 1): 1
separation from gas lines, (Module 4): 6
and transformer safety, (Module 1): 28
vibration, (Module 3): 4, (Module 6): 8–9
Power loss
localized. *See* Current, leakage; Heat production
system-wide outage
causes, (Module 2): 2, (Module 3): 8, (Module 6): 1
correction of imbalances following, (Module 1): 33
customer reporting to operator or computer, (Module 6): 15
instant reporting by smart meter, (Module 5): 9
from system overload in hot weather, (Module 1): 22
Power plant, (Module 1): 2, 24, 30, (Module 6): 20
Power surge. *See* Current, surge
Power system, three-phase, (Module 1): 31, 33, (Module 2): 8, 10, (Module 5): 3–5, 14, 16. *See also* Transformers, three-phase
Power triangle, (Module 1): 22, 23
Puller/pulling machine, (Module 3): 9, 10, 20, 21, 24
Pulley, for conductor installation, (Module 3): 9
Pulling eye, (Module 4): 14, 16, 20, 24, (Module 6): 5
Pulling rope, (Module 3): 19, 20
PVC. *See* Polyvinyl chloride
Pythagorean theorem, (Module 1): 22

Q

Quadruplex, (Module 5): 4, 7, 13, 24

R

Radial-feed design, (Module 4): 3–4, 14, 17, 30, (Module 5): 2, 3, 24
Radio communication, (Module 6): 11, 13
Radio tuner, (Module 1): 1, 19
Ratings
capacitance, (Module 1): 14
current (ampacity; current-carrying capacity), (Module 2): 14, (Module 4): 8, 30
dual-rated cable terminals or connectors, (Module 4): 10, 11, 30
fuse, (Module 2): 12, (Module 6): 22
interrupting, of protective device, (Module 6): 21
temperature, of insulation, (Module 4): 8
temperature-rise, (Module 2): 2
voltage, (Module 1): 14–15, 30, 31, (Module 4): 8
voltage-amperes (VA), (Module 2): 2, (Module 6): 12
Reactance, (Module 1): 11–12, 16–17, 19, 22, 37, (Module 3): 6
Recloser, (Module 2): 18, 19, 20, 30, (Module 6): 10–11, 22
Rectifier, (Module 1): 2
Recycling, (Module 3): 1, (Module 6): 12, 15
Reel, conductor, (Module 3): 9, 10, 16, 20, (Module 4): 6, (Module 6): 7
Reflectometer, time domain (TDR), (Module 6): 17, 18
Regulator. *See* Voltage regulator
Rejacketing procedure, (Module 4): 11
Relay, (Module 1): 30
Remote control, (Module 2): 17
Repair. *See* Maintenance
Residences
cable from utility pole to. *See* Service drop
conductor damage by customer, (Module 6): 19
motor homes/mobile homes, (Module 5): 7, 11, 19
single-phase light and power system, (Module 2): 8
three-phase power system, (Module 1): 31, 33
typical voltage, (Module 2): 1, 15

underground power distribution. *See* Power distribution, underground residential
Resistance
in AC circuits, (Module 1): 7–9
ACSR, (Module 3): 6
due to corrosion, (Module 1): 34
due to friction, (Module 1): 11
effects of temperature, (Module 3): 8
in metal-oxide varistor, (Module 2): 11
in Ohm's law formulas, (Module 1): 7–8
relationship with true power, (Module 1): 19
of a worker's body, (Module 1): 28. *See also* Shock, electrical
Resistor, (Module 1): 20, (Module 6): 12, 20. *See also* Varistor, metal-oxide
Resonance, (Module 1): 19, 37
Ring, pull or hookstick, (Module 2): 19, 22, (Module 6): 12
Riser. *See* Poles (utility), riser
RMC. *See* Conduit, rigid metal
rms. *See* Root-mean-square
Rods
armor (line guard), (Module 3): 14, 16, 17, 24, (Module 6): 6
filler, (Module 3): 17, (Module 6): 7
ground, (Module 2): 24, (Module 4): 20, (Module 5): 8, (Module 6): 2
Root-mean-square (rms), (Module 1): 4, 5, 6, 19, 37
Rope. *See* Finger (pilot) rope; Pulling rope
Rubber
blanket, (Module 2): 1, (Module 3): 1, (Module 5): 1, (Module 6): 1
cable insulation, (Module 4): 8
cover for transformer connection, (Module 4): 20
gloves and sleeves, (Module 2): 1, 2, (Module 3): 1, (Module 5): 1
lightning/surge arrester insulation, (Module 4): 14
tape, (Module 4): 12
Runoff, (Module 4): 7

S

Safety. *See also* Tools, insulated
conductor installation, (Module 3): 20
conductor tie installation, (Module 3): 12
draining transformer oil prior to fuse removal, (Module 3): 20
effects of fatigue, (Module 6): 2–3
hazardous materials, (Module 6): 9
lift operations, (Module 2): 25
PCB toxicity, (Module 2): 6, (Module 6): 10
responsibility for, (Module 6): 3
service drop over a swimming pool, (Module 5): 10
working near traffic, (Module 6): 3
Safety, electrical
AC power, (Module 1): 31
check for voltage prior to maintenance, (Module 5): 14, (Module 6): 2, 15
current transformer, (Module 1): 28
deadbreak elbow, (Module 4): 18
isolating device, (Module 2): 16–20, 21
isolation transformer on equipment chassis, (Module 1): 27, 29
overview, (Module 2): 1–2, (Module 3): 1, (Module 5): 1, (Module 6): 1
pad-mounted transformer kept locked, (Module 5): 17
power line maintenance, (Module 6): 1–2, 3, 10
residual charge on capacitor, (Module 1): 15, (Module 2): 16
rotary tap-changing switch inside shell, (Module 2): 15
training, (Module 1): 31
underground *vs.* overhead power distribution, (Module 4): 3
Safety tag, (Module 6): 1
Sag chart, (Module 3): 10, 13
Sag scope, (Module 3): 10, 14, (Module 6): 8
Sand bedding, (Module 4): 7
Screens. *See* Shields
Sectionalizer, (Module 2): 18, 20, 30, (Module 6): 11, 12
Self-inductance, (Module 1): 9, 28, 37
Semiconductor material, (Module 4): 8, 12, 13
Series. *See* Circuits, series
Service drop
definition, (Module 3): 24, (Module 5): 3
inspection, (Module 5): 13
overhead cable
conductor clearance, (Module 5): 8, 10, 11, 12
installation, (Module 3): 14–15, 16, (Module 5): 3, 12–14
over a swimming pool, (Module 5): 10
use of, (Module 3): 2, 3, (Module 4): 2, (Module 5): 5
three-phase, (Module 5): 5
underground, (Module 5): 6, 7
Service head, (Module 5): 7, 8, 10, 11, 12, 13
Service loop, (Module 5): 4
Sewer lines, (Module 4): 6
Sheath, on cable, (Module 4): 8, 11, 12, (Module 5): 14
Shell, on transformer, (Module 2): 2, 3, 8, 30
Shields (screens)
concentric neutral metallic, (Module 4): 8, 12, 17, 18
protective, (Module 2): 1, 2, (Module 3): 2
semiconductor, (Module 4): 8, 12, 14
on URD cable, (Module 4): 8, 12
Shock, electrical. *See also* Tools, insulated
from equipment chassis, (Module 1): 27, 29
from residual charge on capacitor, (Module 1): 15, (Module 2): 16
from static electricity build up, (Module 3): 9, 10, (Module 6): 1, 2
during transformer work, (Module 1): 28
working with AC power, (Module 1): 31
Shoes, (Module 2): 1, (Module 3): 1, (Module 5): 1
Shotgun stick (clamp stick), (Module 2): 22
Shunt, splice, (Module 3): 18, 19, 24, (Module 6): 7
Silicon, in transformer core, (Module 2): 5
Silicone, lubricant, (Module 4): 17
Silver, (Module 3): 2, (Module 4): 14
Sine value, (Module 1): 2–3
Sine wave, (Module 1): 1–6
Sleeves
aluminum outer, (Module 3): 17, 18
conductor repair, (Module 3): 18–19, 24, (Module 4): 11, (Module 6): 7, 19
protective, (Module 2): 1, 2, (Module 3): 1, 17, (Module 6): 7
service drop conductor, (Module 5): 13
steel, (Module 3): 18
Smart Grid, (Module 2): 12
Smart meter, (Module 5): 9
Snow. *See* Weather conditions, ice and snow
Soil. *See* Ground (soil)
Soldering, (Module 4): 9, 12
Solenoid, (Module 2): 18
SP. *See* Transformers, self-protected
Spacer, (Module 3): 4, 7, 12
Spade terminal
for transformer connection, (Module 2): 8
transformer primary terminal, (Module 4): 24, 25

Spade terminal (*continued*)
transformer secondary terminal, (Module 2): 11, (Module 4): 20, 22, 26, (Module 5): 16, 20
Spark gap, (Module 2): 11
Spider® System, (Module 3): 10, 12
Splices
automatic, (Module 3): 19, 20, (Module 6): 7, 8
conductor, (Module 3): 17, 24, (Module 6): 6, 7, 8
conductor crimp-on, (Module 3): 18, 19, (Module 6): 7, 19
full-tension (non-crimp), (Module 3): 17, 18, 24, (Module 6): 7
full-tension crimp-on, (Module 3): 17–18, 24
on meter loop, (Module 5): 10
procedure, (Module 4): 11–12, (Module 5): 20, (Module 6): 7, 19
on service drop, (Module 5): 13
URD cable, (Module 4): 9, 11–12
Splice shunt, (Module 3): 18, 19, 24, (Module 6): 7
Splice/termination kit, (Module 4): 12, 24, 25, (Module 6): 19
Spoiler, air flow, (Module 3): 7
Standoff, (Module 3): 12, (Module 5): 6
Static electricity, (Module 3): 9, 10, (Module 6): 1, 2
Steel
ACSR. *See* Aluminum conductor steel reinforced
characteristics, (Module 3): 2
in copperweld and alumoweld conductors, (Module 3): 2, 24
galvanized, mast for meter loop, (Module 5): 10
from recycled transformer, (Module 6): 15
in transformer core, (Module 1): 22, 24, (Module 2): 5
Sticks. *See* Clamp stick; Extendo stick; Hook stick; Hot stick
Strap hoist, (Module 5): 12
Street lights, (Module 2): 23–26, (Module 4): 25, (Module 5): 19–20
Strength, tensile, (Module 3): 2, 3, 24
Stripper, cable, (Module 4): 9, 10
Subdivision, residential, (Module 1): 23, (Module 5): 5
Sublets, (Module 3): 17
Substation, (Module 1): 24, (Module 2): 1, (Module 6): 22
Super Beast meter circuit tester, (Module 5): 17
Surge. *See* Current, surge
Swimming pool, (Module 5): 10, (Module 6): 6
Switches
air break, (Module 2): 17–18, 30
bypass, (Module 2): 16, (Module 6): 11
disconnect, (Module 2): 17, 18, 19, 30, (Module 4): 2, 14, (Module 6): 16
electronic, to select winding turns, (Module 2): 16
gang-controlled, (Module 4): 19
inspection, (Module 6): 12
load break sectionalizer, (Module 4): 20, 21
lock out and safety tag, (Module 6): 1
oil, (Module 2): 18, 19, 30
tap-changing, (Module 2): 5, 15, 16, 17, (Module 6): 14, 20
Switchgear, (Module 4): 5, 18–19, 30
Swivel, (Module 2): 21, (Module 3): 9, 11, 20
Symbols, (Module 1): 7, 8, 9, 11, 12, 26, 30

T

Tank. *See* Shell, on transformer
Tap, transformer, (Module 1): 27, 28, (Module 2): 15, 31, (Module 6): 20–21. *See also* Switches, tap-changing; Tap changer
Tap changer, (Module 2): 18, (Module 4): 20, (Module 6): 20–21. *See also* Voltage regulator
Tap percentages chart, (Module 6): 20–21
Tape, to seal cable termination or splice, (Module 4): 9, 12, 24, 26, (Module 5): 20
TDR. *See* Reflectometer, time domain
Telephone service, (Module 4): 6
Television service, cable, (Module 4): 6
Temperature
ambient, (Module 3): 8, 10, 13, (Module 4): 8
effects on conductor sag, (Module 6): 7
high, effects on voltage rating, (Module 1): 14
operating, transformer, (Module 2): 2
Temperature rating, (Module 4): 8
Temperature-rise rating, (Module 2): 2
Tensile strength, (Module 3): 2, 3, 24
Tension stringing method, (Module 3): 9, 20–21, 24
Tensioning and sagging process, (Module 3): 10–11, 13, 21, (Module 5): 12, (Module 6): 7–8
Tensioner, (Module 3): 9, 11, 20, 21, 24
Terminals
compression, (Module 4): 11, 12–13, 24, (Module 5): 19
insulated copper pin, (Module 2): 9
lug, (Module 4): 10, 24, 25, 26, (Module 5): 7, 19–20
primary, (Module 2): 9, 10, (Module 4): 24–25
on secondary cable, (Module 2): 9, 10–11, (Module 4): 25–26, (Module 5): 14
single-phase meter base, (Module 5): 9
spade. *See* Spade terminal
on a transformer, (Module 2): 6–7, 8–9, 10–11, (Module 4): 23–26, (Module 5): 16
Termination methods, (Module 4): 9–13, (Module 5): 14
Tesla, Nikola, (Module 1): 1
Tests
boom leakage, (Module 2): 2
capacitance discharge, (Module 6): 17
capacitor, (Module 6): 12
circuit, prior to meter installation, (Module 5): 14
high potential (hi-pot), (Module 6): 17
infrared scanning of power line, (Module 6): 7
for PCBs, (Module 6): 10
Super Beast meter circuit tester, (Module 5): 17
thermal, (Module 6): 7, 10
transformer, (Module 6): 13–15
utility pole decay, (Module 6): 5
voltage, (Module 6): 2, 14–15, 17–18. *See also* Voltmeter
Thermal expansion/contraction, conductor, (Module 3): 3, 8, 10
Thermal imaging, (Module 3): 18
Thermovision, (Module 6): 10, 26
Thumper, (Module 6): 16, 17
Ties
conductor, (Module 3): 11–14, 15, 16, (Module 6): 8, 9
guy, (Module 6): 5, 6
Western Union, (Module 3): 14, 16, 24
Tie kit, (Module 3): 12, 13, 15
Tin, (Module 2): 8
Tools
banding, (Module 3): 11
cable cutter, (Module 4): 9
cable-spiking, (Module 6): 18–19
cable stripper, (Module 4): 9, 10
to check for pole decay, (Module 6): 5
come-along, (Module 6): 5
compression tool and die, (Module 4): 11, 12, 13, 14
crimping, (Module 3): 17, 18, (Module 4): 12–13, 26
cross-arm extender, (Module 6): 7
disconnect attachments, (Module 2): 22, 23
insulated, (Module 2): 20–23, 24
load break, (Module 2): 17, 22–23, 24, (Module 6): 11

sticks. *See* Clamp stick; Extendo stick; Hook stick; Hot stick
for tree pruning, (Module 6): 7
universal tool head for, (Module 2): 21, 22
wire brush, (Module 2): 17, (Module 4): 25, 26, (Module 5): 14
Total inrush VA, (Module 1): 30
Total steady-state VA (sealed VA), (Module 1): 30
Toxic chemicals, (Module 2): 6, (Module 5): 4, (Module 6): 4–5, 9, 10, 13
Traffic, (Module 4): 22, (Module 6): 3, 16
Training, (Module 1): 31, (Module 3): 20, (Module 6): 19
Transformers
autotransformer, (Module 1): 28, 29
basic principle, (Module 1): 9, (Module 4): 2, 18, 19
completely self-protected (CSP), (Module 2): 2, 4, 6, 30
connections, (Module 2): 7–8, 9–11, (Module 4): 20, 23–26
construction and components, (Module 1): 22–24, (Module 2): 5–9, 10–11, 31, (Module 4): 20–21, 22, (Module 5): 16
current, (Module 1): 28–30, (Module 5): 20–21, (Module 6): 11
dead-front, (Module 4): 24, 30
definition, (Module 1): 22
faults, (Module 6): 15–16
inspections, (Module 6): 14, 15, 16
isolation, (Module 1): 27–28, 29, 30, 31
like-wound, (Module 1): 25
live-front, (Module 4): 20, 24–25, 30
maintenance, (Module 1): 30–31, (Module 6): 12–13
operating properties, (Module 1): 24–25
pad-mounted in URD system
construction, (Module 4): 20–21, 22, (Module 5): 16
installation, (Module 4): 21–23, (Module 5): 16
location, (Module 4): 3–5, 20
overview, (Module 1): 33, (Module 2): 4, 5, (Module 4): 1, 3
sites which use, (Module 1): 33, (Module 2): 23
pole-mounted, for overhead power distribution, (Module 2): 1, 2–4, 5, (Module 4): 2, (Module 6): 12–13, 20
potential, (Module 1): 30
recycling, (Module 6): 15
role in power distribution system, (Module 1): 1, 2, 22, 24, 30
selection, (Module 1): 30–31
self-protected, (Module 2): 2, 30
single-phase
additive polarity, (Module 2): 7
composite pad for, (Module 4): 21
connections, (Module 2): 8, 9, (Module 4): 25
fuse on, (Module 2): 14
live-front pad-mounted, (Module 4): 20
measuring secondary voltage on, (Module 6): 15
in three-phase power distribution, (Module 5): 3, 4
slang terms for, (Module 2): 3
symbol, (Module 1): 26
tapped, (Module 1): 27, 28, (Module 2): 15, 16, 31, (Module 6): 20–21
three-phase, (Module 1): 31–34, (Module 2): 2–3, 15, (Module 4): 20–21, 22, (Module 5): 14, 16
troubleshooting, (Module 6): 13–15
turns and voltage ratios, (Module 1): 25–27, 29, (Module 6): 20
types, (Module 1): 27–30, (Module 2): 5–6
typical dimensions and weights, (Module 2): 3
unlike-wound, (Module 1): 25
vault, (Module 2): 4, (Module 4): 1, 2
Transformer cabinet, (Module 5): 6, 17, 21
Trees, (Module 3): 17, (Module 4): 3, (Module 5): 4, (Module 6): 5, 7
Trencher, (Module 4): 6, (Module 5): 18
Trenching and backfilling methods, (Module 4): 5–7, (Module 5): 17, 18, (Module 6): 19
Triangle, power, (Module 1): 22
Trigonometry, (Module 1): 2–3, 22
Trip, (Module 1): 33, (Module 4): 16
Triplex
definition, (Module 3): 15
flat parallel arrangement, (Module 5): 16
overhead service loop, (Module 5): 7, 10, 12
overview, (Module 4): 9, (Module 5): 3, 4, 17
underground, (Module 5): 15, 17, 18
Troubleshooting
cable fault, (Module 6): 15, 16–18
fuse, (Module 6): 15–16, 22
riser pole, (Module 6): 15–16
transformer, (Module 6): 13–15
underground power distribution, (Module 4): 3, 13, (Module 6): 15–19
use of faulted circuit indicator, (Module 2): 19, (Module 6): 17
Truck, bucket, (Module 3): 15, (Module 6): 7, 13
Tubing, electrical metallic (EMT), (Module 5): 12
Turnbuckle, (Module 6): 5
Turns, (Module 1): 9, 10, 25–27, 29–30, 31, (Module 6): 20

U

U-guard, (Module 5): 5, 6, 24
Underground distribution. *See* Power distribution, underground; Power distribution, underground residential
Uniline®, (Module 3): 19
United States, loading districts, (Module 3): 4, 7, 24
Universal tool head, (Module 2): 21, 22
URD. *See* Power distribution, underground residential
Utility companies
and the common-trench, (Module 4): 5–6, 30
conductor ground clearance policy, (Module 5): 8
data from smart meters, (Module 5): 9
frequency values, commonly used, (Module 1): 4
fuse selection policy, (Module 2): 14
ownership of meters, (Module 5): 14
ownership of street lights, (Module 2): 23
service mast materials policy, (Module 5): 12

V

VA. *See* Volt-amperes
Vacuum, (Module 2): 18, 20, (Module 4): 19, (Module 6): 11
Valves, transformer
oil drain, (Module 4): 21, 22, (Module 6): 12, 13, 14
oil pressure-release, (Module 4): 20, 22, (Module 6): 12, 13
Vandalism, (Module 3): 17, (Module 6): 13, 14
VAR. *See* Volt-amperes-reactive
Varistor, metal-oxide (MOV), (Module 2): 11, 30, (Module 4): 13, 14, 30
Vault, underground, (Module 2): 4, (Module 4): 1, 2
Vector, (Module 1): 7, 17, 19, 22, 30
Vehicles, (Module 6): 3, 5. *See also* Traffic; Truck, bucket
Ventilation, U-guard, (Module 5): 6
Vibration, in the power lines, (Module 3): 4, (Module 6): 8–9, 26
Voltage
in an inductive AC circuit, (Module 1): 9–11
calculation, (Module 1): 3

Voltage *(continued)*
in a capacitive AC circuit, (Module 1): 15–16
in a capacitor, (Module 2): 16, (Module 6): 12
and conductor selection, (Module 3): 4, 5–6
creation, (Module 1): 1–3, 9
effective, (Module 1): 4–5
induced from adjacent power lines, (Module 6): 1, 2
maximum for insulated hand tools, (Module 2): 20
measurement. *See* Multimeter, digital; Voltmeter
in Ohm's law formulas, (Module 1): 7–8, 12
one cycle of alternating, (Module 1): 3
peak, (Module 1): 4–5, 6, 37
range for electrocution, (Module 1): 31
relationship with apparent power, (Module 1): 20
relationship with turns, (Module 1): 25–27, 28
secondary voltage test, (Module 6): 14–15, 17–18
step-down, (Module 1): 2, 24, 26, 28, 30, (Module 4): 1, (Module 6): 1–2
step-up, (Module 1): 2, 24, 26, 28, (Module 6): 1–2
tap-changing calculations, (Module 6): 21
typical
along power distribution system, (Module 1): 24, 33, (Module 2): 1, 2, (Module 6): 20
for appliances and equipment, (Module 4): 1, (Module 6): 19–20
fluorescent lighting, (Module 1): 31
residences, (Module 2): 15, (Module 4): 1, 19, (Module 5): 2
for street lights, (Module 2): 23
in three-phase power distribution, (Module 1): 31, 33, (Module 2): 8, (Module 5): 4
Voltage drop, (Module 1): 29, (Module 2): 17, (Module 6): 20
Voltage rating, (Module 1): 14–15, 30, (Module 4): 8
Voltage ratio, (Module 1): 25–27, (Module 6): 20
Voltage regulator, (Module 2): 15–16, 18, (Module 6): 13, 14, 20
Volt-amperes (VA), (Module 1): 20, 22, 23, 30–31, (Module 6): 12
Volt-amperes-reactive (VAR), (Module 1): 20, 22, 23
Voltmeter, (Module 1): 30, (Module 2): 7, (Module 5): 4, (Module 6): 2, 14–15

W

Waste disposal, (Module 2): 6, (Module 6): 7, 9, 10, 12, 15
Watt, (Module 1): 19
Watt-hour, (Module 5): 14, 15
Waveform, (Module 1): 1–6
Weather conditions. *See also* Temperature, ambient; Wind
conditions for static electricity, (Module 3): 9, (Module 6): 2
extreme, (Module 2): 2, (Module 3): 1, (Module 5): 1
ice and snow, (Module 3): 3, 4, 7, 8, 17, (Module 6): 3
rain, and soil erosion, (Module 4): 7
Westinghouse, George, (Module 1): 1
Wildlife, (Module 2): 2, 5, 6, (Module 5): 4, (Module 6): 5, 9
Winch, capstan, (Module 6): 7, 13
Wind
insulator damage from, (Module 6): 13
static electricity build up from, (Module 3): 9, (Module 6): 2
stress on conductor from, (Module 3): 3, 4, 7, 8, 17
Windings, (Module 1): 10, 22–33, (Module 5): 4, 5
Wire
aluminum, (Module 4): 8
copper, (Module 1): 1, (Module 4): 8
guy, (Module 6): 5, 6
sizes, (Module 3): 4, (Module 5): 20
street lighting circuit, (Module 5): 19, 20
three-wire secondary service, (Module 4): 1, 25
in URD cable, (Module 4): 8
Wire holder, (Module 5): 7, 10, 12, 13
Wire wrap, (Module 3): 10, (Module 6): 5, 6
Wood preservatives, (Module 5): 4, (Module 6): 4–5
Wye-wye arrangement, (Module 1): 31, 32, (Module 2): 8, 10, (Module 5): 4

X

XLPE. *See* Cross-linked polyethylene